Carbon Capture

Jennifer Wilcox

Carbon Capture

Prof. Jennifer Wilcox
Dept. of Energy Resources Engineering
Stanford University
Stanford, CA
USA

ISBN 978-1-4939-0125-8 ISBN 978-1-4614-2215-0 (eBook)
DOI 10.1007/978-1-4614-2215-0
Springer New York Dordrecht Heidelberg London

Printed on acid-free paper

Springer is part of Springer Science+Business Media (www.springer.com)

Preface

The burning of fossil fuels, including coal, oil, and gas meets more than 85% of the world's energy needs today. The associated CO_2 emissions cause detrimental changes to the earth's climate. The scientific community agrees that the solution for mitigating CO_2 emissions lies in a portfolio of strategies, including carbon capture and storage (CCS) and potentially carbon capture and utilization (CCU). The mitigation of CO_2 through its separation from gas mixtures such as power plant emissions, *i.e.*, *Carbon Capture*, will be a critical step toward stabilizing global warming. Amine-based scrubbing for CO_2 capture has been taking place for over 70 years to purify natural gas, but it is unclear whether this technology will be optimal for tackling the scale of CO_2 emitted on an annual basis ($\sim$30 Gt worldwide). The next several decades of engineers, chemists, physicists, earth scientists, mathematicians, and social scientists will advance traditional separation technologies. Thus, this book spans a wide range of disciplines, as will the portfolio of solutions.

The core of the book focuses on the most advanced CO_2 capture technologies including absorption, adsorption, and membranes. The book also includes chapters on algae and electrochemical/photocatalytic CO_2-to-fuel conversion processes. The reduction of CO_2 via photosynthesis or electrochemical/photo catalysis are routes to alternative fuels. Along with climate change, the development of alternatives to crude oil for transportation fuels will be a strong drive of policy. These two drivers are likely to remain present for a long time, with indeterminate relative weights. An additional motivation for including these topics is the vision that one day they may be advanced to the extent that CCS can take place in a single process, rather than the current three-step sequence of capture, compression, and storage. The book does not discuss geological storage of CO_2, but focuses solely on methods of capture.

This book is the first of its kind. As an emerging field, carbon capture ranges many disciplines. No single source exists that explains the fundamentals of gas separation and their link to the design process. The closest predecessor to this book is the 2005 IPCC Special Report on CCS, which provides only a high-level overview of CCS options. This book provides the reader with the skillset needed to recognize the limitations of traditional gas-separation technologies in the context of CO_2 capture, and how they may be advanced to meet the scale challenge required to substantially decrease CO_2 emissions.

This book can be used in the classroom at the undergraduate and graduate levels for students examining separations processes in the context of carbon capture. It can be used as educational material for industrial personnel engaged in carbon capture technologies in addition to gas separation processes. The material in this book will benefit scientists and engineers active in the research and development of carbon capture technologies. Finally, engineers evaluating separation processes for carbon capture will find this book useful.

The three core chapters pay significant attention to the pedagogy of absorption, adsorption, and membrane separation processes for CO_2 capture and include many worked examples and end-of-chapter problems. These examples test fundamental concepts, from the chemical physics associated with a given material that binds CO_2, to the unit operations of the process, closely coupled by mass transfer. In general, metric units (*e.g.*, metric tonne) are primarily used throughout the book, with the exception of several instances where US engineering units are used when they are considered the industry standard.

Cover Design: The cover design was motivated by the most advanced CO_2 capture technology: amine scrubbing. The image illustrates that CO_2 is generated from the oxidation of fossil fuels, such as petroleum and coal, and the chemical form of the CO_2 emitted into the atmosphere from fossil-fuel oxidation is of a fairly stable linear structure. An amine capture process involves a complex combination of carbonate and carbamate chemistries dependent upon pH, temperature, and pressure conditions. The chemical process of oxygen (carbonate) or nitrogen (carbamate) binding with the carbon atom of CO_2 forces the molecular structure to transform from a linear geometry (the molecule depicted on the right) to a bent structure (the molecule depicted on the left). This process has an entropy cost in addition to a significant barrier associated with binding, compared to other gas species that already possess the bent structure, *e.g.*, SO_2. Not surprisingly, the reaction rate is faster with SO_2. In an absorption process, CO_2 must bend in preparation for reaction as illustrated in the cover image. The bent structure represents an intermediate stage between the gas and solution phase, in which CO_2 is ultimately captured. Credit is given to artists, Karen Miller and Deborah Hickey for the design of the cover image.

Acknowledgements

Several years ago, I was invited to be part of a team co-led by Robert Socolow and successively Bill Brinkman, Arun Majumdar, and Michael Desmond. The American Physical Society charged the committee to explore the feasibility of direct air capture of carbon dioxide from the atmosphere with chemicals. The team recognized a strong need for one go-to source that provides the necessary skills and tools to move carbon capture research forward. This book aims to provide this foundation.

In particular, I thank Robert Socolow. His encouragement and contagious passion for teaching and learning motivated me throughout the writing of this book. For their helpful discussions and for providing the motivation to take on this challenge, I thank the other members of the APS study committee: Roger Aines, Jason Blackstock, Olav Bolland, Michael Desmond, Tina Kaarsberg, Nathan Lewis, Marco Mazzotti, Allen Pfeffer, Jeffrey Siirola, Karma Sawyer, and Berend Smit.

The individual chapters in this book were reviewed in draft form by individuals chosen for their technical expertise in specific areas. I thank the following individuals for their participation in the review process:

Shela Aboud, Stanford University
Antonio Baclig, C12 Energy, Inc.
Abhoyjit Bhown, Electric Power Research Institute (EPRI)
Heinz Bloch, P.E.
Gordon Brown, Stanford University
Stefano Consonni, Politecnico di Milano
Christopher Edwards, Stanford University
Brice Freeman, Electric Power Research Institute (EPRI)
Thomas Jaramillo, Stanford University
Marco Mazzotti, Swiss Federal Institute of Technology Zurich (ETH)
Lisa Joss, Swiss Federal Institute of Technology Zurich (ETH)
Anthony Kovscek, Stanford University
Nichola McCann, University of Kaiserslautern
Sean McCoy, International Energy Agency (IEA)
Keith Murphy, Air Products & Chemicals, Inc.
Andrew Peterson, Brown University

Valentina Prigiobbe, University of Texas, Austin
Gary Rochelle, University of Texas, Austin
Edward Rubin, Carnegie Mellon University
Douglas Ruthven, University of Maine
Douglas Way, Colorado School of Mines

Without the assistance of my research group at Stanford University, this project would not have been completed nearly as well nor as efficiently. Abby Kirchofer, Mahnaz Firouzi, Guenther Glatz, and Dong-Hee Lim provided content for several of the chapters. Jasper van der Bruggen with his GIS skills assisted in map design. Keith Mosher and Ekin Ozdogan assisted with the images and tracked down lots of numbers for me to make useful tables in many of the chapters. Yangyang Liu laid out and organized the appendix in addition to proofreading chapters. Erik Rupp assisted in the writing of the worked problems throughout the text and Kyoungjin Lee, Jiajun He, and Feng Feng assisted with the solutions manual for the problems at the end of Chaps. 3, 4, and 5. I am also grateful to Panithita Rochana. Her careful attention to detail in creating the majority of the images and the complete notation list made the book far stronger.

I would also like to acknowledge my husband, Austin, for his support and encouragement, day in and day out. Thank you.

Contents

Abbreviations

Notation

A	Cross-sectional area of the column (3.64), L^2 *or* Cross-sectional area of the adsorbate (4.13), L^2 *or* Membrane surface area (5.29), L^2
a	Interfacial area per unit of packed volume (3.55), L^2/L^3 *or* Core radius or the pore, such that $a = r - l$, where r is the pore radius and l is the pore surface film thickness (4.17), L *or* External surface area of the sorbent particles $= 3(1 - \varepsilon)/r$ for spherical particles of radius r (4.41), $1/L$
$B(T)$	Second virial coefficient corresponding to interactions between pairs of molecules (4.8), $L^3/mole$
b	Langmuir affinity constant (5.10), p^{-1}
C	BET constant (4.14), *dimensionless* *or* Dimensionless concentration $= c/c_0$ where c_0 is the initial concentration (4.21), *dimensionless*
C_{avo}	Cost of CO_2 avoided for a power plant (1.13), \$/ton CO_2
$C_{avo,\ DAC}$	Cost of CO_2 avoided from a direct air capture system (1.16), \$/ton CO_2
C_{cap}	Cost per ton CO_2 captured from a power plant (1.14), \$/ton CO_2
$C_{cap,\ DAC}$	Cost per ton CO_2 captured from a direct air capture system (1.15), \$/ton CO_2
$C(T)$	Third virial coefficient corresponding to interactions between triplets of molecules (4.8), $L^6/mole^2$
c	Speed of light $= 2.998 \times 10^8$ m/s
c_{A1}, c_{A2}	Concentrations of gas A at the high and low pressure side of the membrane, respectively (5.1), $mole/L^3$

Numbers in parentheses refer to equations, sections, or tables in which the symbols are defined or first used. Dimensions are given in terms of mass (M), length (L), time (t), temperature (T), amount of substance (*mole*), electric potential (V), electric current (I), energy ($E = ML^2/t^2$), pressure ($p = M/Lt^2$) and *dimensionless*. Symbols that appear infrequently are not listed. Symbols used frequently are not referenced to a given equation, section, or table.

c_B Concentration of the reactive binding species in aqueous solution (3.17), $mole/L^3$
c_H Concentration of the Henry's law species in the dual-mode model (5.7), $mole/L^3$
c'_L Langmuir capacity constant (5.7), $mole/L^3$
or Atomic concentration of hydrogen (5.11), $mole/L^3$
c_i Concentration of component i in a mixture, $mole/L^3$
or Concentration of ions in the solution with charge z_i(3.4), $mole/L^3$
c_p Concentration of the product species (3.13), $mole/L^3$
c_L Concentration of the Langmuir species in the dual-mode model (5.7), $mole/L^3$
c_{S,CO_2} Concentration of CO_2 at the sorbent surface (4.33), $mole/L^3$
c_0 Initial feed concentration of CO_2 (4.43), $mole/L^3$
c_{i,CO_2} Concentration of CO_2 at the gas–liquid interface (3.34) *or* Concentration of CO_2 at the gas-film interface (4.33), $mole/L^3$
c_{∞,CO_2} Concentration of CO_2 in the bulk gas phase (3.34) or bulk liquid phase (3.35), $mole/L^3$
D Diffusivity, L^2/t
D_A Diffusivity of component A (3.7), L_2/t
D_{AB} Diffusivity of gas A in liquid B (Table 3.3), L^2/t
D_B Liquid diffusivity of binding species, B (Table 3.5), L^2/t
D_b Bulk fluid diffusivity (4.32), L^2/t
D_{G,CO_2} Molecular diffusivity of CO_2 in the gas phase (3.34), L^2/t
D_{L,CO_2} Molecular diffusivity of CO_2 in the liquid phase (Table 3.5), L^2/t
D_c Column diameter (3.77), L
D_{He} Diffusion coefficient of the Henry's law species in the dual-mode model (5.9), L^2/t
D_e Effective (or measured) diffusivity (4.19), L^2/t
D_{eA} Effective diffusivity of component A (5.17), L^2/t
D_f Molecular diffusivity of CO_2 in the film (4.33), L^2/t
D_H Diffusion coefficient of atomic hydrogen (5.12), L^2/t
D_i Pipeline inner diameter (2.13), L
D_K Knudsen diffusivity (4.28), L^2/t
D_L Dispersion coefficient from axial mixing (4.47), L^2/t
D_{La} Diffusion coefficient of the Langmuir species in the dual-mode model (5.9), L^2/t
D_m Molecular diffusivity (4.46), L^2/t
D_{Pois} Poiseuille flow across the cylindrical pore (4.31), L^2/t
D_p Diffusivity in the pore (4.36), L^2/t
D_A/D_B Mobility selectivity (5.6), *dimensionless*
d Channel diameter (4.58), L
d_p Particle diameter (4.34), L
E Enhancement factor for the absorption process (3.41), *dimensionless*
or Electric field (4.5), V/t or MLI/t^3
E_{abs} Binding energy of the atom within the bulk crystal (5.8), $E/mole$

E_d	Dissociation energy of gaseous molecule (5.8), *E/mole*
E_i	Enhancement factor corresponding to an instantaneous reaction (3.42), *dimensionless*
E_{full}	Solar energy at full spectrum (7.2), *E/mole*
E_L	Energy associated with condensation to a liquid phase (4.15), *E/mole*
E_{ph}	Photon energy (7.2), *E/mole*
E_0	Standard potential (8.2), *E/mole*
E_1	Activation barrier to overcome the potential associated with adsorption on layer 1 (4.15), *E/mole*
F	Faraday's constant = 96485.3 C/g-equivalent *or* Ratio of Langmuir and Henry's law diffusion coefficients (5.10) = D_{La}/D_{He}, *dimensionless*
F_{CO_2}	CO_2 feed rate to the adsorption bed (4.50), $mole/L^2t$ or M/L^2t
F_P	Packing factor (3.49), *dimensionless*
f	Friction factor (4.63), *dimensionless*
f_F	Fanning friction factor (2.13), *dimensionless*
f_1	Liquid viscosity correction factor = $(\mu/\mu_W)^{0.16}$ (3.77), *dimensionless*
f_2	Liquid density correction factor = $(\rho/\rho_W)^{1.25}$ (3.77), *dimensionless*
f_3	Surface tension correction factor = $(\sigma/\sigma_W)^{0.8}$ (3.77), *dimensionless*
G_i°	Gibbs free energy of gas species *i* at standard conditions (1.8), *E*
g	Gravitational acceleration, L/t^2
g_c	Conversion factor for the force unit (2.13), *dimensionless*
H	Henry's law constant. The unit of the Henry's law constant depends on the Henry's law equation that is being used (3.2), *p* or $pL^3/mole$ *or* Henry's law adsorption equilibrium constant (4.55), *dimensionless*
H^0	Henry's law constant of substance in pure water (3.3), *p* or $pL^3/mole$
H_{CO_2}	Henry's law constant of CO_2 (Table 3.5), *p* or $pL^3/mole$
H_G	Height of individual vapor film mass-transfer unit (3.72), *L*
H_L	Height of individual liquid film mass-transfer unit (3.72), *L*
H_0	Dimensionless Henry's law adsorption equilibrium constant at some initial temperature condition (4.55), *dimensionless*
H_{OG}, H_{OL}	Height of overall mass-transfer unit based on gas phase and liquid phase, respectively (3.68–3.69), *L*
h	Planck's constant = 6.63×10^{-34} J s *or* Height or elevation change (2.13), *L* *or* Solubility factor (3.3), $L^3/mole$ *or* Overall heat transfer coefficient between particle and fluid (4.45), E/L^2t^2
h_G	Contribution of gas to solubility factor (3.5), $L^3/mole$
h_+	Positive ions associated with the dissociated ionic species in electrolyte solution (3.5), $L^3/mole$
h_-	Negative ions associated with the dissociated ionic species in electrolyte solution (3.5), $L^3/mole$
I	Ionic strength of the electrolyte solution (3.3), M/L^3 *or* Molecular moment of inertia (5.8), ML^2

i	Electrode current density (8.15), I/L^2
i_0	Exchange current density (8.15), I/L^2
J	Rate of absorption per unit area (flux) (3.7), $mole/L^2t$ or M/L^2t
J_{f,CO_2}	Molar flux of CO_2 at the interface of the gas and fluid film on the sorbent (4.33), $mole/L^2t$
J_{H_2}	Flux of molecular hydrogen through membrane (5.12), $mole/L^2t$
J_{L,CO_2}	Molar flux of CO_2 in the solution (3.35), $mole/L^2t$
K	Langmuir adsorption equilibrium constant (4.10), *dimensionless*
K_c	Overall (combined) mass-transfer coefficient (4.42), L/t
K_{eq}	Equilibrium constant (3.15). The unit of equilibrium constant depends upon the number of moles of the reactants and products involved in the reaction
K_{eq}^{VL}	Vapor–liquid equilibrium ratio for CO_2 (3.6), *dimensionless*
K_S	Sieverts' Constant (5.11), $mole/L^3p^{1/2}$
K_x	Overall volumetric mass-transfer coefficient for liquid phase, based on concentration driving force (3.58), $mole/L^2t$
K_y	Overall volumetric mass-transfer coefficient for gas phase, based on concentration driving force (3.57), $mole/L^2t$
K_3	Percentage of flooding correction factor from Fig. I.2 (3.78)
k	Boltzmann's constant $= 1.38066 \times 10^{-23}$ J/K *or* Specific heat ratio, c_p/c_v where c_p is the specific heat at constant pressure and c_v is the specific heat at constant volume, *dimensionless* *or* Rate constant (3.12). The units of the rate constant depend on the global order of reaction *or* reaction rate constant $= sk_s$ where s is the surface area per unit volume and k_s is the surface rate constant (4.19), L/t *or* Interaction constant of fluid molecules (4.27), EL^3
k_D	Henry's law constant (5.10), p, $pL^3/mole$ or *dimensionless*
k_f	External mass-transfer coefficient of the film (4.33), L/t
k_{G,CO_2}	Mass-transfer coefficient of CO_2 or mass-transfer velocity (3.34), L/t
k_{L,CO_2}	Liquid-phase mass-transfer coefficient of CO_2 or Gas-phase mass-transfer velocity (3.35), L/t
k_p	Individual volumetric mass-transfer coefficient for the gas phase, based on partial-pressure driving force (3.59), $mole/L^3tp$ *or* Mass-transfer coefficient associated with diffusion into pores or internal mass-transfer coefficient *i.e.*, for a spherical particle $k_p \approx 10D_e/d_p$ where d_p is the particle diameter (4.39), L/t
k_x	Individual volumetric mass-transfer coefficient for liquid phase, based on concentration driving force (3.55), $mole/L^3t$
k_y	Individual volumetric mass-transfer coefficient for gas phase, based on concentration driving force (3.55), $mole/L^3t$
k_0	Gas-phase rate constant (3.12). The units of the rate constant depend on the global order of reaction *or* Standard rate constant for electrochemical reaction (8.13)

k_1, k_{-1} First-order rate constant of the forward and the reverse reaction, respectively (3.13), $1/t$

k_2 Second-order rate constant (3.17), $L^3/mole{\cdot}t$

L Pipeline length (2.13), L
or Bulk liquid velocity (advection) (3.35), L/t
or Molar flow rate of liquid at any section of the absorption column (3.52), $mole/t$
or Length of the sorbent bed (4.43), L

L_w Liquid mass flow rate per unit cross-sectional area in the column (3.77), M/L^2t

LUB Length of the unused bed (Sect. 4.4.2), L

L_1 Liquid molar flow rate entering the top of the absorption column (3.50), $mole/t$

L_2 Liquid molar flow rate exiting the bottom of the absorption column (3.50), $mole/t$

l Film thickness (Sect. 4.1.3), L

l_c Critical layer thickness (4.17), L

M Molecular weight, $M/mole$

M_i Molecular weight of species i in a mixture, $M/mole$

m Mass, M;
or Slope of the equilibrium curve (3.62)

$\dot{m}$ Mass flow rate of CO_2 (2.14), M/t
or Mass flow rate of entering gas (3.81), M/t

N Number of equilibrium stages (mass-transfer units), *dimensionless*

N_A Avogadro's constant $= 6.0221415 \times 10^{23}$ molecules per mole

N_i Diffusive flux of component *i*, $Mole/L^2t$ or M/L^2t

N_m Number of adsorbate molecules comprising of a complete monolayer per unit area (4.11), $molecules/L^2$

N_{OG}, N_{OL} Number of overall mass-transfer units based on gas phase and liquid phase, respectively (3.68–3.69), *dimensionless*

n Polytropic index (2.3), *dimensionless*
or Number of electrons (8.3)

n_i Number of moles of species i in the mixture, *mole*

$n_i^{CO_2}$ Number of moles of CO_2 in the gas mixture of stream i (1.10), *mole*

$n_i^{i-CO_2}$ Number of moles of the remaining gas in the gas mixture of stream i (1.10), *mole*

$\dot{n}$ Molar flow rate, $mole/t$

P Power, E/t

P_A Permeability of component A in a gas mixture $= D_A S_A$ (Sect. 5.3.1), *Barrer*

$\bar{P}_A$ Permeanace of component A in a gas mixture $= \frac{D_A S_A}{z}$ (5.5), *Gas processing unit* (*GPU*)

P_{ad} Adiabatic single-stage horse power (2.4), p

P_{ave} Average pressure along a CO_2 pipeline $= \frac{2}{3}\left(p_2 + p_1 - \frac{p_2 p_1}{p_2 + p_1}\right)$ (2.14), p

P_{H_2}	Hydrogen permeability through membrane (5.14), *Mole/Ltp*$^{0.5}$
P_{iso}	Isothermal compression power (2.7), *p*
PAR	Photosynthetically active radiation (7.2), *mole/L*^{2}*t*
PFD	Annual photon flux molar density (7.2), *mole/L*^{2}*t*
p	Pressure, *p*
p_{A1}	Partial pressure of *A* on the feed side or the membrane (5.3), *p*
p_{A2}	Partial pressure of *A* on the permeate side of the membrane (5.3), *p*
p_{CO_2}	Partial pressure of CO_2 in equilibrium with solution (3.2), *p*
p_c	Critical pressure, *p*
p_{c1}, p_{c2}	Probability of a fluid particle condensing upon collision with layer 1 and layer 2, respectively (4.15), *dimensionless*
p_i	Partial pressure of species *i* in the mixture, *p*
p_{H_2}	Partial pressure of H_2 in equilibrium with metal (5.11), *p*
p_r	Reduced pressure, *p*
p_v	Vapor pressure of the moisture contained in the gas mixture (2.12), *p*
p_0	Vapor pressure (saturation pressure), *p*
p_1, p_2	Inlet and outlet gas pressure from the compression stage (2.1), *p* *or* Total pressure on the feed and the permeate side of membrane, respectively (5.19), *p*
pK_a	Logarithmic value of an acid dissociation constant $= -\log_{10} K_a$ (Table 3.4), *dimensionless*
Q	Heat supplied to the system of interest (1.6), *E* *or* Volumetric flow rate, L^3/t *or* Quadrupole moment (4.5), L^2I
$q(\rho,\theta)$	Local charge density at the point (ρ, θ) with the origin defined as the center of the system (4.6), I/L, I/L^2 or I/L^3
R	Ideal gas constant *or* Dimensionless radius of the particle $= r/R_p$ (4.21), *dimensionless*
R_p	Radius of particle (4.19), *L*
r	Compression ratio (2.4), *dimensionless* *or* Rate of reaction (3.13). The units of the rate of reaction depend on the global order of reaction *or* Distance between particles (4.3), *L* *or* Pore radius (4.27), *L* *or* Ratio of permeate and feed pressures in membrane $= p_2/p_1$ (5.20), *dimensionless*
r_{12}	Distance between two isolated systems 1 and 2 (4.1), *L*
r_f	Rate of forward electrochemical reaction (8.6)
r_r	Rate of reverse electrochemical reaction (8.7)
r_s	Compression ratio per stage (2.10), *dimensionless*
r_t	Total compression ratio (2.10), *dimensionless*
r, θ, z	Cylindrical coordinates
r, θ, ϕ	Spherical coordinates
S_i	Solubility coefficient of component *i* in the gas mixture (5.2), *mole/L*^{3}*p*
S	Total surface area of the sorbent (4.13), L^2

S_A/S_B	Solubility selectivity (5.6), *dimensionless*
s	Number of compression stages (2.10), *dimensionless* *or* Fractional rate of surface-renewal (3.40), L^2/t
T	Temperature, T
T_{ave}	Average temperature, usually assumed to be constant at ground temperature for a CO_2 pipeline model (2.14), T
T_c	Critical temperature, T
T_g	Glass transition temperature, T
T_r	Reduced temperature, T
t	Time, t
t_b	Breakthrough time (4.53), t
t^*	Ideal time if there is no mass-transfer resistance (4.52), t
$\bar{t}$	Dimensionless time $= t/t^*$ where t^* is the ideal time if there is no mass-transfer resistance (4.43), *dimensionless*
u	Fluid velocity, L/t
u_{fl}	Minimum fluidization velocity (4.66), L/t
u_{max}	Maximum allowable upflow fluid velocity in packed bed (4.66), L/t
u_0	Superficial velocity of fluid, L/t
u_{12}	Interaction potential energy between molecules 1 and 2 (4.9), E
V	Volume, L^3 *or* Molar flow rate of vapor (gas) at any section of the absorption column (3.52), *mole*/t
V_{out}	Volumetric flow rate of the permeate stream (5.28), L^3/t
V_w^*	Mass velocity of gas stream based on total column cross section (3.80), M/L^2t
V_1	Vapor (gas) molar flow rate exiting the top of the absorption column (3.50), *mole*/t
V_2	Vapor (gas) molar flow rate entering the bottom of the absorption column (3.50), *mole*/t
$\tilde{V}$	Molar volume, L^3/*mole*
$\dot{V}_{p1}$	Perfect (ideal) gas volumetric flow rate at the inlet of compressor (2.4), L^3/t
$\dot{V}_{r1}$	Real gas volumetric flow rate at the inlet of compressor (2.6), L^3/t
V_1, V_2	Initial and final volume of gas (2.1), respectively, L^3
W	Work done by the system of interest (1.6), *E/mole*; *or* Adsorbate loading (4.10), *mole*/L^3 or *mole*/M or *dimensionless*
W_e	Electrical energy produced per mole of product from the electrochemical reaction (8.5), *E/mole*
W_m	Maximum sorbate loading based on the monolayer assumption (4.10), *mole*/L^3 or *mole*/M
W_{min}	Minimum theoretical work required for separation process, also refers to the reversible work (W_{rev}) (1.7), *E/mole*
W_{real}	Real (actual) work required for separation process (1.12), *E/mole*
W_{sat}	Equilibrium saturation concentration of CO_2 in the porous sorbent (4.43), *mole*/L^3 or *mole*/M

W_0	Initial concentration of CO_2 in the porous sorbent (4.43), *mole/L*^{3} or *mole/M*
w_b	Work associated with blower, *E/mole*
w_c	Work associated with compressor, *E/mole*
w_f	Work associated with fan, *E/mole*
x	Mole fraction of a given component in the liquid phase, *dimensionless* *or* Mole fraction of more permeable species in the feed (5.18), *dimensionless*
x_{CO_2}	Mole fraction of CO_2 in the liquid phase, *dimensionless*
x_e	Mole fraction of a given component in the liquid phase at equilibrium (3.58), *dimensionless*
x_i, x_j	Mole fraction of components *i* and *j* in the feed stream of a given membrane (5.6), *dimensionless*
x_{1,CO_2}	Mole fraction of CO_2 in the liquid entering the top of the absorption column (3.51), *dimensionless*
x_{2,CO_2}	Mole fraction of CO_2 in the liquid exiting at the bottom of the absorption column (3.51), *dimensionless*
y	Mole fraction of a given component in the gas (vapor) phase, *dimensionless* *or* Mole fraction of more permeable species in the permeate (5.18), *dimensionless*
y_A	Mole fraction of component *A* in gas phase, *dimensionless*
y_{CO_2}	Mole of CO_2 in the gas phase, *dimensionless*
y_e	Mole fraction of CO_2 in the gas phase at equilibrium (3.57), *dimensionless*
y_i, y_j	Mole fraction of components *i* and *j* in the product stream of a given membrane (5.6), *dimensionless*
$y_i^{CO_2}$	Mole fraction of CO_2 in the gas mixture of stream *i* (1.11), *dimensionless*
$y_i^{i-CO_2}$	Mole fraction of the remaining gas in the gas mixture of stream *i* (1.11), *dimensionless*
y_{out}	Mole fraction of more permeable species on the permeate side (5.28), *dimensionless*
y_{1,CO_2}	Mole fraction of CO_2 in the gas (vapor) exiting the top of the absorption column (3.51), *dimensionless*
y_{2,CO_2}	Mole fraction of CO_2 in the gas (vapor) entering at the bottom of the absorption column (3.51), *dimensionless*
y^*	Local composition of the gas leaving membrane surface (5.22), *dimensionless*
y^*_{max}	Highest achievable permeate composition (5.27), *dimensionless*
Z_T	Total height of packed section of the column (3.65), *L*
Z_1, Z_2	Gas compressibilities at the inlet and outlet conditions respectively (2.4)
z	Stoichiometric coefficient associated with the reactive binding species (3.16), *dimensionless*
z_i	Charge associated with ion *i* in the solution (3.4), *dimensionless*

Greek Letters

α Polarizability (4.5), L^3

α Parameter associated with the fluid-wall interaction (4.17)
or Membrane selectivity or permeability selectivity (5.6), *dimensionless*
or Charge transfer coefficient (8.9), *dimensionless*

β Shifting factor related to the adsorptive (4.26), *dimensionless*
or $1/kT$ where k is Boltzmann's constant (5.8)

γ_{AC} Activity coefficient of the activated complex (3.12), *dimensionless*

γ_B Activity coefficient of the reactive binding species B (3.12), *dimensionless*

γ_{CO_2} Activity coefficient of CO_2 in solution (3.12), *dimensionless*

ΔE_b Energy content of biomass (7.5), *E/mol*

ΔE_f Formation energy of CH_2O (7.5), *E/mole*

$\Delta E_{1\to 2}$ Change in total energy of the system from state 1 to state 2 (1.6), *E/mole*

ΔG Change in the Gibbs free energy $= \Delta H - T\Delta S$ where ΔH is change in enthalpy (*E/mole*) and ΔS is change in entropy (*E/mole T*) (4.1), (*E/mole*)

ΔG_i Change in the Gibbs free energy in stream i in the separation process (1.7), *E/mole*

$\Delta G_{f,act}$ Gibbs energy of the activation of the forward reaction (8.6), *E/mole*

ΔG_{rxn} Change in the Gibbs free energy of the reaction (8.4), *E/mole*

$\Delta G_{r,act}$ Gibbs energy of the activation of the reverse reaction (8.7), *E/mole*

ΔG_{sep} Change in the Gibbs free energy in separation (1.7), *E/mole*

ΔG_0 Standard Gibbs energy of activation of the reaction if the over potential is zero (8.11), *E/mole*

ΔH_{ad} Change in heat of adsorption between the adsorbed phase and the corresponding gas phase (4.67), *E/mole*

ΔH_{rxn} Change in the enthalpy of the reaction $= \Delta G_{rxn} + T\Delta S_{rxn}$ where ΔS_{rxn} is the entropy change of the reaction (8.4), *E/mole*

ΔP_{flood} Pressure drop at flooding (3.49), *p/L*

Δp Pressure drop across a given unit (fan, pump or packed bed), *p*

Δy_m Logarithmic mean of $y_1 - y_e$ and $y_2 - y_e$ for an absorption column (3.75), *dimensionless*

δ Thickness of gas (or liquid) film (3.34), *L*
or Film thickness (4.33), *L*

ε Internal surface roughness of the pipe (2.15), *L*
or Efficiency, *dimensionless*
or Void fraction (porosity), *dimensionless*

ε_a Biomass accumulation efficiency (7.5), *dimensionless*

ε_d Pump driver efficiency (3.85), *dimensionless*

ε_{en} Energetic efficiency for electrochemical reaction (8.2), *dimensionless*

ε_i Intrinsic pump efficiency (3.83), *dimensionless*

ε_{Far}	Faradaic efficiency $= \frac{n_e F n}{q}$ where n_e is the numbers of electron, F is Faraday constant, n is the numbers of mole of product and q is charge associated in the reaction (8.2), *dimensionless*
ε_{pt}	Photon transmission efficiency (7.4), *dimensionless*
ε_{pu}	Photon utilization efficiency (7.4), *dimensionless*
ε_{12}	Lennard-Jones well depth= $\sqrt{\varepsilon_1 \varepsilon_2}$, T
φ_n	H_{OL} factor from Fig. I.4 (3.78)
η	Effectiveness factor (4.23) *or* Overpotential (8.1), *E/mol*
η_{2nd}	Second-law efficiency (1.12), *dimensionless*
θ	Time of exposure of liquid to gas (3.37), t; *or* Contact angle of the liquid against the pore wall (4.18), *degree* *or* Extent of micropore filling (4.27) *or* Fraction of the bed saturated (4.53)
λ	Wavelength associated with photon (7.3), L
λ_g	Thermal conductivity of fluid (4.46), *E/tLT*
λ_s	Thermal conductivity of solid (4.45), *E/tLT*
μ	Fluid viscosity, *cP or* M/Lt; *or* Dipole moment (4.5), *IL* *or* Chemical potential, $E/mole$
μ_0	Chemical potential of the boundary (4.17), $E/mole$
ν	Vibrational frequency of the adsorbate normal to the surface in its adsorbed state (4.11), $1/L$
ν_H	Vibrational frequency of atomic hydrogen within the crystal lattice (5.11), $1/L$
ν_{H_2}	Vibrational frequency of H_2 in the gas phase (5.8), $1/L$
ν_1, ν_2	Vibrational frequency of the adsorbate normal to layer 1 and layer 2, respectively (4.15), $1/L$
ρ	Fluid density, M/L^3
ρ_b	Bed density (4.51), M/L^3
ρ_p	Particle density (4.40), M/L^3
σ	Surface tension, E/L^2
σ_{12}	Lennard-Jones collision diameter $= \frac{1}{2}(\sigma_1 + \sigma_2)$, L
τ	Tortuosity $= L_{eff}/L$ where L_{eff} is the traveled distance through a given porous solid and L is the simple straight-line path through the solid (4.32), *dimensionless*
Φ	Overall potential for a charged system such as zeolite or MOF, *E/mole* *or* Sphericity or the surface-volume ratio for a sphere of diameter d_p (4.71), *dimensionless*
Φ_D	Attractive potential contributed by the dispersion forces between two isolated systems (4.1), *E/mole*
Φ_{LJ}	Lennard-Jones potential (4.3), *E/mole*
Φ_P	Energy contribution from polarization interaction (4.5), *E/mole*
Φ_Q	Energy contribution from field gradient-quadrupole interaction (4.5), *E/mole*

Φ_R	Short-range repulsive energy associated with the finite size of the systems (4.2), *E/mole*
Φ_S	Energy contribution from sorbate-sorbate interactions at the surface at high coverage (4.7), *E/mole*
Φ_μ	Energy contribution from field-dipole interaction (4.5), *E/mole*
ϕ	Solvent association parameter (3.11), *dimensionless* *or* Thiele modulus (4.19), *dimensionless*
ψ_n	H_{OG} factor from Fig. I.3 (3.77)
Ω	A function of Lennard-Jones well depth (ε) (3.8), T

Named dimensionless groups designated with two letters

Bi_h	Biot number for heat transfer
Bi_m	Biot number for mass transfer
Nu	Nusselt number
Pe	Péclet number
Re	Reynolds number
Sc	Schmidt number
Sh	Sherwood number

Chapter 1
Introduction to Carbon Capture

The capture of CO_2 is motivated by the forecasted change in climate as a result of the world's dependence on fossil fuels for energy generation. Mitigation of CO_2 emissions is the challenge of the future for stabilizing global warming.The separation of CO_2 from gas mixtures is a commercial activity today in hydrogen, ammonia, and natural gas purification plants. Typically, the CO_2 is vented to the atmosphere, but in some cases, it is captured and used. The current primary uses of CO_2 include enhanced oil recovery (EOR) and the food industry (carbonated beverages). The traditional approach for CO_2 capture for these uses is solvent-based absorption. It is unclear whether this technology will be the optimal choice to tackle the scale of CO_2 emitted on an annual basis ($\sim$30 Gt worldwide). A new global interest in extending CO_2 capture to power plants is producing a dramatic expansion in R&D and many new concepts associated with clean energy conversion processes. The application of CO_2 capture technologies beyond concentrated sources is in view, but less tractable. The first and second laws of thermodynamics set boundaries on the minimum work required for CO_2 separation. Real separation processes will come with irreversibilities and subsequent inefficiencies taking us further from best-case scenarios. The inefficiency of a given process reveals itself in the form of operating and maintenance, and capital costs.

The interconnected nature of multiple fields of science and engineering is essential. Appreciating the physical and chemical properties of the various CO_2 emissions sources serves as a critical first step. Envisioning methods for its capture will occupy the next several decades of engineers, chemists, physicists, earth scientists, mathematicians, and social scientists. Where will the capture technology be applied? What might the technology involve? Will we one day place systems in the desert to capture dilute CO_2 directly from air? Will we install CO_2 separators on automobiles and airplanes so that consumers exchange CO_2 for fuel at refueling stations? CO_2 capture is often considered in the context of point sources such as coal-fired power plants or industrial processes, but breakthroughs in research must lead to applications of CO_2 capture to an even wider realization of its abatement potential.

Process engineering, materials science, catalysis and nanoscience will likely play key roles in hybridizing the known technologies toward an integrative approach to meet the challenge. Figure 1.1 illustrates the multiple scales that a particular

J. Wilcox, *Carbon Capture*,
DOI 10.1007/978-1-4614-2215-0_1,

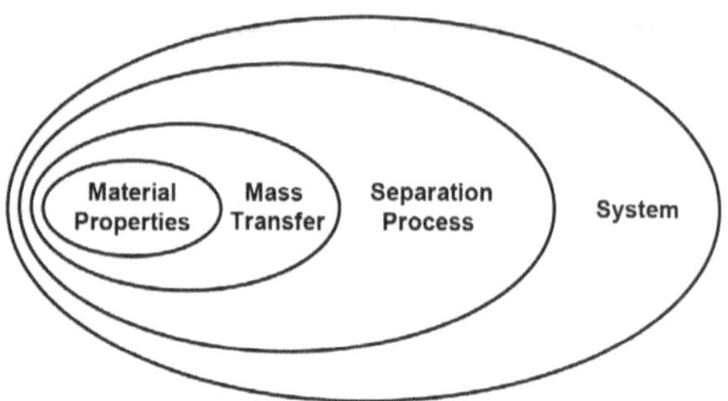

Fig. 1.1 Schematic of the overlap between the scales across the portfolio of solutions to CO_2 capture

solution for CO_2 capture may entail. At the heart of any CO_2 capture technology is the material and its required properties for optimizing mass transfer from a gas mixture to a captured phase. The material may be a solvent, sorbent , membrane or catalyst. Mass transfer acts as the bridge and links the optimal material properties with the overall separation process, whether it is a combined absorber-stripper system, sorption apparatus, membrane module, or catalytic reactor. The material properties, mass transfer, and capture process must be considered as coupled and inherent to the system in total. In this case, the system is defined as the CO_2 capture environment and the local surroundings of the capture assembly.

A CO_2 capture technology may look quite different for the direct air capture of CO_2 (DAC) in which the concentration of CO_2 is quite dilute, *i.e.*, approximately 0.0390 mol% (390 ppm) versus its capture from the exhaust of a power plant, in which its concentration is on average 12 mol%. In addition to the concentration difference, CO_2 capture from a power plant must overcome the challenge associated with the timescale and throughput of emissions. For instance, a 500-MW power plant emits on average 11,000 tons[1] of CO_2 per day. Direct air capture is much less efficient. For example, assuming an air flow rate of 2 m/s, capturing 11,000 tons of CO_2 per day directly from the air requires a surface area of approximately 133,000 m^2 to process 2.31×10^{10} m^3 of air per day.[2]

This chapter provides a brief overview of CO_2 sources, as well as the physical and chemical nature of its environment. This overview motivates further discussion of fuel oxidation and combustion, the heart of CO_2 emissions. Knowledge of the chemical composition of a given exhaust mixture allows for the determination of the minimum work required for CO_2 separation from the mixture. The *2nd-law efficiency* is defined as the ratio between this ideal minimum work and the real work associated with the unit operations of the actual separation process. Known processes relying on absorption, adsorption, membrane, and hybrid processes can then be investigated. Chapter 2 focuses on Compression and Transport of CO_2. Knowledge of CO_2 phase behavior as a function of temperature and pressure throughout compression and transport processes is crucial to determine safe and cost-effective approaches to

[1] Throughout the textbook, "ton" is in reference to a metric ton and is sometimes referred to as tonne, which is equal to 1000 kg.

[2] This assumes 100% capture of the CO_2 at a concentration in air of approximately 390 ppm.

handling CO_2 between the capture and potential-use stages. Once CO_2 is captured and compressed, the question becomes what to do with it. This will be addressed shortly. The phase behavior of CO_2 and fundamental equations associated with its compression also play a role in separations processes. Although compression and transport take place after CO_2 capture, presenting this material prior to the separation processes provides a foundation and set of equations referenced in future chapters.

Chapters 3–6 focus on Absorption, Adsorption, Membrane Technology, and Air Separation, respectively. These broad-reaching chapters are motivated by the operating, maintenance, and capital costs of a given separation process. Costs are bridged in step-wise fashion to the fundamental chemical and physical processes that underlie the mass transfer of CO_2 from a gas to its captured form. In particular, Chapter 6 discusses air separation (*i.e.*, the separation of air into N_2, O_2, and argon) as it may play a key role in CO_2 capture since N_2 is the primary inert gas diluting most CO_2-inclusive gas mixtures.

Chapters 7–9 examine nontraditional separation technologies that in the most ideal sense may be considered carbon-neutral. The topics covered in these chapters include the role that algae plays in CO_2 capture, CO_2 electrocatalysis and photocatalysis to fuel and chemicals, and mineral carbonation, respectively. These are interesting approaches to consider in the CO_2 capture portfolio since currently they are investigated primarily as post-capture processes. For instance, the conversion of CO_2 via algae, electrocatalytic, and photocatalytic processes to a chemical or fuel has mostly been investigated for the conversion of CO_2-rich gas streams as the input. However, with technological advancements it may be possible to use such approaches to combine CO_2 capture and conversion in a single process. Chapter 9 discusses mineral carbonation, focusing on the potential to form mineral carbonates from reacting CO_2 with an alkalinity source. Natural and industrial waste byproduct alkalinity sources are considered, with a particular focus on industrial waste alkalinity sources since the mineral carbonation process would be sequestering CO_2 in addition to other potentially hazardous components of the waste. The potential to reuse the mineral carbonates as aggregate for construction applications may serve as a driver to move this technology forward. These chapters review of each of these processes and their current challenges.

The separation processes considered in Chaps. 3–6 for CO_2 capture may be broken down into their fundamental unit operations with the work requirements determined for each stage of separation, such as blower power for gas compression or pumping power for solvent transport. The reader can work through the application of a capture technology from a gas mixture containing CO_2 and based upon the local environment and composition, determine the feasibility of capture with current separation tools or a hybrid thereof based upon its 2nd-law efficiency.

1.1 Relationship Between CO_2 and Climate

The combustion of fossil fuels, *i.e.*, coal, petroleum, and natural gas, is the major anthropogenic source of CO_2 emissions, with an estimated 30 gigatons (Gt) CO_2 (*e.g.*, 30 billion tons) generated per year as illustrated in Fig. 1.2 [1]. This figure shows the

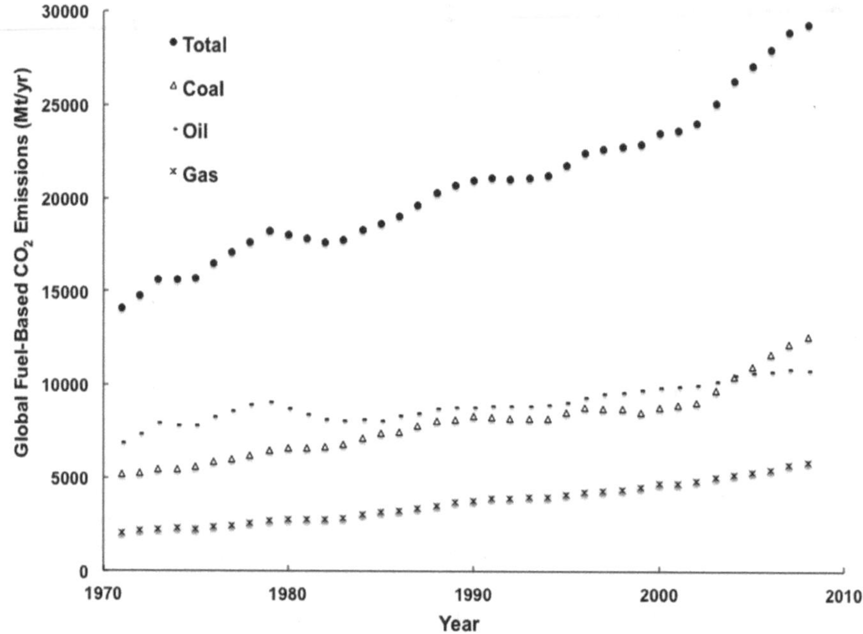

Fig. 1.2 Global fuel-based CO_2 emissions from the period of 1970–2008 (total and by fuel type) [2]

breakdown of CO_2 emissions by fuel type, with coal burning surpassing oil-sourced emissions around 2004. Figure 1.3 shows the breakdown in CO_2 emissions from the top-emitting countries, with China surpassing the U.S. around 2006 [2]. It is clear that human activities over the course of the 20th century, have led to increasing greenhouse gas (GHG) emissions. Studies indicate that increases in CO_2 concentration in the atmosphere leads to irreversible climate changes lasting up to 1,000 years, even after elimination of emissions [3]. Additionally, even if anthropogenic GHG emissions remained constant from today, the world would still experience continued warming for several centuries [4]. For global mean temperature stabilization, GHG emissions would have to cease today [3, 5], which is difficult due to our reliance on the existing fossil-based energy and transportation infrastructure, which is expected to contribute to GHG emissions for tens of years to come [6].

Examples of irreversible climate changes include atmospheric warming, dry-season rainfall reductions in several regions comparable to those of the "dust bowl" era[3], more extreme weather events and inexorable sea level rise [3]. The carbon-climate response, which is defined as the ratio of temperature change to cumulative

[3] The "dust bowl" era from 1930–1936 was a period of dust storms causing major ecological and agricultural damage to the prairie lands of the U.S. (panhandles of Texas and Oklahoma, and neighboring regions of New Mexico, Colorado, and Kansas) and Canada, causing severe drought followed by decades of extensive farming without crop rotation or other techniques to prevent wind erosion. Millions of acres of farmland became useless, with hundreds of thousands of people leaving their homes and migrating to California and other states.

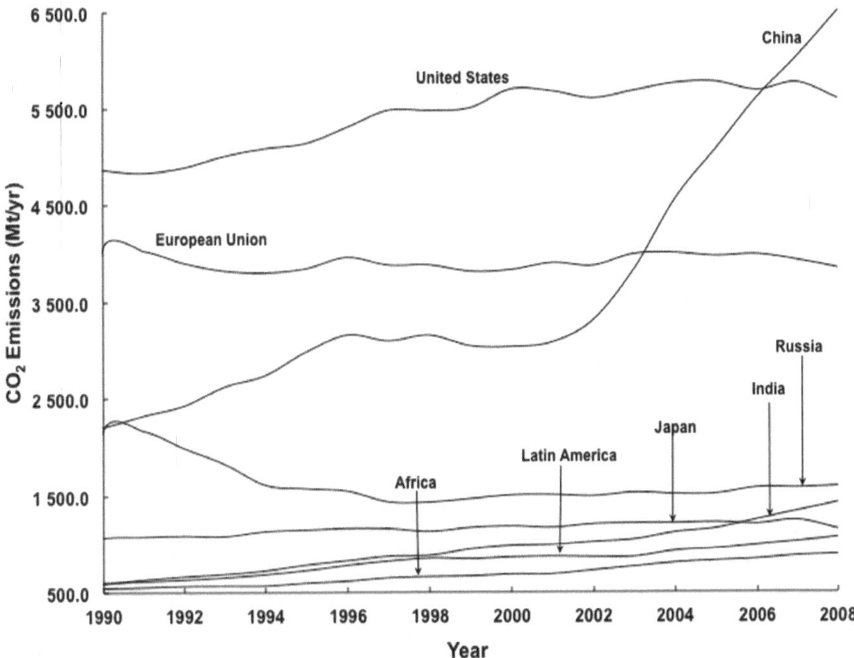

Fig. 1.3 Global fuel-based CO_2 emissions by top-emitting countries from 1990 to 2008 [2]

CO_2 emissions is estimated at approximately 1.0–2.1°C per 3600 Gt of CO_2 emitted into the atmosphere [7]. The natural carbon cycle is responsible for a portion of the removal of CO_2 from the atmosphere, some to the oceans and some to terrestrial vegetation. The natural carbon cycle includes exchange with the land biosphere (photosynthesis and deforestation), the surface layer of the ocean, and a much slower penetration into the depth of the ocean, which is dependent upon vertical transport and the buffering effect of the ocean's chemistry [3]. The carbon flows associated with global photosynthesis and respiration are generally an order of magnitude larger in scale than the extraction and combustion of fossil fuel, but are largely in balance. With updated estimates on CO_2 emissions from both deforestation and coal combustion, estimated emissions from deforestation and forest degradation represent approximately 12 to 13% of global anthropogenic CO_2 emissions [8], different from previous estimates of 20% [9]. In addition to CO_2, other GHGs include methane (CH_4), nitrous oxide (N_2O), hydrofluorocarbons (HFCs), perfluorocarbons (PFCs), and sulfur hexafluoride (SF_6). Researchers predict that non-CO_2 GHG emissions will constitute approximately[4] one-third of total CO_2 equivalent emissions over the 2000–2049 period [10]. This textbook, however, focuses primarily on the separation of CO_2, which remains the primary anthropogenic contributor to climate change [11].

[4] Based on 100-year global warming potentials.

Table 1.1 Estimated 2011 U.S. CO_2 exhaust emissions [2]

Source	Emissions sector (Mt CO_2/year)						Exhaust mixture
	Trans	Elec.	Comm.	Ind.	Res.	*Total*	
Petroleum	1952.68	102.30	54.93	416.83	100.91	2627.64	CO_2 (3–8 vol%) SO_2; NO_X; PM; CO
Natural gas	33.1	319.11	163.06	397.54	262.42	1175.24	CO_2 (3–5 vol%) NO_x; CO; PM
Coal	~0	1983.81	9.28	187.99	0.81	2181.89	CO_2 (12–15 vol%) PM; SO_2; NO_X; CO; Hg
Total	1985.79	2405.22	227.26	1002.6	364.14	5984.77	

Table 1.2 Estimated 2008 U.S. and worldwide CO_2 process emissions [15]

Non-Fossil	Emissions (Mt CO_2/year)	CO_2 content (vol %)
Cement production	50 (world 2000)	14–33
Refineries	159 (world 850)	3–13
Iron and steel production	19 (world 1000)	15
Ethylene production	61 (world 260)	12
Ammonia processing	7 (world 150)	100
Natural gas production	30 (world 50)	5–70
Limestone consumption	19	50
Waste combustion	11 (electricity)	20
Soda ash manufacture	4	
Aluminum manufacture	4 (world 8)	

1.2 CO_2 Sources and Sinks

The combustion of fossil fuels produces an estimated 30 Gt of CO_2 per year. Deforestation of tropical regions accounts for an additional 4 Gt CO_2 per year [8]. Due to the natural carbon cycle and associated terrestrial and ocean CO_2 sinks, the annual increase in CO_2 emissions averages approximately 15 Gt CO_2 per year, which is equivalent to 2 ppm per year. Fossil fuel-based emissions of CO_2 may be sourced from both stationary (*e.g.*, power plant) and non-stationary systems (*e.g.*, automobile). There are approximately 13 Gt CO_2 per year on average from large stationary sources globally, and approximately 2.5 Gt in the U.S. as illustrated in Table 1.1 [2]. In addition to CO_2 emissions generation from the oxidation of fossil fuels, emissions may also be sourced as a result of a chemical process. Although these emissions sources represent a minor portion of total anthropogenic emissions, the chemical processing method currently used may be required for the formation of a useful product, such as cement or steel. Therefore, due to the difficulty of replacing CO_2-generating chemical processes with others that are absent of CO_2, it is crucial that these emissions sources are not disregarded.

Chemical Processes Although fossil-fuel oxidation processes produce the majority of emissions, there is a fraction of emissions generated from chemical processes. Examples presented in Table 1.2 [15] include cement manufacturing, iron and steel production, gas processing, oil refining, and ethylene production. Mitigation associated with the capture of CO_2 from these industrial-based processes is small compared to that of the transportation and electricity sectors; however, there may not be alternatives to the materials created from these processes, such as cement, iron and steel production, etc. Several of these processes are discussed briefly.

Cement manufacturing results in CO_2 emissions sourced from *calcination* in addition to the fuel combustion emissions of the cement kilns. Estimated 2008 emissions worldwide from cement production were approximately 2 Gt of CO_2 with approximately 52 and 48% associated with the calcination process and cement kilns, respectively. Portland cement is a mixture of primarily di- and tricalcium silicates ($2CaO \cdot SiO_2$, $3CaO \cdot SiO_2$) with lesser amounts of other compounds including calcium sulfate ($CaSO_4$), magnesium, aluminum, and iron oxides, and tricalcium aluminate ($3CaO \cdot Al_2O_3$). The primary reaction that takes place in this process is the formation of calcium oxide and CO_2 from calcium carbonate, which is highly endothermic and requires 3.5–6.0 GJ per ton of cement produced.

Within the steel-making process a combination of emissions and the chemical processes comprise the estimated 1 Gt of CO_2 emitted worldwide. Steel-making, generates CO_2 as a result of carbon oxidation to carbon monoxide, which is required for the reduction of hematite ore (Fe_2O_3) to molten iron (pig iron). The CO_2 is also sourced from a combination of coal-burning and limestone calcination. In the second stage of the steel-making process, the carbon content of pig iron is reduced in an oxygen-fired furnace from approximately 4–5% down to 0.1–1%, and is known as the basic oxygen steelmaking (BOS) process. Both of these steps produce a steel-slag waste high in lime and iron content.

In 2008, roughly 630 refineries emitted on average 1.25 Mt each, resulting in approximately 850 Mt of CO_2 emitted in the atmosphere. In an oil refining process, crude oil, a mixture of various hydrocarbon components ranging broadly in molecular weight, is fractionated from lighter to heavier components. In a second stage, the heavier components are catalytically "cracked" to form shorter hydrocarbon chains. In addition to producing CO_2 as a byproduct of the distillation and catalytic cracking processes, the heat and electricity required for the methane reforming process used in H_2 production for hydrocracking and plant utilities produce additional CO_2.

Recovered natural gas from gas fields or other geologic sources often contains varying levels of nonhydrocarbon components such as CO_2, N_2, H_2S, and helium. Natural gas (primarily methane and ethane) and other light hydrocarbons such as propane and butane, to less extent, are the valuable products in these cases, and often the generated CO_2 is a near-pure stream. Concentrations of approximately 20% CO_2 are not uncommon in large natural gas fields. A unique case is Indonesia's Natuna field, which commercially produces natural gas containing approximately 70% CO_2 [12].

Exhaust Emissions Comparing the sectors (electricity, transportation, industrial, commercial, and residential), currently the electricity sector is the largest, representing 40% of total U.S. CO_2 emissions. Among all the sectors, comparing the different fossil fuel sources, *i.e.*, coal, petroleum, and natural gas, petroleum constitutes the majority of the emissions at approximately 43%. Carbon capture technologies and the appropriateness of their application are highly dependent upon the following four factors: 1) the nature of the application, *i.e.*, a coal-fired power plant, an automobile, air, etc., 2) the concentration of CO_2 in the gas mixture, 3) the chemical environment of CO_2, *i.e.*, the presence of water vapor, acid species (SO_2, NO_x), particulate matter (PM), etc., and 4) the physical conditions of the CO_2 environment, *i.e.*, the temperature and pressure. A brief discussion of each of these factors highlights their importance.

The concentration of CO_2 plays a role in that the work required for separation decreases as the CO_2 concentration increases. The greater the CO_2 content in a given gas mixture, the easier it is to carry out the separation. This concept will be revisited toward the end of the chapter. If the CO_2 concentration in a gas mixture is too low then certain separation processes may be ruled out under their current design. For instance, in order for membrane technologies to be effective, a sufficient driving force is required. One of the many benefits of membrane technology is reduced capital cost. Membranes are a once-through technology in that the gas mixture enters the membrane in one stream and leaves the membrane as two streams, with one of the streams concentrated in CO_2. But since most of the sources of CO_2 are fairly dilute as shown in Table 1.1, this technology currently has limited application unless novel measures are taken. The chemical-process-based sources of CO_2 tend to have higher concentrations making these processes targets for membrane technology application. Examples include ammonia, hydrogen, and ethanol production facilities.

In addition, the chemical environment of CO_2 is important when considering the separation technology since some technologies may have a higher selectivity to other chemical species in the gas mixture. For instance, in coal-fired flue gas water vapor and acid gas species such as SO_2 and NO_x compounds may compete with CO_2 for binding in solution or on a material.

The final factor to consider is the temperature and pressure of the potential CO_2 application. If a process occurs at high temperature or pressure it may be possible to take advantage of the work stored at the given conditions for use in the separation process. For instance, a catalytic reaction involving CO_2 may be enhanced at high temperature. An example of a catalytic approach for flue gas scrubbing is the case of NO_x reduction to water vapor and N_2 from the catalytic reaction of NO_x with ammonia across vanadia-based catalysts. This approach to NO_x reduction in a power plant is referred to as selective catalytic reduction. Noncatalytic NO_x reduction, in which ammonia is injected directly into the boiler is also practiced, but is not as effective as the catalytic approach [13]. Membrane technology is another example, in that separation may be enhanced at high pressure.

Carrying out an exergy analysis [14] on a given CO_2 generation source and capture process is a useful exercise, which can aid in determining the potential irreversibilities associated with each step. Exergy is defined as the amount of energy in a process that is potentially available to do work. For instance, if one were to design a CO_2 separation process that effectively used the thermal energy available at the high temperature of the flue gas then this would be maximizing the exergy in the system. The flue gas temperature ranges from approximately 650°C at the exit of the boiler

Table 1.3 Current CO_2-EOR projects taking place in the U.S.[a] [17]

Location	CO_2 sources	CO_2 supply (MMcfd[b])	
		Natural	Anthropogenic
W. Texas/New Mexico/Arizona	Natural CO_2 (Colorado/New Mexico) Gas processing plants (W. Texas)	1670	105
Colorado/Wyoming	Gas processing plant (Wyoming)	–	230
Mississippi/Louisiana	Natural CO_2 (Mississippi)	680	–
Michigan	Ammonia plant (Michigan)	–	15
Oklahoma	Fertilizer plant (Oklahoma)	–	30
Saskatchewan	Coal gasification plant (North Dakota)	–	150
Total (MMcfd CO_2)		*2350*	*530*
Total (Mt CO_2)		*45*	*10*

[a]MIT EOR report estimates 65 Mt CO_2 for EOR annually [18]
[b]MMcfd stands for one million cubic feet per day of CO_2 and may be converted to Mt CO_2 per year by dividing by 18.9 Mcf per metric ton and multiplying by 365

down to approximately 40–65 °C at the stack. Current technologies such as absorption and adsorption are exothermic processes that are enhanced at low temperatures and in a traditional sense are not effective strategies for taking direct advantage of the exergy of the high-temperature flue gas. For instance, the capture of CO_2 is most effective at low temperature for absorption and adsorption processes with the regeneration of the solvent or sorbent most effective at elevated temperatures. Membrane separation and catalytic-based technologies, may however, be enhanced at the elevated temperatures (and pressures) available at exit boiler or gasifier conditions, depending on the specific technology.

CO_2 Usage and Sinks The primary use of CO_2 is for EOR, which has been taking place for several decades, beginning with the Permian Basin located in West Texas and neighboring area of southeastern New Mexico, underlying an area of approximately 190,000 km^2. The primary source of CO_2 for these activities include the natural CO_2 reservoirs in Sheep Mountain and the McElmo dome, both located in Colorado, and Bravo dome in New Mexico. The CO_2 gas is transported to the fields of the Permian Basin through pipeline networks. The existing networks of CO_2 pipeline in the U.S. are discussed in further detail in Chap. 2, Compression and Transport of CO_2. Currently, on average, CO_2-EOR provides the equivalent of 5% of the U.S. crude oil production at approximately 280,000 barrels of oil per day [16]. A limitation of reaching higher EOR production is the availability of CO_2. Natural CO_2 fields provide approximately 45 Mt CO_2 per year, with anthropogenic sources slowly increasing (currently 10 Mt CO_2 per year) as illustrated by the list of current activities in Table 1.3 [17].

Although the discussion of CO_2-EOR thus far has been centered on the usage of CO_2, EOR may also be a viable approach to potentially store CO_2. For instance, the CO_2 used for EOR is not completely recovered with the oil. In fact, only 20–40% of the CO_2 injected for EOR is produced with the oil, separated, and reinjected for additional production [18]. To date, EOR has not had any financial incentive to maximize

CO_2 left below ground. In fact, since the cost of oil recovery is closely coupled to the purchase cost of CO_2, extensive reservoir design efforts have gone into minimizing the CO_2 required for enhanced recovery. If, on the other hand, the objective of CO_2 injection is to increase the amount of CO_2 left underground while recovering maximum oil, then the approach to the design question changes. If there were such an incentive, likely an even larger fraction would stay below ground, via modifications of EOR. Researchers at Stanford University have investigated the co-optimization of CO_2 storage with enhanced oil recovery [19]. Their investigations conclude that traditional EOR techniques such as injecting CO_2 and water in a sequential fashion (*i.e.*, water-alternating-gas process) are not conducive to CO_2 storage. A suggested modified approach includes a well-control process, in which wells producing large volumes of gas are closed and only allowed to open as reservoir pressure increases. In addition to co-optimization of CO_2 storage with EOR, ongoing efforts exist for coupled CO_2 storage with enhanced coalbed methane recovery (ECBM) [20] and potentially enhanced natural gas recovery from gas shales [21].

As previously discussed, postcombustion capture of CO_2 has taken place commercially for decades, primarily for the purification of gas streams other than combustion products. Amine use for CO_2 capture was first patented in 1930 [22] and later in 1952 [23] for the purification of hydrocarbons. Since these times, its primary use has been to meet purity specification requirements for natural gas distribution and the food and beverage industry. Table 1.4 lists some selected power plant and industrial facilities that capture, transport, and store (temporarily in the case of the food industry) CO_2 in an integrated system. Comparing Tables 1.3 and 1.4, it is interesting to notice the difference in scale of CO_2 usage for EOR versus other industries. The Bellingham Cogeneration Facility located in Massachusetts generates electricity and uses a Fluor Econamine FG^{SM} regenerable solvent process and is capable of recovering 85–95% of the flue gas CO_2 for food-grade CO_2 at 95–99% purity. It is important to recognize that usage of CO_2 in the food industry is not a mitigation option as the CO_2 is subsequently reemitted into the atmosphere, yet the usage of CO_2 continues to drive the advancement of the separation technologies.

One might consider the option of converting CO_2 into additional useful products such as carbonates as previously discussed, but taking a look into the current scale of the worldwide chemical industry, one quickly recognizes that even if CO_2 was converted to useful products on the scale of every chemical produced worldwide, this would still constitute less than 5% of the current fossil-based CO_2 emissions. Consider the scale of CO_2 emissions associated with each of the primary energy resources. The annual emissions[5] generated from coal, petroleum, and gas are on the order of 13, 11, and 6 Gt CO_2, respectively [2]. Collectively, the emissions associated with the oxidation of fossil-based energy resources are on the order of 30 Gt CO_2 per year. Now, consider the top chemicals produced worldwide [24], which are shown in Table 1.5.

Lime, sulfuric acid, ammonia, and ethylene production are on the order of 283, 200, 154, and 113 million tons in 2009, respectively. If one could capture the CO_2

[5] Data from 2008, IEA: Coal 12595 Gt CO_2; Oil 10821 Gt CO_2; Gas 5862 Gt CO_2.

Table 1.4 Selected commercial postcombustion capture processes at power plants and industrial facilities [16]

Project name and location	Plant type	Startup year	Capacity (MW)	Solvent	CO_2 captured (Mt/yr)
Projects located in the U.S.					
IMC Global Inc. Soda Ash Plant (Trona, CA)[a]	Coal and petroleum coke-fired boilers	1978	43	Amine (Lummus)	0.29
Bellingham Cogeneration Facility (Bellingham, MA)	Natural gas-fired power plant	1991	17	Amine (Fluor)	0.11
Projects located outside the U.S.					
Soda Ash Botswana Sua Pan Plant (Botswana)	Coal-fired power plant	1991	17	Amine (Lummus)	0.11
Statoil Sleipner West Gas Field (North Sea, Norway)	Natural gas separation	1996	NA	Amine (Aker)	1.0
BP Gas Processing Plant (In Salah, Algeria)	Natural gas separation	2004	NA	Amine (multiple)	1.0
Mitsubishi Chemical Kurosaki Plant (Kurosaki, Japan)	Natural gas-fired power plant	2005	18	Amine (MHI)	0.12
Snøhvit Field LNG and CO_2 Storage Project (North Sea, Norway)	Natural gas separation	2008	NA	Amine (Aker)	0.7

[a] CO_2 is captured from boiler flue gases and is used to carbonate brine for soda ash (sodium carbonate) production; soda ash is primarily used as a water softener, but has other uses associated with pH regulation

Table 1.5 Approximate production of top 10 chemicals produced U.S. and worldwide in 2009 [24]

Chemical	U.S. (Mt)	World (Mt)
1. Sulfuric acid	38.7	199.9
2. Nitrogen[a]	32.5	139.6
3. Ethylene	25.0	112.6
4. Oxygen[a]	23.3	100.0
5. Lime	19.4	283.0
6. Polyethylene	17.0	60.0
7. Propylene	15.3	53.0
8. Ammonia	13.9	153.9
9. Chlorine	12.0	61.2
10. Phosphoric acid	11.4	22.0

[a] N_2 and O_2 are both sourced from air and each have unique markets

Table 1.6 World CO_2 storage capacity estimates for several geologic formations [15b, 26]

Geologic formation	Worldwide (Gt CO_2)
Deep saline aquifers	1,000–10,000[a]
Depleted oil and gas fields	200–900
Unmineable coalbeds	100–300

[a] Source: IPCC (2007), deep saline aquifer capacity is noted as uncertain

at scale and turn it into useable chemicals equating to the current market of these top chemicals produced, this would mitigate a mere 2.5% of the CO_2 generated worldwide annually. This provides insight into the staggering scale of the fossil-based CO_2 emissions.

Although this book is focused solely on CO_2 capture, it is important to recognize the uses of CO_2 that might aid in advancing capture technologies. Additionally, if CO_2 is captured on the scale that is anticipated required for minimizing negative climate change impacts, the amount of CO_2 captured will be far greater than the current market usage. Carbon dioxide capture and storage (CCS) is expected to be a primary component of the portfolio of options for mitigating CO_2 emissions at the required scale [15b, 25].

Storage possibilities include the geologic formations: deep saline aquifers, depleted oil and gas fields, and unmineable coalbeds. Table 1.6 lists the CO_2 storage capacity estimates for these primary types of geologic formations considered. The accurate determination of the storage estimates presented in Table 1.6 is difficult, which is why the estimates range so broadly [27]. The geological nature of a storage site can be quite complex and each site can vary considerably.

The market for CO_2 use and the storage potential for CO_2 exist. The key is to determine the most effective portfolio of solutions [28] for the variety of emissions scenarios that exist to capture CO_2 for the existing markets. Steam-based electricity generation was not invented with the anticipation of capturing CO_2; therefore, it is reasonable to return to the beginning and to understand the formation pathway of CO_2 from the oxidation of fossil fuel and to possibly consider alternative approaches to energy generation that could result in higher plant efficiencies with the inclusion of CO_2 capture.

1.3 Formation Pathways of CO_2

As discussed previously, the primary sources of CO_2 release into the atmosphere are listed in Table 1.1, with the majority of the CO_2 produced from the oxidation of fossil fuels. Understanding CO_2 capture technologies and envisioning solutions to this challenge requires knowledge of how the CO_2 is originally generated. This involves some background on fuel oxidation mechanisms, combustion science, and in general fuel-to-energy conversion processes. A more detailed description of the basic fundamentals of combustion processes is available in Appendix A.

Coal Oxidation Coal is a sedimentary rock formed of fossilized vegetation of different types. Coal may be sourced deep underground (average depths of approximately 600 ft) or close to the surface, having taken up to 400 million years to form under varying temperature and pressure conditions. The four major classes of coal, also known as *rank*, are lignite, subbituminous, bituminous, and anthracite, in order of youngest to oldest. The lowest ranked coal (lignite) has the lowest carbon content (~60 mass%) and is highest in volatiles and moisture content, compared to the highest rank coal, anthracite (carbon content ~90 mass%) [29]. Coal is a complex amorphous mixture of carbon, hydrogen, oxygen, sulfur, nitrogen, moisture, ash, and trace metals. The ideal chemical composition of coal's three principal elements can be written as CH_mO_n. Usually, $m < 1$ and $m < n$. Coal primarily consists of carbon, hydrogen, and oxygen, but may also consist of nitrogen and sulfur, which lead to the formation of NO_x and SO_x, respectively. Additionally, coal contains volatile trace metals that evolve from the coal at the high temperature conditions of the boiler. The trace metals, mercury, selenium, and arsenic are present at ppb levels in most flue gases [30]. However, it is important to note that depending on the type of coal and source from which the coal was mined, the chemical constituents may vary significantly. The chemical composition of coal varies greatly depending upon its rank and origin. Although coal combustion is a complex process since all coal is unique, in general it is consistent in that initially moisture and volatiles are driven off, followed by direct carbon oxidation.

Coal combustion is a heterogeneous reaction involving the oxidation of coal through the transport to and subsequent reactivity of O_2 with its surface. Heterogeneous implies that two phases are taking part in the reaction, *i.e.*, the solid coal surface and oxygen in the gas phase. The coal is comprised of pulverized porous particles ranging in size from 75–300 μm depending upon the boiler type. Gas-phase species reacting with the coal surface must diffuse through the intricate pore network within the coal particles as depicted by the scanning electron microscopy image in Fig. 1.4. In reality, the coal surface may be oxidized by a combination of species present in the gas phase, that is, O_2, CO_2, and H_2O proceeding by the following

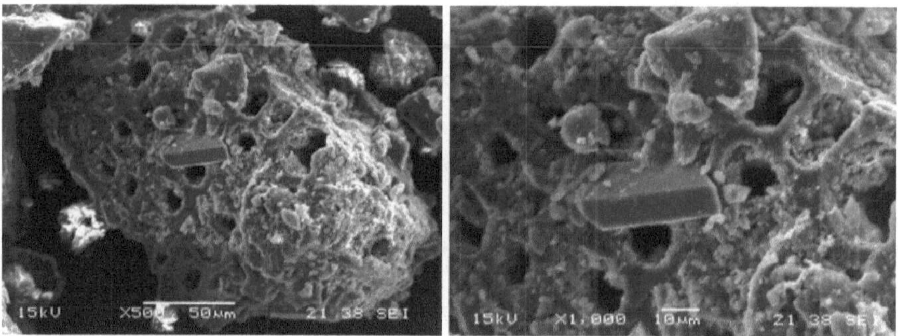

Fig. 1.4 Scanning electron microscopy images of subbituminous coal. The image on the right is a magnification of the particle on the left, both on the micron scale. (Courtesy of [1])

global reactions:

$$C + O_2 \rightarrow CO_2 \tag{1.1}$$

$$2C + O_2 \rightarrow 2CO \tag{1.2}$$

$$C + CO_2 \rightarrow 2CO \tag{1.3}$$

$$C + H_2O \rightarrow CO + H_2 \tag{1.4}$$

A global reaction is one that is comprised of a series of *elementary reactions*, which are reactions that proceed as they are written. For instance, Reaction (1.1) is not elementary since the formation of gas-phase CO_2 in this case would involve a series of elementary steps such as O_2 adsorption, O_2 bond stretching, CO bond forming, etc. At the high temperatures of coal combustion, a dominating reaction is heterogeneous carbon oxidation to the formation of CO, followed by homogeneous CO oxidation by O_2 to CO_2.

Coal-to-Electricity Conversion Just under 50% of the electricity generated in the U.S. is powered by coal-fired steam power plants. Figure 1.5 is a simplified version of a power plant with the basic components responsible for generating power. Understanding the coal-to-electricity conversion process allows one to appreciate the life cycle of CO_2 from the oxidation of the coal's carbon surface to the exit of the stack. Figure 1.5 also shows how other scrubbing technologies have been arranged downstream of the boiler exit. For instance, NO_x reduction (*i.e.*, mitigation) is taking place here using a high-temperature catalyst so its placement is at the boiler exit to maximize the thermal energy required for enhanced conversion. The removal of SO_x with an absorption set-up is placed at the lowest temperature available since absorption is an exothermic process. For CO_2 capture, an additional scrubbing unit would be placed in this plant configuration, with its exact location dependent upon the type of separation process used (*e.g.*, absorption, adsorption, membrane, or catalytic).

The steps of coal conversion to electricity are complex in reality, but are simplified here for a general understanding of the process:

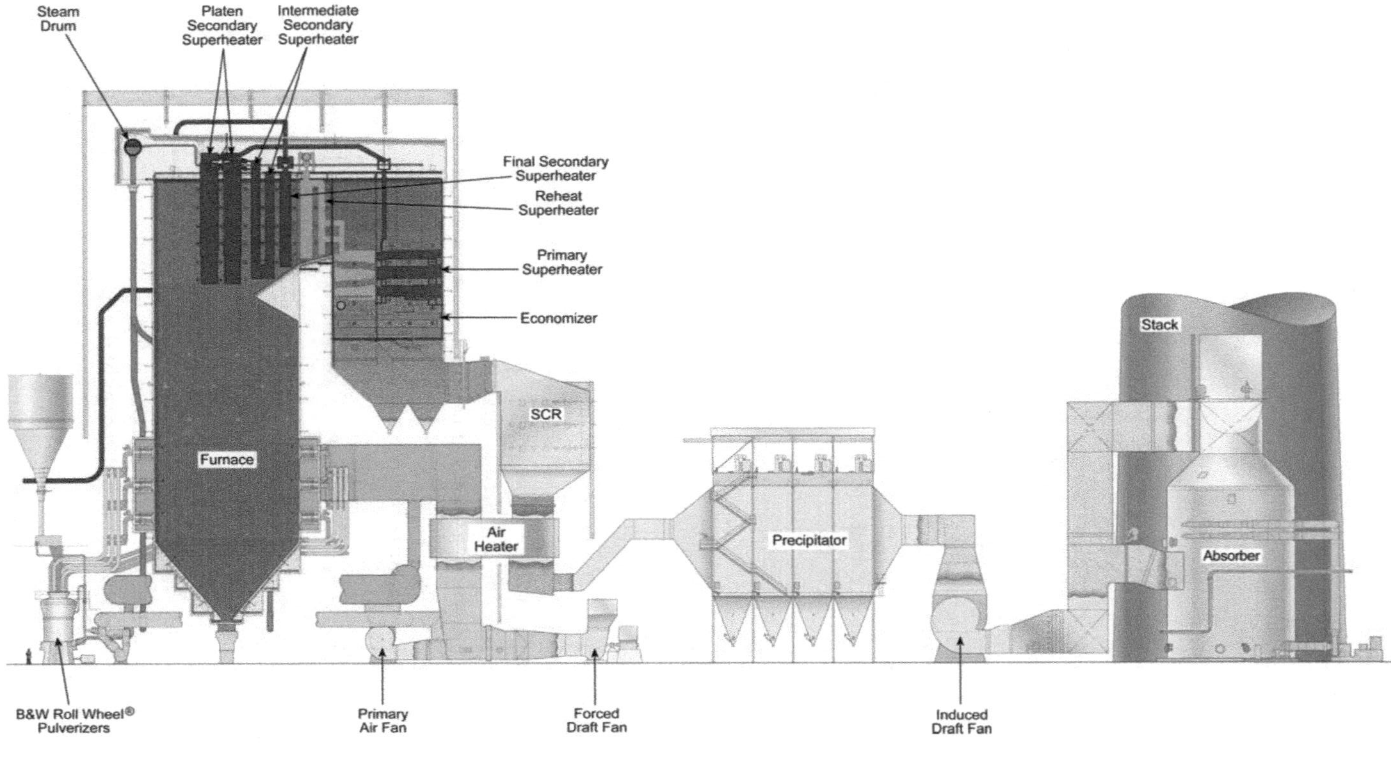

Fig. 1.5 Coal-fired power plant (Courtesy of The Babcock & Wilcox Company)

1. During startup, the furnace is pre-heated by combustion of auxiliary fuel such as natural gas or oil;
2. Pulverized coal powder is blown with air into a combustion chamber (boiler or furnace) through a series of nozzles; combustion of the coal particles creates hot combustion products;
3. Heat is transferred from the hot combustion products to water circulating in tubes along the boiler walls; this produces superheated steam, which is the working fluid for the steam turbines;
4. Pumps are used to increase the pressure of the working fluid;
5. Energy from the hot and pressurized steam is extracted in steam turbines that then transmit the energy to electric generators;
6. The electric generators convert the shaft work of the turbines into alternating current electricity;
7. Heat exchangers are used to condense the energy-drained steam from the turbines;
8. Pumps are used to return the condensed water to the boiler, where the cycle is then repeated; and
9. Pollution control devices are used to scrub the flue gas of NO_x (selective catalytic reduction), particulate matter (electrostatic precipitators and/or fabric filters), and SO_2 (calcium-based flue gas desulfurization units or lime spray dryers). Some power utilities are also equipped with activated carbon injection processes to capture mercury emissions; currently there are no full-scale CO_2 capture methods in place.

Liquid Fuel and Natural Gas Oxidation Converting petroleum to power requires evaporation and burning of a liquid fuel. Applications include diesel, rocket, and gas-turbine engines, oil-fired boilers and furnaces. In liquid fuel oxidation, the fuel is first vaporized and then combusted. Gas combustion may occur with or without flame, and flames are usually characterized as premixed or diffusion flames. In flame combustion, a reaction zone (flame) propagates through an air-fuel mixture where the hot combustion products are left behind the flame with temperature and pressure rising in the unburned fuel. In flameless combustion, rapid oxidation reactions occur throughout the fuel leading to very rapid combustion. The volumetric exothermicity that takes place in an engine is called autoignition. Premixed versus diffusion flames are characterized by the level of mixing that takes place between the fuel and oxidizing agent. A spark-ignition engine is an example in which the fuel and oxidizing agent are mixed prior to any combustion activity. On the other hand, within a diffusion flame the fuel is initially isolated from the oxidizing agent and the combustion reaction takes place simultaneous to mixing with flame propagation at the interface of the fuel and oxidant.

There are clear distinctions between the oxidation processes of coal, petroleum, and gas and subsequent differences between the mechanisms of CO_2 generation in each case. Understanding more thoroughly the nature by which CO_2 is formed may lead to advancements in fuel-to-energy conversion processes that minimize its generation. The capture of CO_2 in a traditional pulverized coal combustion process is termed *postcombustion capture* (PCC), since capture is taking place after the

Table 1.7 Approximate efficiencies of various plants with and without CO_2 capture (CC) [31]

Plant type	Plant efficiency w/out CC (%)[a]	Plant efficiency w/CC (%)
Coal, subcritical	33–39	23–25
Coal, supercritical	38–44	29–31
Coal, ultrasupercritical	43–47	34–37
NGCC[b]	45–51	38–43
IGCC[c]	37–44	32–39

[a]Plant efficiency is based upon the high heating value (HHV) of the fuel
[b]NGCC refers to a natural gas combined cycle power plant
[c]IGCC refers to an integrated gasification combined cycle power plant

combustion process. Another possibility is to modify the fuel-to-energy conversion process to maximize the concentration of CO_2 as to minimize the work associated with its separation.

1.4 Advanced Coal Conversion Processes

When fuel oxidation was first carried out for energy generation there was no intention to capture the CO_2 generated from the process. Air being the primary source of fuel oxidation is approximately 78% N_2, 21% O_2, and 0.95% Ar by volume on a dry basis, *i.e.*, excluding moisture content (plus trace amounts of CO_2, Ne, He, Kr, and Xe) [30]. The N_2 is predominantly an inert gas throughout the combustion process, thereby diluting the CO_2 generated in the flue gas stream and increasing the work required for CO_2 separation. Table 1.7 shows the difference in the efficiency with and without CO_2 capture for various plants.

Advanced coal conversion processes[6] are currently under development that reduce the work required for separation by creating CO_2-concentrated gas outlet streams. These include coal gasification, oxycombustion, and chemical looping combustion. These processes in fact, could be carried out on any fossil-based energy resource. Coal is primarily discussed since this is the most common energy resource available worldwide. Figure 1.6 shows the pathway of each advanced energy conversion process with the energy resource options as coal, biomass, waste, petroleum coke/residue or natural gas.

Gasification Coal gasification is a process in which the oxidation of coal is kept at a minimum, with just enough exothermicity to provide the required energy for driving the gasification reactions. The heat is controlled through the control of air

[6] Although not discussed specifically, another energy conversion option is electrochemical conversion in a direct carbon fuel cell. Challenges are associated with the accessibility of the oxidizer to the electrochemical reaction sites, but progress continues to be made in this field. [51, 52] Electrochemical conversion processes are described in more detail in Chapter 8, but are focused on CO_2 reduction toward fuel synthesis, in which energy (renewable) is required as an input, rather than direct carbon (*e.g.*, coal, biomass, etc.) oxidation toward energy production.

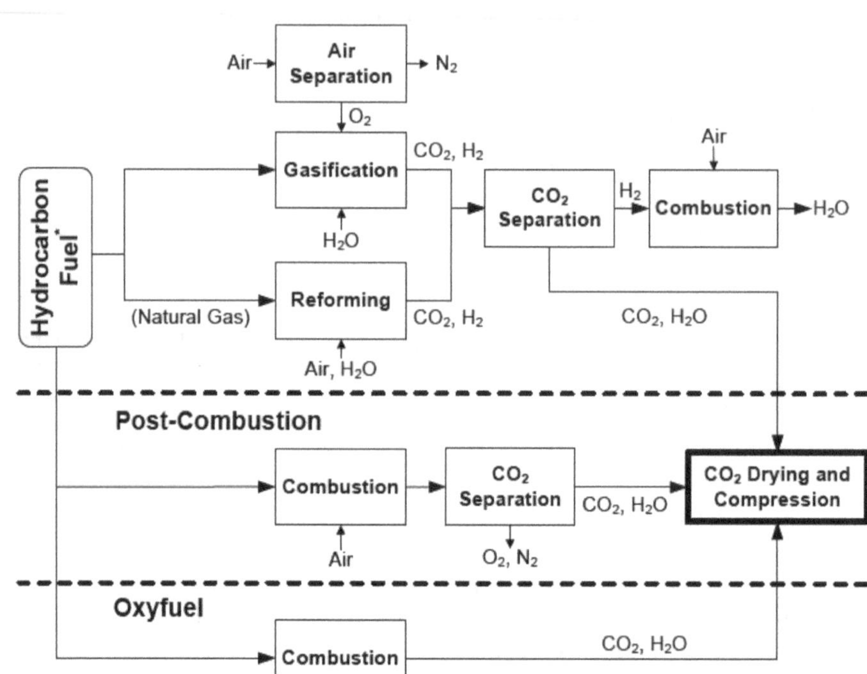

Fig. 1.6 Schematic of primary options for CO_2 capture from various hydrocarbon-based energy conversion processes

or more often, oxygen input into the gasifier. A limitation of gasification is the need for an air separation unit (ASU) for the generation of high-purity O_2 as a feed gas to the gasifier. The gasification process suppresses the formation of water and instead produces primarily CO and hydrogen gas (H_2). IGCC systems operate at high pressures (*e.g.*, 500–700 psia) and require the oxidant stream to also be pressurized. Gasification takes place rapidly at temperatures above 1260°C, which is greater than the ash fusion temperature, allowing ash to become molten and separating easily from the gas. In addition to H_2 and CO, CH_4 is generated in small amounts and H_2S is also generated depending upon the extent of sulfur present in the energy resource. The hot and pressurized synthesis gas exits the gasifier and a particulate control device then removes particulate matter, after which steam is added to the fuel gas (also known as synthesis or syngas) to promote the conversion of H_2 and CO_2, which is called the water gas shift reaction, *i.e.*,

$$CO + H_2O \leftrightarrow CO_2 + H_2 \tag{1.5}$$

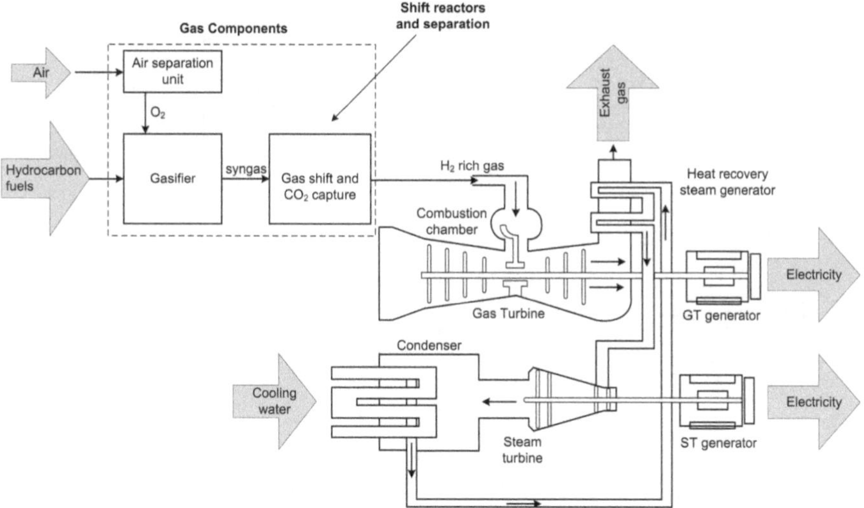

Fig. 1.7 Detailed schematic of an integrated gasification combined cycle (IGCC) plant

and is exothermic by approximately 41 kJ/mol CO_2 generated. The reaction is equilibrium-limited, leading to an increase in conversion as temperature decreases. The current industrial approach is to cool and clean the fuel stream before entering a high-temperature shift reactor at approximately 315–445°C.This step is sometimes followed by a low-temperature shift reactor that operates at approximately 200–250°C. The equilibrium can be shifted using a catalytic metallic membrane reactor selective to the removal of H_2 on the right hand side of Reaction (1.5). These types of reactors will be discussed in more detail in Chap. 5 on Membrane Technology. Shifting the equilibrium to optimize conversion is preferred, since this limits the cooling required of the gas stream.

The concentration of CO_2 in this process is substantially greater than in coal combustion making its separation from the fuel gas mixture easier. The capture of CO_2 in an IGCC process is termed *precombustion capture* since the fuel combustion takes place after the capture process as demonstrated in Fig. 1.7. For electricity generation, the synthesis gas (largely hydrogen) is burned directly and then passed through a gas turbine for electricity generation. The heat recovered from this process is used to generate steam, which is passed through a steam turbine for additional electricity generation, resulting in a combined cycle. The efficiency of an IGCC plant is on the order of 37–44%, compared to a newer existing ultrasupercritical pulverized coal-fired power plant, which is on the order of 43–47%. It is important to recognize the competition that will likely exist between steam and gas processes as technologies are advanced (*i.e.*, turbine technology) toward the handling of gas. Although the cost of electricity of a traditional coal combustion power plant is lower than that of an IGCC plant, with the inclusion of CO_2 capture, the efficiency is higher (see Table 1.7) and hence the cost is less in the gasification case [33].

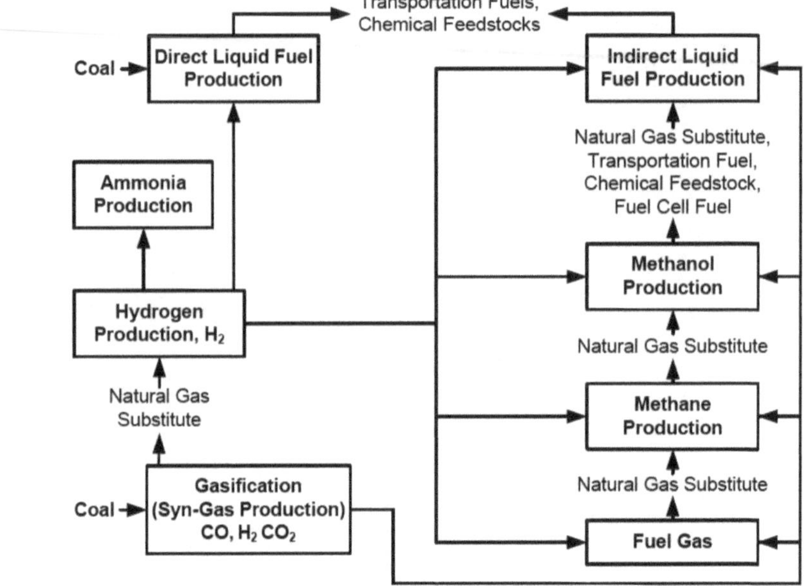

Fig. 1.8 Schematic of fuel and chemical products produced from a synthesis gas via the Fischer-Tropsch (FT) process

Fischer-Tropsch Process Rather than directly combusting the synthesis gas in an IGCC process, the syngas may also be converted to fuel and/or chemicals via Fischer-Tropsch (FT) synthesis. The original inventors of the process were Franz Fischer and Hans Tropsch, who worked at the Kaiser Wilhem Institute during the 1920s. The first commercialization took place in 1934, with the first industrial plant in operation in Germany in 1936, with annual production of over 1 million tons of FT liquid fuel at that time. The largest scale FT operation, based on a coal-to-liquid (CTL) conversion process, is operated by Sasol in South Africa, which is high in coal reserves and low in petroleum [34]. Due to oil embargos in opposition to Apartheid, South Africa advanced their CTL process in an attempt to gain independence from petroleum [35]. Figure 1.8 demonstrates the flexibility of the FT process in terms of the variety of petroleum-based products that can result.

An important parameter for controlling the FT conversion and selectivity is the hydrogen-carbon ratio (H/C). The required H/C ratio for commercial production of hydrocarbon fuels, *i.e.*, Diesel and gasoline is approximately 2 on a molar basis, while the ratio ranges from 1.3–1.9 for petroleum crude oil and 0.8 for typical bituminous coals [36]. A critical challenge for FT conversion is increasing the H/C ratio or the H_2/CO ratio. The hydrogen content in the syngas enhances the conversion efficiency toward the production of hydrocarbon products, with the additional hydrogen sourced from steam during the gasification step. Steam reacts with CO resulting in the water-gas shift reaction (*i.e.*, $CO + H_2O \leftrightarrow CO_2 + H_2$). The effect of the water-gas shift reaction can be varied depending on the type of FT catalyst, with the direction of the water-gas shift equilibrium dependent on the syngas composition [37]. Though the H/C or the H_2/CO ratio for the FT process is increased through the addition of steam, there is a subsequent efficiency penalty associated with the use of steam generated from the gasifier, in addition to the condensation of unused steam after the FT reactor. The future of coal-to-liquid processes may be limited due to: 1) its competition with existing "alternative liquid fuels," such as biofuels, as well as 2) its inevitable result of net CO_2 emissions since the H:C ratio is higher in liquids than in coal.

Oxyfuel Combustion Sometimes referred to as oxycombustion, oxyfuel combustion involves an ASU to allow coal to be burned in an oxygen-enriched environment, which leads to minimal dilution of CO_2 in the flue gas stream. The flue gas stream in an oxyfuel combustion process includes primarily CO_2 and water vapor, which can be condensed out fairly easily. To prevent the temperatures in the boiler from getting too large, up to 70% of the flue gas stream is recycled back to dilute the oxygen-enriched environment and maintain temperatures similar to conventional air-blown designs. Major challenges associated with the application or retrofit of oxyfuel combustion on an existing power plant is the cost of the ASU and potential leakage of air into the flue gas stream. It is important to keep in mind, however, that a typical stream from an ASU may contain approximately 3% N_2 and 2% Ar depending upon the method used for separation. In general, major oxyfuel developers now see the primary application in new plants, or at existing plants that are "repowered" with modern and more efficient boilers. The following demonstration plants are currently in place: Vattenfall in Sweden [38], IHI in Japan [39], and Alstom Power in the U.S. [40], with 2×900, 1000, and 450-MW capacities, respectively. Although the cost of electricity is greater from an oxyfuel combustion plant compared to a coal-fired plant with air, it has been shown that the cost may be lower with the inclusion of CO_2 capture [41].

Chemical Looping Combustion The chemical looping combustion process involves the injection of metal oxides into the boiler that act as oxygen transporters (rather than air) for coal of natural gas oxidation, similar to the other processes previously discussed, minimizing the dilution of CO_2 in the flue gas stream [42]. Within the chemical looping process, the solid oxygen carrier circulates between two fluidized bed reactors to transport oxygen from the combustion air to the fuel. Typical metal oxides include iron and nickel. Iron is low cost and nonhazardous making it an optimal choice. The metal oxide is reduced in a fuel reactor (*i.e.*, reduction reactor) while oxidizing the fuel and is then transported into an air reactor (*i.e.*, oxidation reactor) where it is reoxidized by air. The carrier particles are then transported back into the fuel reactor after passing through a cyclone for separation from the hot N_2 and O_2 gases. Benefits of this process are similar to those of oxyfuel combustion, *i.e.*, the flue gas is not diluted with N_2 and this process leads to lower NO_x formation. However, it is important to note that the majority (~70%) of the NO_x formed is not from the N_2 in the combustion air, but rather, from the surface functional groups on the coal itself. A simple schematic demonstrating the principle of chemical looping combustion is shown in Fig. 1.9. Additional details regarding these advanced coal conversion processes are available in the literature [43].

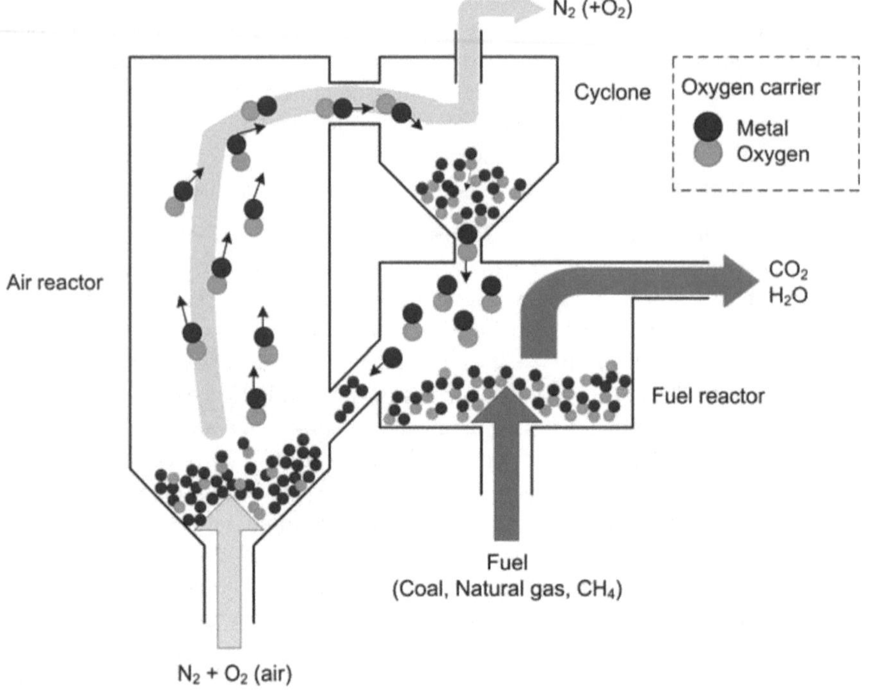

Fig. 1.9 Schematic demonstrating the principle of chemical looping combustion

1.5 Minimum Thermodynamic Work for CO_2 Separation

The first law of thermodynamics is concerned with the conservation of energy. The change in total energy of the system going from state 1 to state 2 is equal to the heat added to the system minus the work done, and can be expressed as,

$$Q - W = \Delta E_{1\to 2}, \tag{1.6}$$

in which Q and W may be positive or negative with the sign indicating the direction of energy flow. Examples of energies include internal energy, kinetic energy, and potential energy. The internal energy is comprised of the molecular energies within a given system, more specifically these include the energy contributions from the translational, vibrational, and rotational degrees of freedom within a molecule. Kinetic energy can be expressed as $\frac{1}{2}mv^2$, where m is the mass of a moving object and v is its velocity. Potential energy can be expressed as mgz, where m is the mass of a stationary object, g is the acceleration due to gravity, and z is elevation. Additional energies exist such as nuclear, electromagnetic, etc., but are not necessary for discussion in the context of CO_2 capture.

The second law of thermodynamics is concerned with entropy, which is generated during irreversible processes. Examples that include such irreversibilities are heat transfer across a temperature gradient, friction, mixing or stirring processes, and

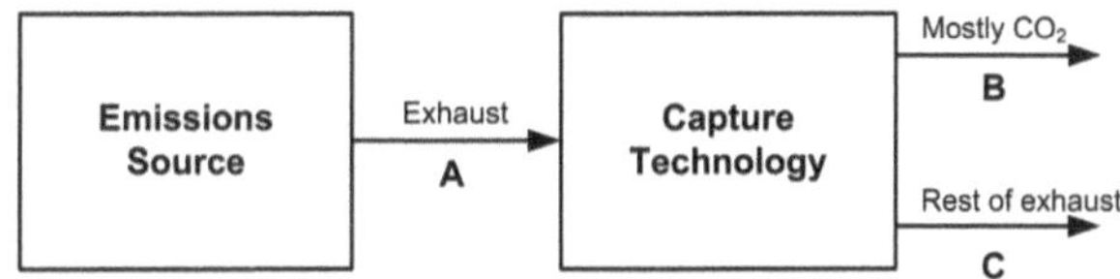

Fig. 1.10 Schematic of carbon capture

many chemical reactions. Irreversible processes prevent a system from returning to its original state and additionally, they reduce the available work (*i.e.*, exergy) of a given system.

The minimum work required to separate CO_2 from a gas mixture can be calculated based upon the combined first and second laws of thermodynamics. Figure 1.10 is a representation of the a generic CO_2 separation process along with the general emissions source and capture technology with corresponding gas streams. Stream A represents a CO_2-inclusive gas stream mixture (not limited to a combustion exhaust) while stream B contains mostly CO_2 depending upon the process purity, and stream C contains primarily the remainder of gas stream A. Ideally, stream B would be mostly (or all) CO_2 and stream C would be very low (or zero) in CO_2.

The minimum work required for separating CO_2 from a gas mixture for an *isothermal* (constant temperature) and *isobaric* (constant pressure) process is equal to the negative of the difference in Gibbs free energy of the separated final states (streams B and C in Fig. 1.10) from the mixed initial state (stream A in Fig. 1.10). For an *ideal gas* (minimal gas species interactions), the Gibbs free energy change between stream A to streams B and C is:

$$W_{min} = \Delta G_{sep} = \Delta G_B + \Delta G_C - \Delta G_A \tag{1.7}$$

For an ideal mixture, the partial molar Gibbs free energy for each gas is [44, 45]:

$$\frac{\partial G}{\partial n_i} = G_i^\circ + RT \ln\left(\frac{p_i}{p}\right) \tag{1.8}$$

such that p_i is the partial pressure of the ith gas and p is total pressure. Therefore, the total Gibbs free energy of an ideal gas mixture is:

$$G_{TOTAL} = \sum_i n_i \frac{\partial G}{\partial n_i} \tag{1.9}$$

The minimum work required to go from state 1 to states 2 and 3 is associated with the free energy difference between the product and reactant states, which can be calculated by combining Eqs. (1.8) and (1.9) as:

$$\begin{aligned} G_A &= n_A^{CO_2} G^\circ_{CO_2} + n_A^{A-CO_2} G^\circ_{A-CO_2} + RT\left(n_A^{CO_2} \ln\left(y_A^{CO_2}\right) + n_A^{A-CO_2} \ln\left(y_A^{A-CO_2}\right)\right) \\ G_B &= n_B^{CO_2} G^\circ_{CO_2} + n_B^{B-CO_2} G^\circ_{B-CO_2} + RT\left(n_B^{CO_2} \ln\left(y_B^{CO_2}\right) + n_B^{B-CO_2} \ln\left(y_B^{B-CO_2}\right)\right) \\ G_C &= n_C^{CO_2} G^\circ_{CO_2} + n_C^{C-CO_2} G^\circ_{C-CO_2} + RT\left(n_C^{CO_2} \ln\left(y_C^{CO_2}\right) + n_C^{C-CO_2} \ln\left(y_C^{C-CO_2}\right)\right), \end{aligned} \tag{1.10}$$

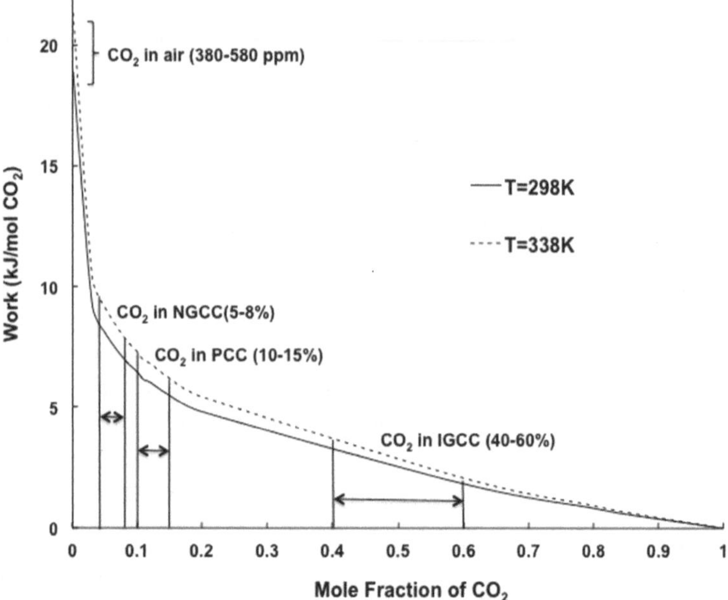

Fig. 1.11 Minimum thermodynamic work for various coal or gas-to-electricity conversions

And thus the minimum work is:

$$\begin{aligned} W_{\min} = RT\left[n_B^{CO_2}\ln\left(y_B^{CO_2}\right) + n_B^{B-CO_2}\ln\left(y_B^{B-CO_2}\right)\right] \\ + RT\left[n_C^{CO_2}\ln\left(y_C^{CO_2}\right) + n_C^{C-CO_2}\ln\left(y_C^{C-CO_2}\right)\right] \\ - RT\left[n_A^{CO_2}\ln\left(y_A^{CO_2}\right) + n_A^{A-CO_2}\ln\left(y_A^{A-CO_2}\right)\right] \end{aligned} \tag{1.11}$$

where R is the ideal gas constant (8.314 J/mol K), T is the absolute temperature, $y_i^{CO_2}$ is the mole fraction of CO_2 in the gas mixture, i, such that i can represent either stream A, B, or C in Fig. 1.10, and $y_i^{i-CO_2}$ represents the remainder of a given gas stream A, B, or C. The quantity of greatest interest is the minimum work per mole of CO_2 removed, that is, $W_{min}/(n_{CO_2})$. For $T = 25°\text{C}$ (298 K), the minimum work when beginning with coal flue gas at 12% CO_2, is 172 kJ/kg CO_2; when beginning with air at 0.04% CO_2, is 497 kJ/kgCO_2, which is approximately a factor of three greater. The minimum work required for separation is highly dependent upon the starting concentration of CO_2 in a given gas mixture.

Figure 1.11 is a plot of the minimum work, W_{min} at varying temperatures for CO_2 separation as a function of the molar concentration of CO_2 in the initial gas mixture. As the concentration of CO_2 decreases, the minimum work required for separation increases. Additionally, an increase in temperature leads to an increase in the thermodynamic minimum work required for separation.

Example 1.1 Assume a 500-MW coal-fired power plant emits a flue gas containing 4 kmol CO_2/s, 5 kmol H_2O/s, 1 kmol O_2/s and 20 kmol N_2/s. What is the minimum work for the isothermal and isobaric separation of CO_2 from the flue gas mixture for 90% capture and 98% purity at 45°C?

Solution

Given: $\dot{n}_{CO_2} = 4 \text{ kmol } CO_2/\text{s}$ $\quad \dot{n}_{H_2O} = 5 \text{ kmol } H_2O/\text{s}$

$\dot{n}_{O_2} = 1 \text{ kmol } O_2/\text{s}$ $\quad \dot{n}_{N_2} = 20 \text{ kmol } N_2/\text{s}$

Capture = 0.90 $\quad$ Purity = 0.98

Using Eq. (1.11),

$$\begin{aligned} W_{\min} =& RT\left[n_B^{CO_2}\ln\left(y_B^{CO_2}\right) + n_B^{B-CO_2}\ln\left(y_B^{B-CO_2}\right)\right] \\ &+ RT\left[n_C^{CO_2}\ln\left(y_C^{CO_2}\right) + n_C^{C-CO_2}\ln\left(y_C^{C-CO_2}\right)\right], \\ &- RT\left[n_A^{CO_2}\ln\left(y_A^{CO_2}\right) + n_A^{A-CO_2}\ln\left(y_A^{A-CO_2}\right)\right] \end{aligned}$$

where stream A is the flue gas mixture entering the separator, stream B at 98% purity contains 90% of the CO_2 contained in stream A, and stream C contains the remaining 10% of the CO_2 contained in stream A. Performing a mole balance on the separator yields:

Stream A: $\dot{n}_A^{CO_2} = 4 \text{ kmol } CO_2/\text{s}$ $\quad \dot{n}_A^{A-CO_2} = 26 \text{ kmol A} - CO_2/\text{s}$

$\dot{n}_A = 30 \text{ kmol A gas/s}$

$y_A^{CO_2} = 0.13$ $\quad y_A^{A-CO_2} = 0.87$

Stream B: $\dot{n}_B^{CO_2} = (4 \text{ kmol } CO_2/\text{s})(0.90) = 3.6 \text{ kmol } CO_2/\text{s}$

$\dot{n}_B = (3.6 \text{ kmol } CO_2/\text{s})/(0.98)$

$= 3.67 \text{ kmol B gas/s to ensure 98\% purity}$

$\dot{n}_B^{B-CO_2} = 3.67 \text{ kmol B gas/s} - 3.6 \text{ kmol } CO_2 \text{ gas/s}$

$= 0.07 \text{ kmol B} - CO_2/\text{s}$

$y_B^{CO_2} = 0.98$ $\quad y_B^{B-CO_2} = 0.02$

Stream C: $\dot{n}_C^{CO_2} = (4 \text{ kmol } CO_2/\text{s})(0.1) = 0.4 \text{ kmol } CO_2/\text{s}$

$\dot{n}_C = \dot{n}_A - \dot{n}_B$

$= 26.33 \text{ kmol C gas/s}$ from the system (separator) molar balance

$$\dot{n}_C^{C-CO_2} = 26.33 \text{ kmol C gas/s} - 0.4 \text{ kmol } CO_2 \text{ gas/s}$$
$$= 25.93 \text{ kmol C} - CO_2/\text{s}$$

$$y_C^{CO_2} = 0.015 \qquad y_C^{C-CO_2} = 0.985$$

The flue gas is entering the separator at 45°C or 318 K. Substituting these values into Eq. (1.11) yields a thermodynamic minimum work of 24,756 kJ/s or 6.88 kJ/mol CO_2 captured, with a CO_2 capture rate of 3.6 kmol CO_2/s.

2nd-Law Efficiency Real systems will always use more energy than the thermodynamic minimum since the minimum is derived for a reversible isothermal process. The capture of CO_2 or more specifically, separating CO_2 from a gas mixture, takes significant work for dilute mixtures of CO_2 and additional work for increased capture and purity. The separation process may depend upon current technologies such as absorption, adsorption, membranes, or some hybrid approach that has yet to be developed. Take for example absorption, which is outlined in Chap. 3; within this process blowers are used to drive a flue gas upward through a packed bed or spray tower, in which a liquid solvent is driven downward countercurrently using pump work. Additional work is required through the addition of heat for the solvent regeneration process, which drives the CO_2 off in a pure stream for compression, which also takes work. Each of these processes have associated efficiencies based upon irreversibilities, such as friction, heat transfer, gas expansion, gas mixing, etc.; therefore, the actual work required for CO_2 separation from a gas mixture deviates from the thermodynamic minimum work based upon the unit operations of the process and the extent of their individual inefficiencies. The ratio of the reversible or thermodynamic minimum work to the real work is termed the 2nd-Law efficiency [46] and is defined as:

$$\eta_{2nd} = \frac{W_{min}}{W_{real}} \tag{1.12}$$

The chemistry and physics of the underlying mechanisms of the various unit operations are discussed in the following chapters and analyzed in addition to the design and process in which the underlying chemistry and physical principles exist. Through the investigation of each of the steps in the given separation process there will be the opportunity to question and probe the actual work required of a given process and to assess the sensitivity of potentially tuning the related parameters to maximize the process's 2nd-law efficiency.

The 2nd-law efficiency has been investigated for a variety of gas scrubbing processes that span a wide range of concentrations. In Fig. 1.12, the 2nd-law efficiency is plotted as a function of decreasing concentration for 90% CO_2 capture from coal-fired power and NGCC plants, as well as for varying levels of NO_x, SO_x and mercury (Hg) scrubbing. For all "actual" work calculations, the Integrated Environmental Control Module developed by Rubin et al. [47] from Carnegie Mellon University was used. In the case of the postcombustion capture of CO_2, SO_x, NO_x, and Hg it was assumed

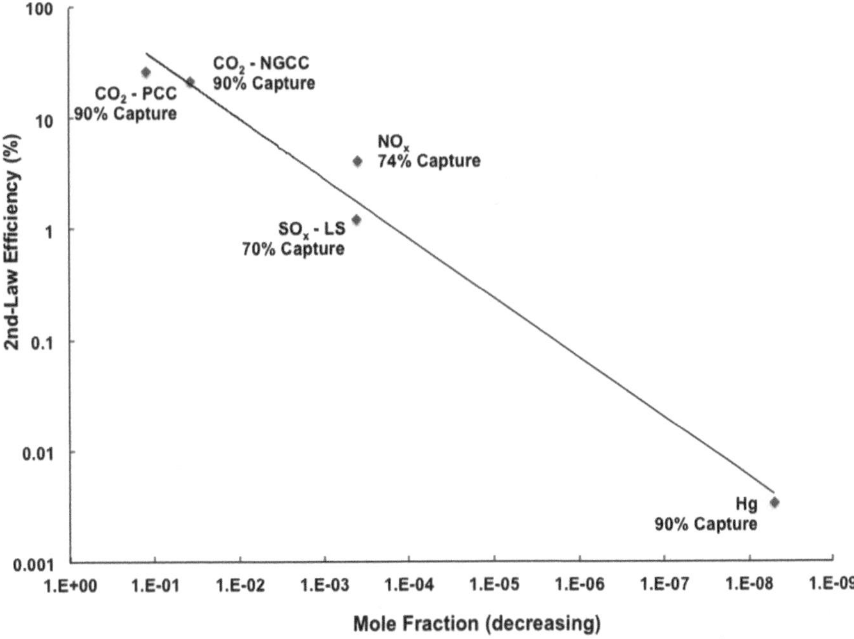

Fig. 1.12 Plot showing the relationship between the 2nd-law efficiency and concentration of initial gas mixture

that low-sulfur Appalachian bituminous coal was burned in a 500-MW utility boiler. The capture technologies assumed for CO_2, SO_x, NO_x, and Hg consisted of amine scrubbing, wet flue gas desulfurization, selective catalytic reduction, and activated carbon injection, respectively. In the case of NGCC, precombustion separation based upon amine scrubbing is assumed for a 477-MW plant. It is interesting to note that the 2nd-law efficiency decreases with decreasing CO_2 concentration [48]. This implies that there are still efficiencies to gain in SO_x, NO_x, and Hg capture since these processes do not include regeneration, yet they still follow the trend.

1.6 Cost of CO_2 Capture

Costs of separation may be divided into technical versus non-technical costs. Examples of non-technical costs may include account depreciation and return on investment, interest rate, labor, etc. Costs associated with the technology include the cost of the equipment, chemicals used, power consumption to operate the separation process, power cost, etc. Additional factors that may affect the cost of CO_2 capture include the choice of power plant and capture technology. For instance, will an existing plant be retrofitted or will the capture technology be applied to a new plant? The process design and variables associated with plant operation, such as plant

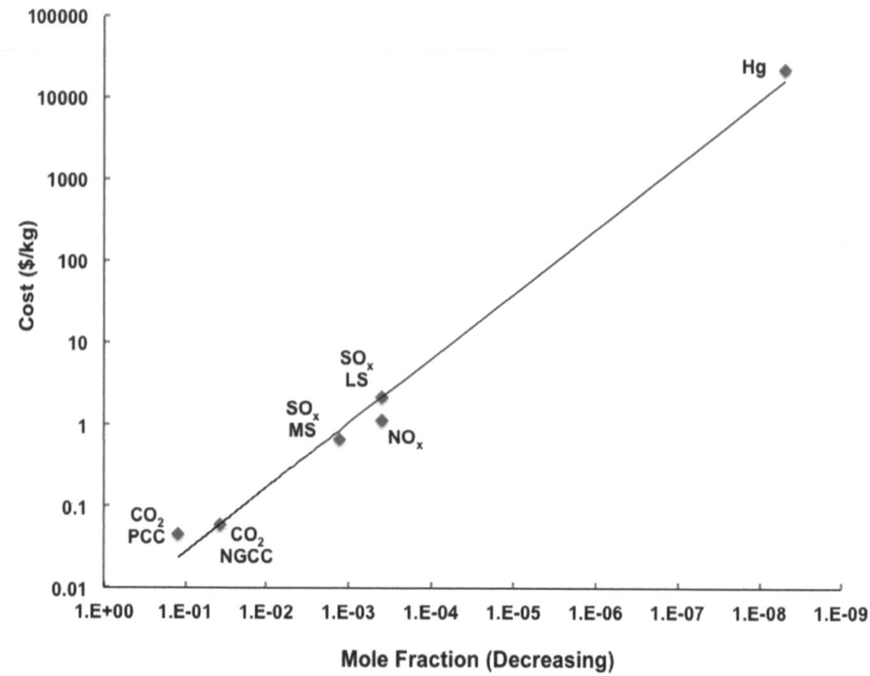

Fig. 1.13 Sherwood plot of various gas scrubbing processes demonstrating the cost increase with decreasing concentration

capacity, capture rate, and CO_2 concentration in and out will also influence costs. The system boundaries may influence the cost, for instance whether the capture technology is implemented on one facility or multiple plants or whether there will be energy resource integration to power the capture process, such as the implementation of natural gas, nuclear, wind, or solar. Whether a plant is a first-of-a-kind or *nth* plant may also influence the cost as technology is learned and advanced with time.

The Sherwood plot [49] has proved to be a useful correlation for estimating the separation cost as a function of starting concentration. Figure 1.13 provides a Sherwood plot demonstrating the relationship between cost increase with increasingly dilute gas mixtures. Based upon the previous assumptions from the data plotted in Fig. 1.12, the Sherwood correlation is illustrated in Fig. 1.13. The concentration of each of these species in the flue gas is also listed in Table 1.8. It is clear that an increase in dilution in the flue gas is correlated with an increase in the unit cost of capture. The cost-to-concentration relationship for medium-sulfur coal compared to the low-sulfur coal, *i.e.*, 1270 versus 399 ppm, is also consistent with this trend. Similarly, amine scrubbing for a 477-MW NGCC power plant generating approximately 3.7 mol% CO_2 follows the trend in that it is higher in cost compared to PCC in which the CO_2 concentration is approximately 12 mol% CO_2. Lightfoot and Cockrem [49] described this concept well in a publication titled, What Are Dilute Solutions? Their

Table 1.8 Cost and scale of various coal and natural gas oxidation processes

Process	Price ($/kg)	Concentration (mole fraction)	Emissions (kg/day)	Cost (1000s $/day)
CO_2–PCC	0.045	0.121	8.59×10^6	392
CO_2–NGCC	0.059	0.0373	3.01×10^6	178
SO_x (MS)	0.66	0.00127	8.94×10^4	59.6
SO_x (LS)	2.1	0.000399 (399 ppm)	2.32×10^4	50.4
NO_x	1.1	0.000387 (387 ppm)	1.11×10^4	12.5
Hg	22000	5×10^{-9} (5 ppb)	0.951	21.6

assessment was that the "recovery of potentially valuable solutes from dilute solution is dominated by the costs of processing large masses of unwanted materials." It is interesting to consider DAC in this context as CO_2 is present in the air at similar concentrations of NO_x in flue gas. However, the scale of the CO_2 emissions compared to NO_x as shown in Table 1.8, makes this approach to CO_2 mitigation less desirable than capture from more concentrated sources. Similar to the results of Table 1.8, recent investigations [48, 50] have concluded that cost estimates may be as high as $1000 per ton of CO_2 for DAC.

Defining the Cost of CO_2 Avoided and Captured Imagine that a CO_2 capture system is installed at a fossil-fuel power plant, and that the energy required to operate the capture system is provided by fossil fuels. In principle, nothing prevents the capture system from emitting more CO_2 than it captures. Such a system would be counterproductive, however. One needs a vocabulary that distinguishes the gross CO_2 removed by a capture device and the net CO_2 that does not enter the atmosphere, which is the gross CO_2 removed minus the CO_2 emitted by the capture system itself. The concepts of gross and net CO_2 prevented from entering the atmosphere are called "captured CO_2" and "avoided CO_2." Avoided CO_2 is always less than captured CO_2 since any capture system will emit some CO_2. Equivalently, the cost of avoided CO_2 (in dollars per ton of CO_2) is always greater than the cost of CO_2 captured.

Figure 1.14 provides a schematic demonstrating the differences between CO_2 captured versus CO_2 avoided for a typical power plant application, assuming that the capture plant captures and stores 90% of the total generated CO_2 emissions and that some CO_2 emissions will be associated with the energy needed to operate the capture process itself. Thus, the cost of CO_2 avoided is the cost of delivering a unit of useful product (in this case, electricity) while avoiding a ton of CO_2 emissions to the atmosphere. Avoidance costs should always include the cost of CO_2 compression, transport and storage since "avoided" means not emitted into the atmosphere.

The cost of CO_2 avoided for a power plant, C_{avo}, is calculated on a net kWh basis by the equation,

$$C_{avo}\left(\$/\text{ton CO}_2\right) = \frac{\left(^{\$}/_{\text{kWh}}\right)_{cap} - \left(^{\$}/_{\text{kWh}}\right)_{ref}}{\left(^{CO_2}/_{\text{kWh}}\right)_{ref} - \left(^{CO_2}/_{\text{kWh}}\right)_{cap}} \tag{1.13}$$

such that $\left(^{\$}/_{\text{kWh}}\right)_{ref}$ and $\left(^{\$}/_{\text{kWh}}\right)_{cap}$ represent the cost per net kWh of electricity produced by the reference and capture plant, respectively and $\left(^{CO_2}/_{\text{kWh}}\right)_{ref}$ and

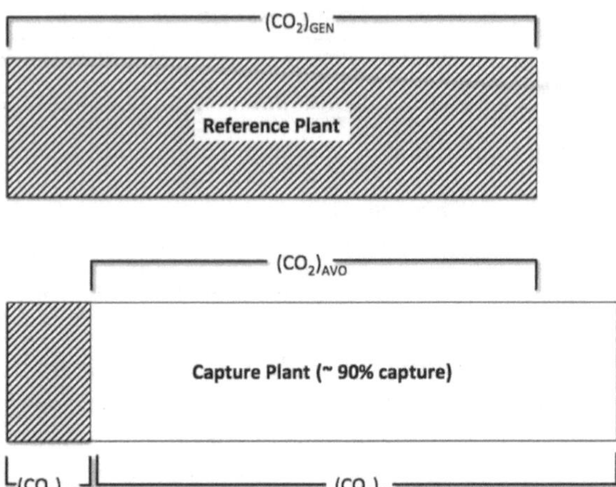

Fig. 1.14 Schematic demonstrating the difference in CO_2 captured versus CO_2 avoided for a point-source capture scenario. Here, the capture plant produces the same useful product output as the reference plant, so additional capacity is needed to operate the capture system. For a power plant all values are typically normalized on the net kWh generated

$\left({}^{CO_2}/_{\mathrm{kWh}}\right)_{cap}$ represent the tons of CO_2 emitted to the atmosphere per net kWh of electricity produced by the reference and capture plant, respectively.

In contrast, the cost per ton of CO_2 captured, C_{cap}, can be calculated by the equation,

$$C_{cap}\left(\$/\mathrm{tonCO_2}\right) = \frac{\left({}^{\$}/_{\mathrm{kWh}}\right)_{cap} - \left({}^{\$}/_{\mathrm{kWh}}\right)_{ref}}{\left({}^{CO_2}/_{\mathrm{kWh}}\right)_{cap}} \tag{1.14}$$

such that the numerator again represents the incremental cost of the capture system, while the denominator is the quantity of CO_2 captured, with all values again normalized on the net plant output. The key difference from Eq. (1.13) is that the capture cost does not include the cost of energy to operate the capture system; nor does it typically include the costs of CO_2 transport and storage. Thus, the cost per ton captured is always less than the cost per ton avoided.

For a co-generation power plant that produces both electricity and heat, costs and emissions can be normalized on the total equivalent thermal energy output (in kJ). For other types of point sources, such as an oil refinery stack or a cement plant, costs and emissions would be normalized on the relevant measure of useful output (*e.g.*, barrels of oil or tons of cement) when calculating the cost of CO_2 avoided or captured.

Cost for Direct Air Capture For CO_2 removed directly from the atmosphere (a concept being developed, but not yet practiced on a commercial scale) there is no reference plant or specific product associated with the capture system, as with point sources. Nonetheless, the concepts of gross and net CO_2 removal from the

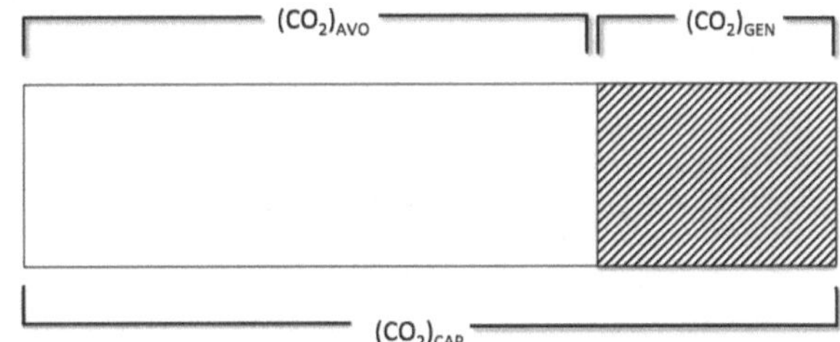

Fig. 1.15 Schematic demonstrating the difference between CO_2 captured versus CO_2 generated for a DAC scenario

atmosphere are analogous to those associated with emission prevention. The two analogous concepts are again called CO_2 captured (gross) and CO_2 avoided (net). The amount and cost of CO_2 captured in the case of DAC is defined in the same way as the previous method, namely the incremental cost of the capture system divided by the amount of CO_2 captured. Thus,

$$C_{cap,DAC}\left(\$/\text{ton CO}_2\right) = \frac{\left(^{\$}/_{\text{yr}}\right)_{DAC}}{\left(^{CO_2}/_{\text{yr}}\right)_{DAC}} \tag{1.15}$$

where the numerator is the levelized annual cost of the capture system and the denominator is the annual average amount captured.

Figure 1.15 shows a schematic demonstrating the difference between avoided versus captured. In the case of DAC there is only a capture plant since the CO_2 emissions are captured directly from air, compressed, transported, and stored (or sequestered). Similar to the point source definition, however, allowance is made for the generation of emissions associated with the purchased energy resource(s) used to fuel the DAC plant and the energy required to capture a given number of tons of CO_2 from the air. Effectively, the cost of CO_2 avoided is the cost of the capture plant (plus transport and storage costs) divided by the *net* amount captured, rather than the total amount.

The cost of CO_2 avoided, $C_{avo,DAC}$, in the case of DAC can be calculated by,

$$C_{avo,DAC}\left(\$/\text{ton CO}_2\right) = \frac{\left(^{\$}/_{\text{yr}}\right)_{DAC}}{\left(^{CO_2}/_{\text{yr}}\right)_{DAC} - \left(^{CO_2}/_{\text{yr}}\right)_{gen}} = \frac{\left(^{\$}/_{\text{yr}}\right)_{DAC}}{\left(^{CO_2}/_{\text{yr}}\right)_{net}} \tag{1.16}$$

such that the numerator represents the total annualized cost of building and operating the DAC plant and the denominator represents the net number of tons captured per year from the DAC plant. Energy to operate the plant typically will be some combination of electricity and heat sourced by fossil fuels or renewables.

References

1. Image courtesy of Yangyang Liu, using a JEOL JSM 5600 Scanning Electron Microscope, (2009) School of Earth Sciences, Stanford University

2. EIA (2011) Annual Energy Outlook Energy Information Administration. U.S. Department of Energy, Washington, DC
3. Solomon S, Plattner GK, Knutti R, Friedlingstein P (2009) Irreversible climate change due to carbon dioxide emissions. Proc Natl Acad Sci U S A 106(6):1704–1709
4. (a) Ramanathan V, Feng Y (2008) On avoiding dangerous anthropogenic interference with the climate system: Formidable challenges ahead. Proc Natl Acad Sci U S A 105(38):14245; (b) Wigley TML (2005) The climate change commitment. Science 307(5716):1766; (c) Friedlingstein P, Solomon S (2005) Contributions of past and present human generations to committed warming caused by carbon dioxide. Proc Natl Acad Sci U S A 102(31):10832
5. (a) Matthews HD, Weaver AJ (2010) Committed climate warming. Nat Geosci 3(3):142–143; (b) Matthews HD, Caldeira K (2008) Stabilizing climate requires near-zero emissions. Geophys Res Lett 35(4):L04705
6. Ha-Duong M, Grubb MJ, Hourcade JC (1997) Influence of socioeconomic inertia and uncertainty on optimal CO_2-emission abatement. Nature 390(6657):270–273
7. Matthews HD, Gillett NP, Stott PA, Zickfeld K (2009) The proportionality of global warming to cumulative carbon emissions. Nature 459(7248):829–832
8. Van Der Werf GR, Morton DC, DeFries RS, Olivier JGJ, Kasibhatla PS, Jackson RB, Collatz GJ, Randerson JT (2009) CO_2 emissions from forest loss. Nature Geosci 2(11):737–738
9. (a) Denman KL (2007) In: Climate Change 2007: the physical basis. contribution of working group i to the fourth assessment report of the intergovernmental panel on climate change. Cambridge University Press, Cambridge; (b) Gibbs HK, Herold M (2007) Tropical deforestation and greenhouse gas emissions. Environ Res Lett 2:045021; (c) Schrope M (2009) When money grows on trees. Nat Rep Climate Change 3:101–103
10. Meinshausen M, Meinshausen N, Hare W, Raper SCB, Frieler K, Knutti R, Frame DJ, Allen MR (2009) Greenhouse-gas emission targets for limiting global warming to 2°C. Nature 458(7242):1158–1162
11. (a) Houghton JT, Ding Y, Griggs DJ, Noguer M, Van der Linden PJ, Dai X, Maskell K, Johnson CA (2001) IPCC, Climate Change 2001: The scientific basis. contribution of working group I to the Third assessment report of the Intergovernmental Panel on climate change. Cambridge University Press, Cambridge; (b) Houghton JT (eds) (1996) Climate change 1995: the science of climate change. Cambridge University Press, Cambridge
12. Best D, Mulyana R, Jacobs B, Iskandar UP, Beck B (2011) Status of CCS development in Indonesia. Energy Procedia 4:6152–6156
13. Srivastava RK, Hall RE, Khan S, Culligan K, Lani BW (2005) Nitrogen oxides emission control options for coal-fired electric utility boilers. J Air Waste Manage Assoc 55(9):1367–1388
14. Amrollahi Z, Ertesvag IS, Bolland O (2011) Thermodynamic analysis on post-combustion CO_2 capture of natural-gas-fired power plant. Int J Greenh Gas Control 5:422–426
15. (a) Marland G, Boden TA, Andres RJ (2006) Global, Regional, and National Annual CO_2 Emissions from Fossil-Fuel Burning, Cement Manufacture, and Gas Flaring: 1751–2003. U.S. Department of Energy (DOE), Carbon Dioxide Information Analysis Center, Environmental Sciences Division, Oak Ridge; (b) Metz B (2005) IPCC special report on carbon dioxide capture and storage. Cambridge University Press, Cambridge, p 431; (c) IEAGHG (2008) Global IEA GHG CO_2 Emissions Database. IEA Greenhouse Gas R&D Programme, Cheltenham
16. Rubin ES, Mantripragada H, Marks A, Versteeg P, Kitchin J (2011) The outlook for improved carbon capture technology. Prog Energy Combust Sci (in press)
17. Kuuskraa VA (2010) Challenges of implementing large-scale CO_2 enhanced oil recovery with CO_2 capture and storage In: symposium on the role of enhanced oil recovery in accelerating the deployment of carbon capture and storage. Advanced Resources International Inc., Cambridge
18. Moniz EJ, Tinker SW (2010) Role of enhanced oil recovery in accelerating the deployment of carbon capture and sequestration; an MIT Energy Initiative and Bureau of Economic Geology at UT Austin Symposium July 23, 2010
19. (a) Kovscek AR, Cakici MD (2005) Geologic storage of carbon dioxide and enhanced oil recovery. II. Cooptimization of storage and recovery. Energy Convers Manag 46(11–12):1941–1956; (b) Jessen K, Kovscek AR, Orr FM (2005) Increasing CO_2 storage in oil recovery. Energy Convers Manag 46(2):293–311

20. (a) Ross HE, Hagin P, Zoback MD (2009) CO_2 storage and enhanced coalbed methane recovery: reservoir characterization and fluid flow simulations of the Big George coal, Powder River Basin, Wyoming, USA. Int J Greenh Gas Con 3(6):773–786; (b) Jessen K, Tang GQ, Kovscek AR (2008) Laboratory and simulation investigation of enhanced coalbed methane recovery by gas injection. Transport Porous Med 73(2):141–159
21. (a) Liu Y, Wilcox J (2011) CO_2 adsorption on carbon models of organic constituents of gas shale and coal. Environ Sci Technol 45:809–814; (b) Liu Y, Wilcox J (2012) Effects of surface heterogeneity in the adsorption of CO_2 in microporous carbons. Environ Sci Technol 46:1940–1947; (c) Nuttall BC, Eble CF (2003) Analysis of Devonian black shales in Kentucky for potential carbon dioxide sequestration and enhanced natural gas production; NASA Center for AeroSpace Information, 7121 Standard Dr Hanover, Maryland, pp 21076–1320; (d) Vermylen J, Hagin P, Zoback M (2008) In: Feasibility assessment of CO_2 sequestration and enhanced recovery in gas shale reservoirs. Americal Geophysical Union, p 990
22. Bottoms R (1930) Process for separating acid gases, U.S. Patent Office, Girdler Corporation, 1783901
23. Riesenfeld FC, Frazier HD (1952) Separation of Acidic constituents, U.S. Patent Office, Fluor Corporation, Ltd., 2600328
24. Bhown AB, Freeman BC (2011) Analysis and status of post-combustion Carbon Dioxide capture technologies. Environ Sci Technol 45:8624–8632
25. (a) http://fossil.energy.gov/programs/sequestration; (b) Carbon capture and storage, assessing the economics. McKinsey and Company, London, (2008)
26. Parson EA, Keith DW (1998) Fossil fuels without CO_2 emissions. Science 282(5391):1053
27. Bachu S, Bonijoly D, Bradshaw J, Burruss R, Holloway S, Christensen NP, Mathiassen OM (2007) CO_2 storage capacity estimation: Methodology and gaps. Int J Greenh Gas Control 1(4):430–443
28. Pacala S, Socolow R (2004) Stabilization wedges: Solving the climate problem for the next 50 years with current technologies. Sci 305(5686):968–972
29. van Krevelen DW (1993) Coal: typology, physics, chemistry, constitution, 3rd edn. Elsevier, Amsterdam
30. (a) Vejahati F, Xu Z, Gupta R (2010) Trace elements in coal: associations with coal and minerals and their behavior during coal utilization.-A review Fuel 89(4):904–911; (b) Senior C, Otten BV, Wendt JOL, Sarofim A (2010) Modeling the behavior of selenium in Pulverized-Coal Combustion systems. Combust Flame 157(11):2095–2105; (c) Wilcox J (2011) A kinetic investigation of unimolecular reactions involving trace metals at post-combustion flue gas conditions. Environ Chem 8(2):207–212; (d) Wilcox J (2009) A kinetic investigation of high-temperature mercury oxidation by chlorine. J Phys Chem A 113(24):6633–6639; (e) Wilcox J, Rupp E, Ying SC, Lim D.-H., Negreira AS, Kirchofer A, Feng F, Lee K (2012) Mercury adsorption and oxidation in coal and gasification processes. Int J Coal Geol 90–91:4–20
31. Frayne C (2002) Boiler water treatment: principles and practice. Chemical Pub Co, New York
32. (a) Chen C, Rubin ES (2009) CO_2 control technology effects on IGCC plant performance and cost. Energy Policy 37(3):915–924; (b) Rubin ES, Chen C, Rao AB (2007) Cost and performance of fossil fuel power plants with CO_2 capture and storage. Energy Policy 35(9):4444–4454; (c) van den Broek M, Hoefnagels R, Rubin E, Turkenburg W, Faaij A (2009) Effects of technological learning on future cost and performance of power plants with CO_2 capture. Prog Energy Combust Sci 35(6):457–480; (d) Yeh S, Rubin ES (2007) A centurial history of technological change and learning curves for pulverized coal-fired utility boilers. Energy 32(10):1996–2005; (e) Zhai H, Rubin ES (2010) Performance and cost of wet and dry cooling systems for pulverized coal power plants with and without carbon capture and storage. Energy Policy 38(10):5653–5660
33. White CM, Strazisar BR, Granite EJ, Hoffman JS, Pennline HW (2003) Separation and capture of CO_2 from large stationary sources and sequestration in geological formations: coalbeds and deep saline aquifers. J Air Waste Manage Assoc 53(6):645–715
34. Leckel D (2009) Diesel production from Fischer-Tropsch: the past, the present, and new concepts. Energy Fuels 23(5):2342–2358

35. Conlon P (1985) The Sasol coal liquefaction plants: economic implications and impact on the south Africa's ability to withstand an oil cut-off. United Nations, New York, p 59
36. Williams RH, Larson ED (2003) A comparison of direct and indirect liquefaction technologies for making fluid fuels from coal. Energy Sustain Dev 7(4):103–129
37. (a) Steynberg AP (2004) Introduction to fischer-tropsch technology. Stud Surf Sci Catal 152:1–63; (b) Dry ME (2004) Chemical concepts used for engineering purposes. Stud Surf Sci Catal 152:196–257
38. Birkestad H (2002) Separation and compression of CO_2 in an O_2/CO_2-fired power plant. Department of Energy Conversion, Thesis (MSc), Chalmers University of Technology, Gothenburg
39. Okawa M, Kimura N, Kiga T, Takano S, Arai K, Kato M (1997) Trial design for a CO_2 recovery power plant by burning pulverized coal in O_2/CO_2. Energy Convers Manag 38:123–127
40. Marion J, Nsakala N, Bozzuto C, Liljedahl G, Palkes M, Vogel D, Gupta JC, Guha M, Johnson H, Plasynski S (2001) In: Engineering feasibility of CO_2 capture on an existing US Coal-Fired Power Plant. 26th International Conference on Coal Utilization & Fuel Systems, Clearwater
41. Singh D, Croiset E, Douglas PL, Douglas MA (2003) Techno-economic study of CO_2 capture from an existing coal-fired power plant: MEA scrubbing vs. O_2/CO_2 recycle combustion. Energy Convers Manag 44(19):3073–3091
42. (a) Fan LS (2010) Chemical looping systems for fossil energy conversions. Wiley Online Library, Hoboken; (b) Li F, Fan LS (2008) Clean coal conversion processes – progress and challenges. Energy Environ Sci 1(2):248–267
43. (a) Wall TF (2007) Combustion processes for carbon capture. P Combust Inst 31(1):31–47; (b) Rackley SA (2010) Carbon capture and storage. Butterworth-Heinemann, Burlington, p 392; (c) Folger P (2010) In: Carbon capture and sequestration (CCS). BiblioGov
44. Gaskell D (1995) Introduction to the thermodynamics of materials. Taylor & Francis, Washington D.C., pp. 219–264
45. House KZ, Harvey CF, Aziz MJ, Schrag DP (2009) The energy penalty of post-combustion CO_2 capture & storage and its implications for retrofitting the U.S. installed base. Energy Environ Sci 2:193–205
46. Turns SR (2006) Thermodynamics: concepts and applications. Cambridge University Press, Cambridge, p 736
47. Rubin ES Integrated Environemental Control Model, Version 6.2.4. http://www.cmu.edu/epp/iecm/index.html
48. House KZ, Baclig AC, Ranjan M, van Nierop EA, Wilcox J, Herzog HJ (2011) Economic and energetic analysis of capturing CO_2 from ambient air. Proc Natl Acad Sci U S A 108(51):20428–20433
49. Lightfoot EN, Cockrem MCM (1987) What are dilute solutions? Separ Sci Technol 22(2):165–189
50. Direct Air Capture of CO_2 with Chemicals (2011) The American Physical Society: College Park, MD. http://www.aps.org/policy/reports/popareports/loader.cfm?csModule=security/getfile&PageID=244407, Accessed 10 June 2011
51. Gür TM (2010) Mechanistic modes for solid carbon conversion in high temperature fuel cells. J Electrochem Soc 157(5):B751–B759
52. Lee AC, Mitchell RE, Gür TM (2009) Thermodynamic analysis of gasification-driven direct carbon fuel cells. J Power Sources 194(2):774–785

Chapter 2
Compression and Transport of CO_2

Following capture and separation of CO_2 from a gas mixture, the CO_2 must be compressed for transport, regardless of whether it is destined for storage in underground geologic reservoirs or used as a feedstock in chemical processing, aggregate formation, or EOR. There are various modes of transport such as rail, ship, truck, and pipeline. On average, a single 500-MW coal-fired power plant would need to transport on the order of 2–3 Mt per year of compressed CO_2, which makes pipeline transport a feasible option due to the potential large-scale application [5]. For instance, the installed capacity in the U.S. is approximately 315 gigawatts (GW) [6], China's installed capacity of coal-fired power is approximately 600 GW and India's approximately 100 GW. The installed capacity in China and India will undoubtedly grow since as of 2009 these regions had populations of 186 million (China and East Asia) and 612 million (South Asia) without electricity [7]. China, U.S., and India are the top three coal producers, with production of 2 716, 993, and 484 million tons produced in 2008, respectively [7]. With this heavy reliance on coal for advancing electrification in these regions, there will be an inevitable increase in CO_2 emissions. Mitigation of CO_2 via carbon capture will require CO_2 transport. This chapter focuses on the compression and transport of CO_2 after capture.

2.1 Thermodynamic Properties of CO_2

Pipeline transport of CO_2 requires CO_2 to be compressed and then cooled to the liquid state, since CO_2-gas transport would result in a higher pressure drop per unit of pipeline and lower throughput for a given pipe diameter [8]. By operating the pipelines above 7.38 MPa, the supercritical pressure of CO_2, the formation of CO_2 gas and subsequent complexities of multiphase flow are avoided. A phase diagram of CO_2 is shown in Fig. 2.1, illustrating the critical temperature and pressure conditions of CO_2. Typical transport conditions are approximately 10-15 MPa (110 bar) and 35°C [9]. Typical ship conditions are approximately 0.7 MPa (7 bar) and −50°C. Due to the different properties of CO_2 compared to natural gas, accurate representation of the phase behavior, density, and viscosity of CO_2 and CO_2-containing mixtures is crucial for the safe and effective design of the pipeline.

J. Wilcox, *Carbon Capture*,
DOI 10.1007/978-1-4614-2215-0_2, © Springer Science+Business Media, LLC 2012

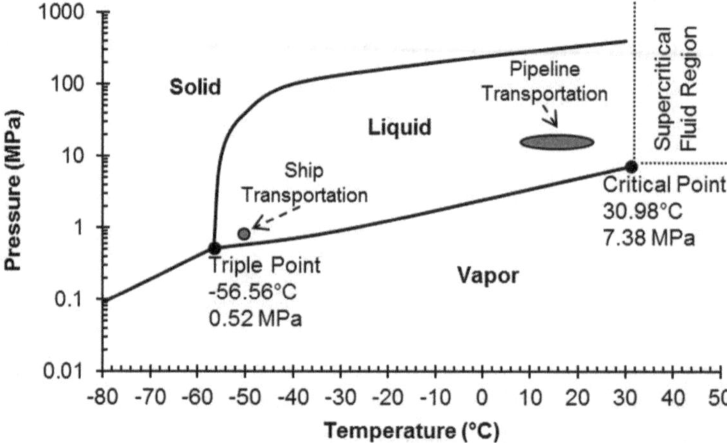

Fig. 2.1 Phase diagram of CO_2 indicating the *triple* and *critical* points. Pipeline transport takes place in the liquid phase above the critical pressure, 7.38 MPa

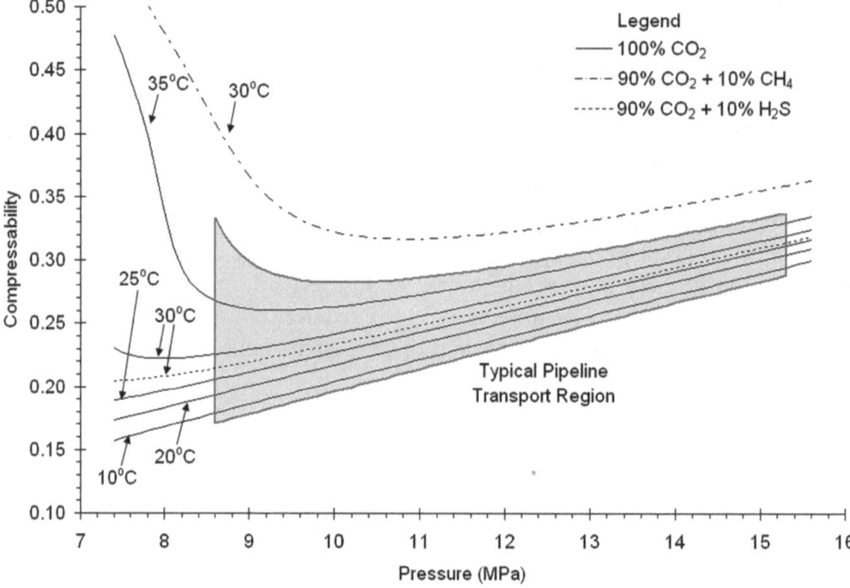

Fig. 2.2 CO_2 compressibility predicted by the Peng-Robinson equation of state showing sensitivity to CH_4 and H_2S. (With permission from Elsevier [2])

McCoy and Rubin [10] used a cubic equation of state with Peng-Robinson parameters and mixing rules based upon a binary interaction parameter [11] to determine the CO_2 compressibility as a function of pressure, including the range applicable for pipeline transport, as shown in Fig. 2.2. The data plotted in Fig. 2.2 illustrates the significant effect that the inclusion of impurities such as H_2S or CH_4 have on the

CO_2 compressibility. In addition, over the typical pipeline transport region shown, the CO_2 density ranges between approximately 800 and 1000 kg/m^3. The operating temperature of the pipeline is assumed equivalent to the surrounding soil temperature, which can vary seasonally from several degrees below zero to 6–8°C, with extremes in tropical regions peaking at 20°C [12].

Gases, unlike most liquids, are compressible and the following compression devices enable gas transport:

- Compressors move gas at high differential pressures, ranging from 35 to 65,000 psi (0.2 to 450 MPa),
- Blowers move large volumes of gas at pressures of up to 50 psi (0.3 MPa), and
- Fans move gas at pressures of up to approximately 2 psi (15kPa).

The compression of a gas results in an increased number of collisions per unit area of the walls of the enclosed system. Gas compression to higher pressures results in increased temperature, which can influence the design and performance of a compressor. To circumvent these potential limitations compression often occurs in multiple steps, or stages.

2.2 How a Compressor Works

The most common type of compressor is the *reciprocating positive displacement*, in which an inlet volume of gas is confined in a given space and then compressed by a reduction in the confined space. At the elevated pressure, the gas is subsequently expelled into a discharge piping or vessel system. Other types include *centrifugal* and *axial flow compressors*. Performance characteristics of the different types of compressors are shown in Fig. 2.3.

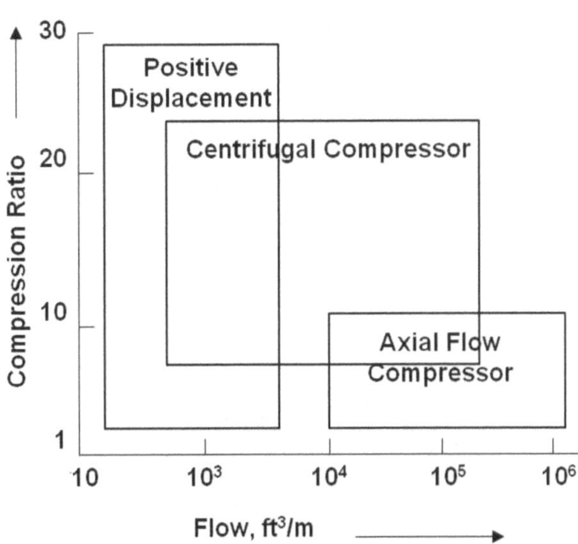

Fig. 2.3 Performance characteristics of different types of compressors

The *compression ratio* is defined as the ratio between the absolute discharge pressure (psia) and the absolute inlet pressure (14.696 psi). As an example, a compressor operating on air with a discharge pressure of 100 psi at sea level would have a compression ratio of (100 psig + 14.7)/14.7 = 7.8. *Gauge pressure* (psig) is defined as the pressure, in pounds per square inch, above local atmospheric pressure. *Absolute pressure* (psia) is defined as the gauge pressure plus the local atmospheric pressure. At sea level, this is equivalent to the gauge pressure plus 14.7 psi. As elevation increases above sea level, the atmospheric pressure decreases. At 5,000 ft, for example, the pressure is approximately 12.2 psia.

Figure 2.4 illustrates the steps of gas compression and expansion for the simple case of a single-cylinder reciprocating compressor with only one side of the piston being acted upon, to illustrate the steps of the compression process [13]. Within these types of compressors, spring-loaded valves open in response to a specific pressure change across the valve. For instance, the inlet valve opens when the cylinder pressure is slightly below the intake pressure, and the discharge valve opens when the cylinder pressure is slightly above the discharge pressure. Figure 2.4a shows the cylinder full of gas with point 1 on the theoretical *p-V* diagram representing the start of the compression process, with both valves initially closed. In Fig. 2.4a, the *compression stroke*, in which the piston moves to the right leads to a decrease in gas volume with

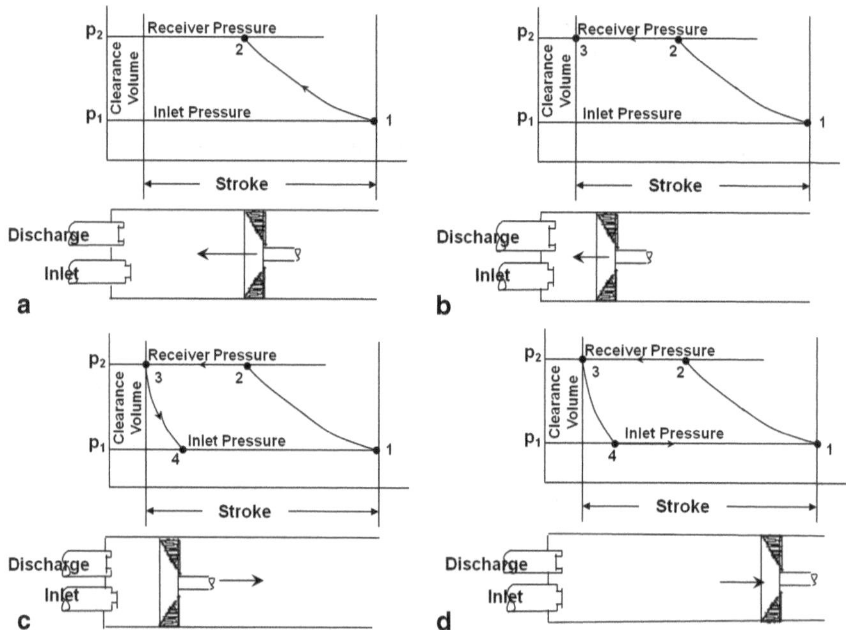

Fig. 2.4 **a** Compression stroke, with both valves closed and a pressure increase from point 1 to 2, **b** Delivery stroke, with discharge valve opening just beyond point 2, **c** Expansion stroke, with both valves closed and a pressure reduction from point 3 to 4, and **d** Inlet valve opens at point 4, with stroke ending at point 1. (Based on original material by [1])

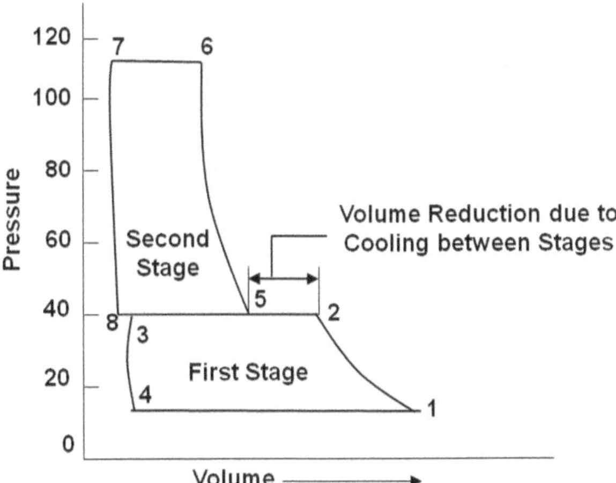

Fig. 2.5 The *p-V* diagram for a 2-stage compressing process. (Based on original material by [1])

a corresponding increase in pressure. The compression from point 1 to 2 on the *p-V* diagram is shown with both valves remaining closed during this step. Figure 2.4b shows the piston completing the *delivery stroke* with the discharge valve opened just beyond point 2. Once the piston reaches point 3, the discharge valve will close and the clearance space will be filled with gas at the discharge pressure. Figure 2.4c shows the *expansion stroke*, in which the air trapped in the clearance space expands with the valves remaining closed, leading to a decrease in pressure. As the piston moves to the right, the cylinder pressure decreases below the inlet pressure at point 4, with the inlet valve opening allowing for inflow of gas until the end of the reverse stroke is reached at point 1, as illustrated in Fig. 2.4d. At point 1 on the *p-V* diagram of Fig. 2.4d, the inlet valves will close, with the cycle repeating on the next crank revolution.

In a two-stage reciprocating compressor, the cylinders are proportional with respect to the compression ratio. Since the gas is partially compressed and cooled in the first compression stage it occupies less volume in the second stage of compression. This is demonstrated in the *p–V* diagram of Fig. 2.5. In general, multiple staging involving positive displacement exhibits this pattern.

2.3 Compression Cycles

The two theoretical compression cycles applicable to positive displacement compressors are isothermal and near-adiabatic (isentropic) compression. If the temperature is constant while the pressure increases, the compression is isothermal. This process requires continuous heat removal and compression is represented by:

$$p_1 V_1 = p_2 V_2 = \text{constant} \tag{2.1}$$

such that $p_1,\ V_1,\ p_2,\ V_2$ represent the inlet and outlet pressures and volumes, respectively. When there is no heated added or removed from the gas during compression, the process is termed near-adiabatic with compression represented by:

$$p_1 V_1^k = p_2 V_2^k \tag{2.2}$$

such that k is the ratio of *specific heats*. For an ideal gas, the relationship between the specific heat at constant volume, c_v, and the specific heat at constant pressure, c_p, on a molar basis is $c_p = c_v + R$ such that R is the ideal gas constant.

Example 2.1 Consider the syngas mixture of 30 mol% H_2, 20 mol% CO_2 and 50 mol% CO. Determine the ratio of specific heats of this mixture at 338.5 K.

Solution: The average temperature for compressor work is 338.5 K (65.5°C). To calculate the specific heat for the mixture, a mole-weighted average of the specific heats of each component should be taken. The molar heat capacity at 338.5 K is provided in Appendix B. Details of the calculation are shown in the table below:

Gas	Mol%	Mol gas/Mol mixture	c_p at 338.5 K of component (J/K mol)	Product (J/K mol)
H_2	30	0.3	29.04	8.71
CO_2	20	0.2	38.93	7.79
CO	50	0.5	29.14	14.57
	100	1.0		31.07

Therefore, the molar specific heat of the gas mixture is 31.07 J/K mol. The ratio of specific heat is determined from

$$k = \frac{c_p}{c_v} = \frac{c_p}{c_p - R} = \frac{31.07}{31.07 - 8.314} = 1.37$$

Figure 2.6 shows the various cycles on a p-V diagram, with the area ADEF representing the work required for an isothermal compression process, while the area represented by ABEF represents the work required for an adiabatic process. Neither the isothermal, nor the adiabatic cycles are generally exactly achievable and an actual compression process usually takes place via a polytropic cycle, in which compression is represented by:

$$p_1 V_1^n = p_2 V_2^n \tag{2.3}$$

such that the exponent, n is obtained from experiment for a given piece of equipment at a given set of conditions, and is typically less than k.

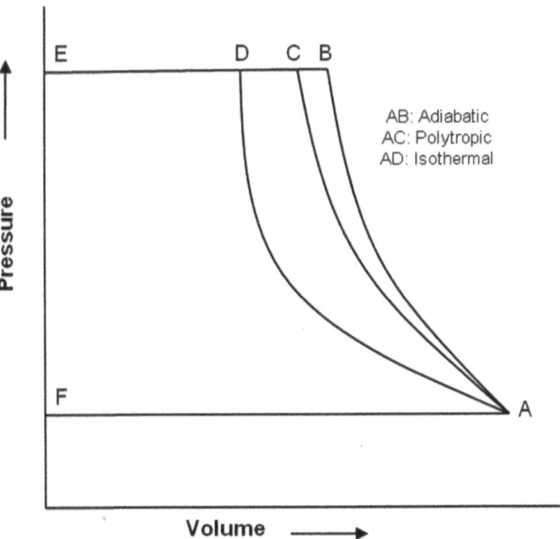

Fig. 2.6 Representation of theoretical compression cycles on a *p-V* diagram. (Based on original material by [1])

2.4 Compressor Power

The power requirement for a real compressor is related to the theoretical cycle through the compression efficiency. Positive displacement compressors have typical efficiencies between 88 and 95% depending on the size and type [13]. In general, compressor efficiencies are reported between 85 and 90% [14]. The compression efficiency is equal to the ratio of the theoretical power to the real power; therefore, the real power can be obtained by dividing the theoretical power by the efficiency. The theoretical adiabatic single-stage power is:

$$P_{ad} = p_1 \dot{V}_{p1} \frac{k}{k-1} \left(r^{(k-1)/k} - 1 \right) \frac{Z_1 + Z_2}{2Z_1} \tag{2.4}$$

such that p_1 is the inlet pressure, $\dot{V}_{p1}$ is the ideal gas volumetric flow rate at the inlet, Z_1 and Z_2 are the gas compressibilities at the inlet and outlet conditions, respectively, and r is the compression ratio. In later chapters, an alternate, but identical, version of Eq. (2.4) will be used to determine the power requirements of gas blowing/compression for overcoming pressure drop for absorption, adsorption, and membrane separation processes:

$$P_{ad} = \frac{\dot{m}RT}{M} \frac{k}{k-1} \left(r^{(k-1)/k} - 1 \right) \frac{Z_1 + Z_2}{2Z_1} \tag{2.5}$$

such that $\dot{m}$ is the inlet mass flow rate of gas, T is the inlet temperature and M is the molecular weight of the inlet gas. Equations (2.4) and (2.5) are related directly using the ideal gas law. In addition, since power is defined as the work done per unit time, dividing Eq. (2.4) and Eq. (2.5) by the gas molar flow rate results in the work per mole of gas compressed.

The real gas volumetric flow rate is related to the ideal gas volumetric flow rate by the compressibility, *i.e.*, $\dot{V}_{r1} = \dot{V}_{p1} Z_1, = \dot{n}_{p1} RT/P$ such that $\dot{V}_{p1}$ and $\dot{n}_{p1}$ are the volumetric and molar flow rates, respectively, of the perfect or ideal gas at the inlet conditions. In this situation, Eq. (2.4) reduces to:

$$P_{ad} = \frac{p_1 \dot{V}_{r1}}{Z_1} \frac{k}{k+1} \left(r^{(k-1)/k} - 1 \right) \frac{Z_1 + Z_2}{2Z_1} \tag{2.6}$$

The isothermal compression power for any number of stages provided r is the overall compression ratio is expressed by:

$$P_{iso} = \frac{p_1 \dot{V}_{r1} \ln r}{Z_1} \frac{Z_1 + Z_2}{2Z_1} \tag{2.7}$$

As pressure increases and temperature decreases, intermolecular forces between gases play an increasing role in describing their occupied volume. The compressibility factor Z, unique from the gas compressibility, is typically determined by experimental data is often used as a multiplier to improve the accuracy of the ideal gas law, *i.e.*, $P\dot{V} = \dot{n}ZRT$. Compressibility data can be obtained from generalized compressibility charts based upon reduced conditions. Reduced pressure and temperature are defined as:

$$p_r = \frac{p}{p_c} \tag{2.8}$$

and

$$T_r = \frac{T}{T_c} \tag{2.9}$$

such that p_c and T_c are the critical pressure and temperature of the gas, respectively. Generalized gas compressibility charts are available in Appendix C.

Example 2.2 Find the compressibility factors at inlet and discharge conditions for the gas mixture in Example 2.1 when compressed from 2172 kPa and 65.5°C (338.5 K) to 6653 kPa. Also determine the volume or power required.

Solution: Note that temperatures and pressures ***must*** be in absolute values in this calculation.

Step 1 Calculate pseudocritical temperature and pressure for a given gas mixture. Recall that the properties of a gas mixture can be determined by the properties for the components through a mole-weighted average.

Gas	Mol %	T_c (K)	Pseudo T_c (K)	p_c (kPa)	Pseudo p_c (kPa)
H_2	30	46	13.8	2254	676.2
CO_2	20	304	60.8	7398	1479
CO	50	134	67.0	3496	1748
	100		*141.6*		*3903*

Step 2 For adiabatic compression, calculate the discharge temperature (T_2) from the following equation.

$$T_2 = T_1 r^{(k-1)/k} \quad \text{where } T_1 \text{ is the inlet condition}$$

The compression ratio (r) is $6653/2172 = 3.06$ and the ratio of specific heat (k) is 1.37 from Example 2.1. Therefore, the theoretical discharge temperature is

$$T_2 = (338.5\ \text{K})\left(3.06^{(1.37-1)/1.37}\right) = 458\ \text{K}$$

Step 3 Determine the reduced temperature and pressure for the inlet and discharge conditions using Eqs. (2.7) and (2.8).

	Inlet	Discharge
Pressure, kPa	2172	6653
Temperature, K	338.5	458
Reduced pressure (p_r)	0.55	1.70
Reduced temperature (T_r)	2.39	3.23

Step 4 Read the compressibility factor, Z from the applicable generalized chart in Appendix C.

	Inlet	Discharge
Compressibility	0.99	1.03

Step 5 Use the compressibility factor in Eq. (2.5) to determine the theoretical power for an adiabatic single-stage compressor. A basis for $\dot{V}_{r1}$ of 1 m^3/min (0.0167 m^3/s) at inlet conditions is assumed.

$$P_{ad} = \frac{(2.2 \times 10^6)(0.0167)}{0.99}\left(\frac{1.37}{1.37-1}\right)\left(3.06^{(1.37-1)/1.37} - 1\right)\left(\frac{0.99 + 1.03}{(2)(0.99)}\right)$$

$$= 49.1\ \text{kW}$$

Note that this is the required theoretical power based on 1 m^3/min at inlet conditions for a single-stage compressor. In reality, other parameters such as mechanical losses, and line leakage should be considered.

In compression processes involving high gas throughputs, multiple stages are desired for controlling the discharge temperature, limiting the pressure differential, and the potential power savings. A reciprocating compressor typically requires an individual cylinder for each stage with gas cooling in between. An example of the p-V diagram

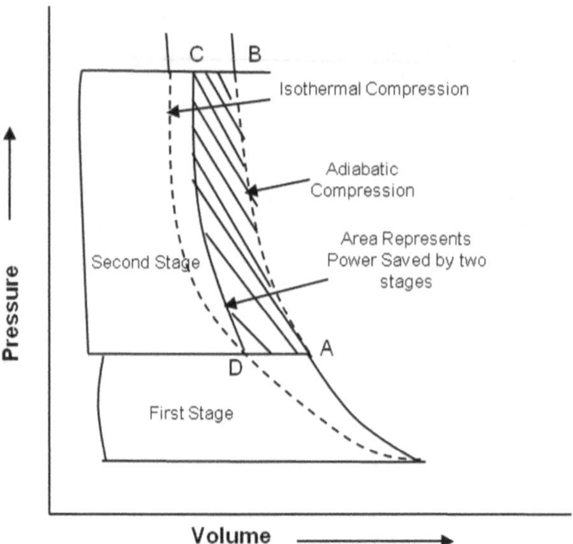

Fig. 2.7 Combined *p-V* diagram for a 2-stage air compressor. (Based on original material by [1])

for a 2-stage 100-psig air compressor is shown in Fig. 2.7. To minimize power with ideal intercooling between each stage, a theoretical relation exists between the intake pressures and succeeding stages, which is derived by equating the ratios of compression and temperatures of each stage. The formula for the compression ratio per stage r_s, based upon the total compression ratio r_t is:

$$r_s = \sqrt[s]{r_t} \tag{2.10}$$

such that s is the number of stages. In this way, the compression ratio per stage for a 2-stage compression process is $\sqrt[2]{r_t}$, and for a 3-stage, $\sqrt[3]{r_t}$, etc. Each stage is considered as a separate compressor, with the capacity (V_1) of each stage calculated separately from the first-stage real intake volume. Corrections are made to account for actual pressure and temperature conditions existing at the higher-stage cylinder inlet in addition to any moisture content in case condensation occurs during inter-stage cooling.

The initial inlet volume and those of subsequent stages may be expressed in terms of scfm (14.7 psia, 60°F, dry) as:

$$\dot{V}_1 = \frac{\dot{n} Z_1 R T_1}{p_1} \tag{2.11}$$

The volume can be corrected for a wet gas by multiplying by the following correction factor:

$$\frac{p_1}{p_1 - p_0} \tag{2.12}$$

such that p_0 is the vapor pressure of the moisture contained in the gas mixture.

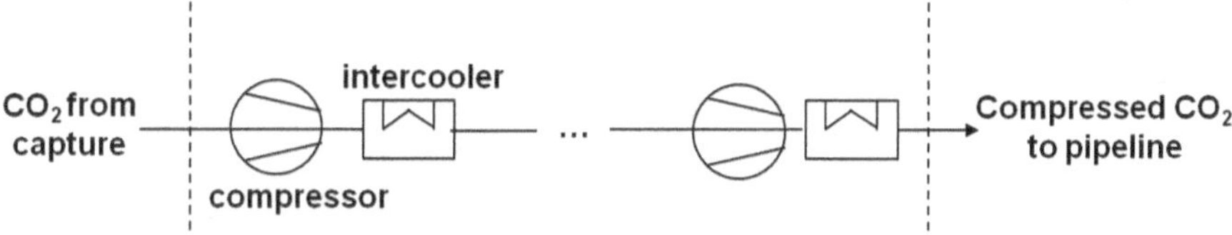

Fig. 2.8 Schematic of the multiple-stage CO_2 compression process

There are various methods available to determine the total number of stages recommended for a given desired compression ratio. However, as a rule-of-thumb, in modern centrifugal compressors, compression ratios are typically limited to 4 at each stage to avoid excessive heating of the gas and allow for cooling of the gas between stages, which is termed intercooling [13]. Compression ratios are typically higher in reciprocating compressors than centrifugal compressors. Figure 2.8 shows a schematic representing the steps required following CO_2 capture, *i.e.*, multiple-stage compression with intercooling prior to pipeline transport. A quick selection method for multistage compressors is available in Appendix D.

2.5 Advanced Compression

To decrease the cost associated with compression of CO_2, whether for a separation process or for pipeline transport, advanced compression technologies must be pursued. For instance, estimates sourced from DOE/NETL indicate that centrifugal compression of CO_2 from the pressure conditions of the regeneration column (*i.e.*, 20–25 psi or 0.14-0.17 MPa) to the desired pipeline pressure (*i.e.*, 1450-2200 psi or 10-15 MPa) costs on the order of 0.1 MW of electricity per ton of CO_2 captured for a new 667-MW_{gross} supercritical pulverized coal-fired power plant [15]. In addition to multi-stage compression with interstage cooling, as previously discussed, another route may include a combination of compression, cooling, and pumping to supercritical pressures. This approach is being investigated by Southwest Research Institute [16], in which refrigeration is used to liquefy the CO_2 so that pumping rather than compression is used for increasing pressure. In this case, the dominant power requirements are for an initial compression stage that increases the CO_2 pressure to 250 psi (1.7 MPa) in addition to the cooling required to liquefy gaseous CO_2. Once liquefied, the subsequent pumping power required to increase the pressure to the desired pipeline condition is minimal.

Another approach developed by Lawlor of Ramgen Power Systems, Inc., [17] relies on a supersonic shock-wave compression technology that uses a rotating disk operating at high speeds to generate shock waves that lead to the compression of CO_2. This technology allows for a resulting high-efficiency compressor capable of single-stage compression ratios greater than what is possible with existing axial or centrifugal compressors, thereby reducing the capital costs associated with the overall compression process. An additional benefit is the potential recovery of

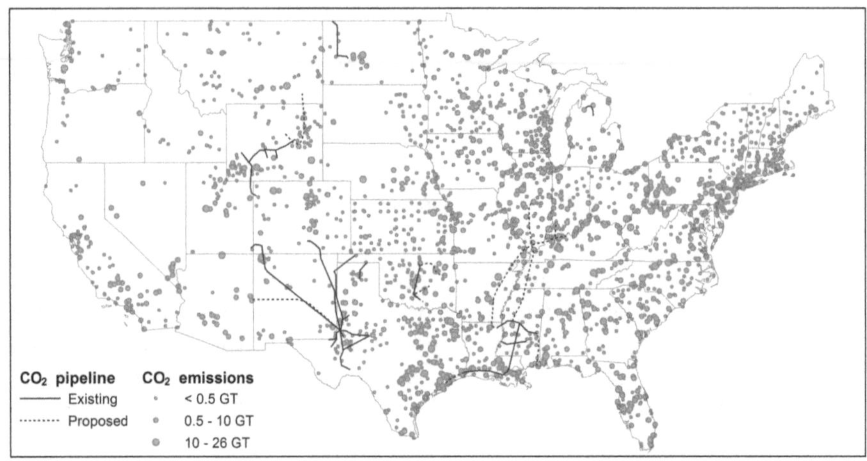

Fig. 2.9 Major CO_2 pipelines in the U.S. (September 2010) [3]

compressed CO_2 at higher temperatures due to the increased compression ratio, which may provide opportunities for additional heat recovery from the system.

2.6 Pipeline Transport

Pipelines are the most common method for transporting large volumes of CO_2 over long distances. Figure 2.9 shows a map of the U.S. with existing CO_2 pipelines, which are primarily used for EOR. The Canyon Reef Carriers Pipeline located in Texas is the oldest long-distance pipeline in the U.S. The pipeline is 225 km long and began service in 1972 for EOR in the regional oil fields. The pipelines are operated at ambient temperature and high pressure with primary compression stations located at the injection sites and booster compression stations located intermittently along the pipeline [18]. In total, approximately 3,900 miles [3, 19] of CO_2 pipeline have been constructed in the U.S. as of September 2010, compared to the approximate 300,000 miles [20] of natural gas pipeline. As of 2008, approximately 50 Mt/y of CO_2 is transported over land for EOR operations [18], which is the equivalent of the CO_2 generated by approximately twenty 500-MW coal-fired power plants. The largest of the existing CO_2 pipelines, completed in 1983, is the 30-inch Cortez pipeline that extends approximately 500 miles from McElmo Dome in Southwestern Colorado to the EOR fields in West Texas [18]. For climate stabilization, a study by Dooley et al. [19] estimates that between 11,000 and 23,000 additional miles of CO_2 pipeline will be added to the existing network in the U.S. before 2050.

The pipeline diameter is the primary design constraint when considering CO_2 pipeline transport, and the parameters that must be considered in its estimation are pressure drop, elevation change, intended CO_2 mass flow rate, CO_2 compressibility and viscosity. The differential form of the energy balance is:

$$\frac{\rho u}{g_c \nu} du + \frac{1}{\nu} dp + \frac{g}{g_c \nu^2} dh + \frac{2 f_F (\rho u)^2}{g_c D_i} dL = 0 \tag{2.13}$$

such that the terms on the left hand side of Eq. (2.13) represent the change in kinetic energy, pressure-volume work, change in potential energy, and energy loss due to surface roughness in the flow system, respectively; u is the fluid velocity, ρ is the fluid density, g is the acceleration due to gravity, ν is the specific volume of the fluid, p is the pressure, h is the height, g_c is the conversion factor for the force unit, which is unity in SI units, f_F is the Fanning friction factor, L is the pipeline length, and D_i is the inner pipeline diameter. Integration over each term and solving for the pipe diameter as a function of pressure drop yields [10]:

$$D_i = \left\{ \frac{-64 Z_{ave}^2 R^2 T_{ave}^2 f_F \dot{m}^2 L}{\pi^2 \left[M Z_{ave} R T_{ave} \left(p_2^2 - p_1^2 \right) + 2 g P_{ave}^2 M^2 (h_2 - h_1) \right]} \right\}^{1/5} \tag{2.14}$$

such that $\dot{m}$ is the maximum annual mass flow rate of CO_2, Z_{ave} is the average fluid compressibility, and R is the ideal gas constant. The Fanning friction factor [21] may be approximated by:

$$\frac{1}{2\sqrt{f_F}} = -2.0 \log \left\{ \frac{\varepsilon / D_i}{3.7} - \frac{5.02}{Re} \log \left[\frac{\varepsilon / D_i}{3.7} - \frac{5.02}{Re} \log \left(\frac{\varepsilon / D_i}{3.7} + \frac{13}{Re} \right) \right] \right\} \tag{2.15}$$

such that ε is the internal surface roughness of the pipe and Re is the Reynolds number. A reasonable estimate for surface roughness is 0.0457 mm [22]. The fluid velocity is well within the turbulent region in pipeline transport with Reynolds numbers typically on the order of 10^6.

Using Eq. (2.14), the pipeline diameter is plotted as a function of pipeline segment length with a given soil temperature of 12°C and a pressure change from 10 to 14 MPa in Fig. 2.10 for various CO_2 mass flow rates. An obvious trend is that the pipe diameter increases with increasing mass flow rate. Another trend that can be seen from Fig. 2.10 is the step-wise increase in the pipe diameter as the pipeline length increases for a given mass flow rate. It is important to recognize that if the CO_2 mass flow rate does not consist of pure CO_2, that the fluid properties and subsequent transport properties will be influenced [23]. Goos et al. [24] investigated the phase behavior and fluid densities of CO_2–N_2 gas mixtures based upon the Soave-Redlich-Kwong equation of state. The phase diagrams for mixtures of CO_2 and N_2, ranging from near-pure CO_2 to a mixture containing 80 mol.% CO_2 with the balance N_2 are shown in Fig. 2.11. As the purity of CO_2 in the gas mixture decreases, a 2-phase region develops, and grows as can be seen from Fig. 2.11 b to c.

In the work of Goos et al. [24], an 8-stage compressor was assumed in modeling the compression of a 1000 kmol/h flow rate. The density predictions and corresponding compression energy requirements are listed in Table 2.1. Impurities such as SO_x, NO_x, HCl, and H_2 are also present in both flue and fuel (H_2, in particular) gases.

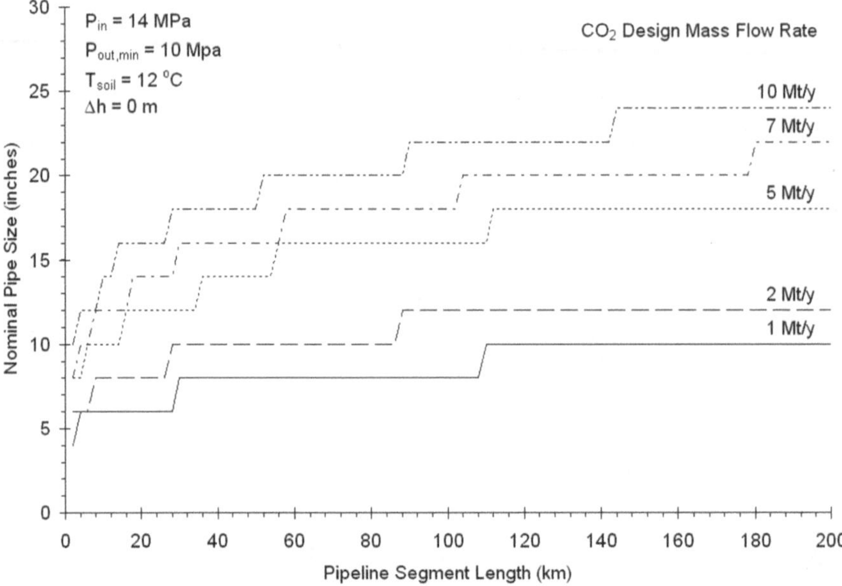

Fig. 2.10 Pipeline diameter as a function of pipeline segment length for a series of mass flow rates. (With permission from Elsevier [2])

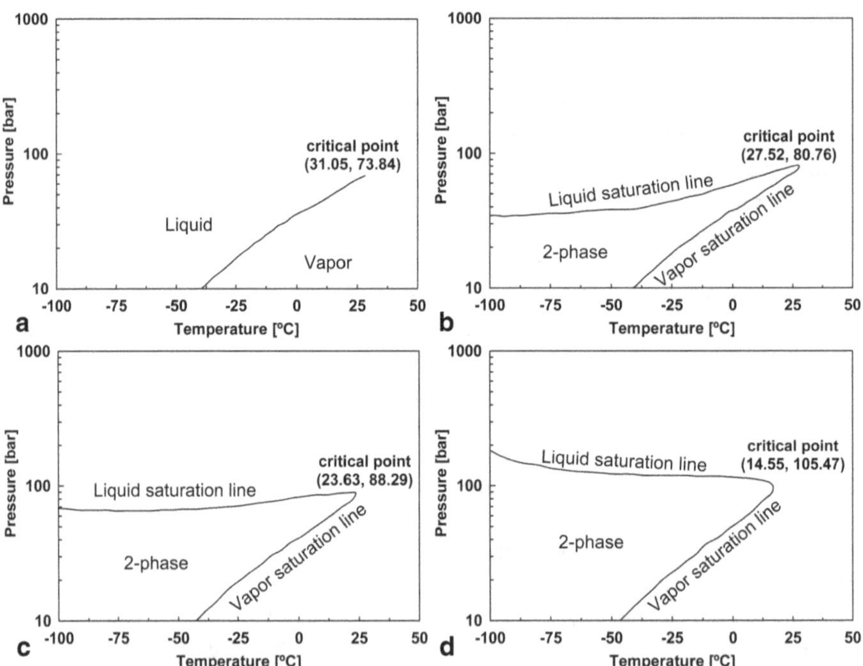

Fig. 2.11 Phase diagrams generated by PRO/II for different CO_2–N_2 gas mixtures using the Soave-Redlich-Kwong (SRK) equation of state at **a** 99.99 mol% CO_2, 0.01 mol% N_2, **b** 95 mol% CO_2, 5 mol% N_2, **c** 90 mol% CO_2, 10 mol% N_2, and **d** 80 mol% CO_2, 20 mol% N_2. The two-phase region (envelope) becomes apparent with CO_2 purity of 95% and less. (with permission from Elsevier [4])

Table 2.1 Compression energy and corresponding fluid density [24] at 110 bar, 30°C

Composition	Energy (kWh/t CO_2)	Density (kg/m^3)
99.99 mol% CO_2, 0.01 mol% N_2	86	792
95 mol% CO_2, 5 mol% N_2	87	681
90 mol% CO_2, 10 mol% N_2	88	536
80 mol% CO_2, 20 mol% N_2	89	343

Table 2.2 European Union's recommended quality specifications for pipeline transport of CO_2

Component	Concentration limit	Application
H_2O	300–500 ppm	Free water minimization
H_2S	200 ppm	Health and safety
CO	2000 ppm	Health and safety
SO_x	100 ppm	Health and safety
NO_x	100 ppm	Health and safety
O_2	<4 vol%	Aquifer storage
	<1000 ppm	EOR technical limit
CH_4	<4 vol%	Aquifer storage
	<2 vol%	EOR technical limit
$N_2 + Ar + H_2$	<4 vol% total	

How the density changes as a function of the presence of these impurities is still unknown.

The extent of impurities in the CO_2 stream will also influence the extent of corrosion of the pipeline materials. This is clear for the case of acidic species such as H_2S, but can also be a problem for water. For instance, if the water concentration in the stream is greater than the solubility limit of the operating conditions (*i.e.*, temperature and pressure) of the pipeline, the presence of free water will lead to the formation of carbonic acid (H_2CO_3), which will lead to the corrosion of carbon steel pipelines. An additional challenge associated with the presence of water is the potential for *hydrate* formation. In general, *clathrates* are chemical frameworks in which a gas is trapped within a cage or lattice formed by molecules of a second type, with a hydrate as an example in which the molecular lattice is formed by water molecules. Hydrates of CO_2 ($CO_2 \cdot 6H_2O$) can be formed in a pipeline with free water present down to 300 ppm if the operating temperature drops to below approximately 15°C. This results in the formation of an ice-like solid that can plug the pipeline and is extremely difficult to remove once formed.

In the case of precombustion CO_2 separation from the fuel gas, the presence of H_2 can be a challenge. If the concentration of H_2 gas is not below a certain level, hydrogen migration into the carbon steel can lead to embrittlement of the material. The mechanism by which H_2 can dissociate and diffuse into the material is identical to that of the H_2-selective membranes discussed in Chap. 5. Recommended quality specifications for CO_2 pipeline transport published by the European Union's Dynamis project are summarized in Table 2.2.

In addition to the initial compression, booster pumps are positioned along the length of the pipeline. Figure 2.12 demonstrates how the compression power changes

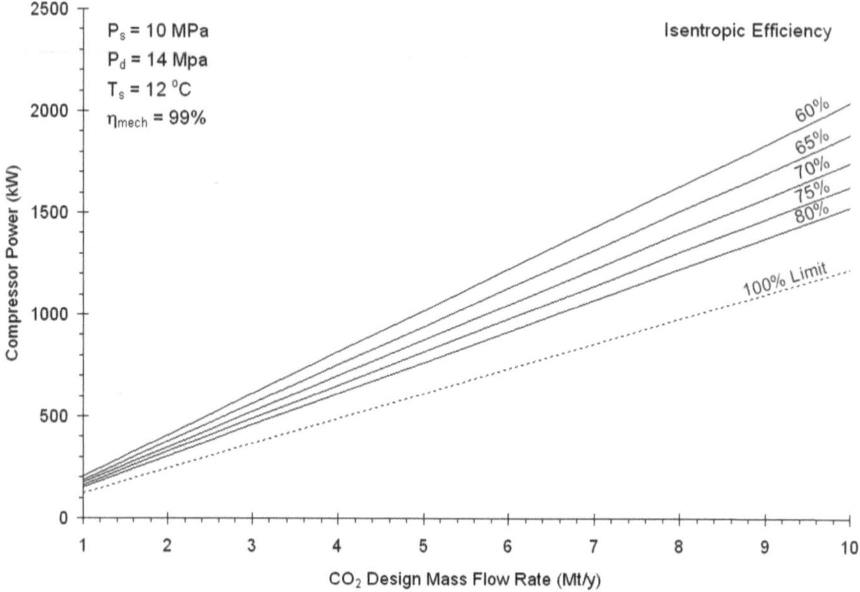

Fig. 2.12 Results of a booster compression model showing the linear relationship between the compression size and the mass flow rate for various isentropic efficiencies. (With permission from Elsevier [2])

linearly with the mass flow rate, with the slope steepening as the isentropic efficiency decreases [10].

For pipeline design, the choice of appropriate material selection for the prevention of fracture propagation from a puncture due to the low temperatures caused by decompression in the line is also important [25]. Other important considerations in the pipeline design include having appropriately spaced isolation valves (*e.g.*, closer together in populated areas and low-lying areas) to limit leakage in case of a break.

References

1. Dresser-Rand and Ingersoll-Rand as reported in A Practical Guide to Compressor Technology, 2nd Ed., 2006, John Wiley & Sons, Inc.
2. McCoy ST, Rubin ES (2008) An engineering-economic model of pipeline transport of CO_2 with application to carbon capture and storage. Int J. Greenh. Gas Control, 2(2):219–229
3. Bliss K, Eugene D, Harms RW, Carrillo VG, DCoddington K, Moore M, Harju J, Jensen M, Botnen L, Marston PM, Louis D, Melzer S, Dreschsel C, Moody J, Whitman L (2010) Amann, R. (ed), A policy, legal, and regulatory evaluation of the feasibility of a national pipeline infrastructure for the transport and storage of carbon dioxide. Interstate Oil and Gas Compact Commission, Oklahoma City, OK
4. Goos E, Riedel U, Zhao L, Blum L (2011) Phase diagrams of CO_2 and CO_2-N_2 gas mixtures and their application in compression processes. Energy Procedia, 3778–3785

5. Svensson R, Odenberger M, Johnsson F, Strömberg L (2004) Transportation systems for CO_2-application to carbon capture and storage. Energy Convers Manag 45(15–16):2343–2353
6. EIA (2008) Existing electric generating units by energy source; Energy Information Administration (EIA). Department of Energy (DOE), Washington, D.C.
7. World Energy Outlook (2009) International Energy Agency, Access to Electricity. (accessed August 13, 2011)
8. Zhang ZX, Wang GX, Massarotto P, Rudolph V (2006) Optimization of pipeline transport for CO_2 sequestration. Energy Convers Manag 47(6):702–715
9. Amrollahi Z, Ertesvag IS, Bolland O (2011) Thermodynamic analysis on post-combustion CO_2 capture of natural-gas-fired power plant. Int J Greenh Gas Control 5:422–426
10. McCoy ST (2008) The economics of CO_2 transport by pipeline and storage in saline aquifers and oil reservoirs. Carnegie Mellon University, Pittsburgh PA, Thesis (PhD)
11. Reid RC, Prausnitz JM, Poling BE (1987) The properties of gases and liquids, 4th edn. McGraw Hill Book Co., New York, NY, p 598
12. Skovholt O (1993) CO_2 transportation system. Energy Convers Manag 34(9–11):1095–1103
13. Bloch HP (2006) A practical guide to compressor technology. Wiley-Interscience, Hoboken, NJ
14. (a) van der Sluijs J, Hendriks C, Blok K (1992) Feasibility of polymer membranes for carbon dioxide recovery from flue gases. Energy Convers Manag 33(5–8):429–436;(b) Favre E (2007) Carbon dioxide recovery from post-combustion processes: can gas permeation membranes compete with absorption? J Membrane Sci 294(1–2):50–59
15. Haslbeck JL, Black J, Kuehn N, Lewis E, Rutkowski MD, Woods M, Vaysman V (2008) Volume 1: Bituminous Coal to Electricity; National Energy Technology Laboratory (NETL). U.S. Department of Energy, p 315
16. Moore JJ, Nored MG, Gernentz RS, Brun K (2007) Novel concepts for the compression of large volumes of Carbon Dioxide. Oil Gas J
17. Lawlor SP, Baldwin P (2005) In: Conceptual design of a supersonic CO_2 compressor. ASME TURBO EXPO 2005, June 6–9, 2005, Reno NV
18. Metz B (2005) IPCC special report on carbon dioxide capture and storage. Cambridge University Press, Cambridge, p 431
19. Dooley JJ, Dahowski RT, Davidson CL (2009) Comparing existing pipeline networks with the potential scale of future US CO_2 pipeline networks. Energy Procedia 1(1):1595–1602
20. EIA (2007) About U.S. natural gas pipelines – transporting natural gas; Energy Information Administration. U.S. Department of Energy, Washington, DC, p 76
21. Zigrang DJ, Sylvester ND (1982) Explicit approximations to the solution of Colebrook's friction factor equation. AIChE J 28(3):514–515
22. Green DW (eds) (2008) Perry's chemical engineers' handbook. McGraw-Hill, New York
23. Farris C (1983) Unusual design factors for supercritical CO_2 pipelines. Energy Progress 3(3):150–158
24. Goos E, Riedel U, Zhao L, Blum L (2011) Phase diagrams of CO_2 and CO_2-N_2 gas mixtures and their application in compression processes. Energy Procedia 4:3778–3785
25. Mohitpour M, Golshan H, Murray MA (2007) Pipeline design & construction: a practical approach. ASME Press, New York

Chapter 3
Absorption

In absorption and stripping processes mass transfer takes place between gas and liquid phases at each stage throughout a column. In the absorption process the *solute*, or component to be absorbed (*e.g.*, CO_2), is transferred from the gas phase to the liquid phase. In the stripping process, the opposite occurs, *i.e.*, mass transfer occurs from the liquid to the gas phase. These two units are traditionally coupled, as shown in Fig. 3.1, for the solvent to be recovered and recycled, and for an effective separation of CO_2 from a gas mixture to produce a somewhat pure stream of CO_2.

The columns used for absorption and stripping most often use structured packing, but may use trays or random packing; spray towers and bubble columns may also be used as shown in Fig. 3.2. Structured packing is usually preferred in the absorber to minimize gas-side pressure drop. The acid gas treatment industry uses both structured packing and trays in amine strippers.

The rate of mass transfer in an absorption process is complex and dependent upon a number of factors, which include the length of the mass-transfer zone, the solubility of CO_2 in a given solvent, the diffusivity of CO_2 through the solvent, and the chemical reactivity between CO_2 and the reactive component of the solvent. Before considering the details required to design an absorption column for a specific CO_2 separation application it is important to first consider each of these aspects that influence the mass transfer individually.

Since absorption is a complex separation process, a schematic showing how the chapter is organized is presented in Fig. 3.3. As discussed in Sect. 1.6 of Chap. 1, the costs can be divided into non-technical and technical. The factors that influence the cost of the technology associated with the absorption separation process are discussed in this chapter and hence, the compression of CO_2 required for transport is not included in the schematic. Figure 3.3 is a simplistic view of the relationship between cost, equipment, material, and process parameters to emphasize that these aspects are not disconnected, but rather, highly dependent and intertwined in a complex fashion. This schematic is not meant to be inclusive of all connections and components of absorption, but rather a general overview of the primary constituents. At the start of the chapter, the fundamental aspects of solubility, diffusion and kinetics (rate constant) are discussed as they relate to the rate of absorption, since the rate of absorption determines the ratio of liquid-to-vapor flow rates, which then determines the number of stages required for a desired separation. Next, parameters associated with process

J. Wilcox, *Carbon Capture*,
DOI 10.1007/978-1-4614-2215-0_3, © Springer Science+Business Media, LLC 2012

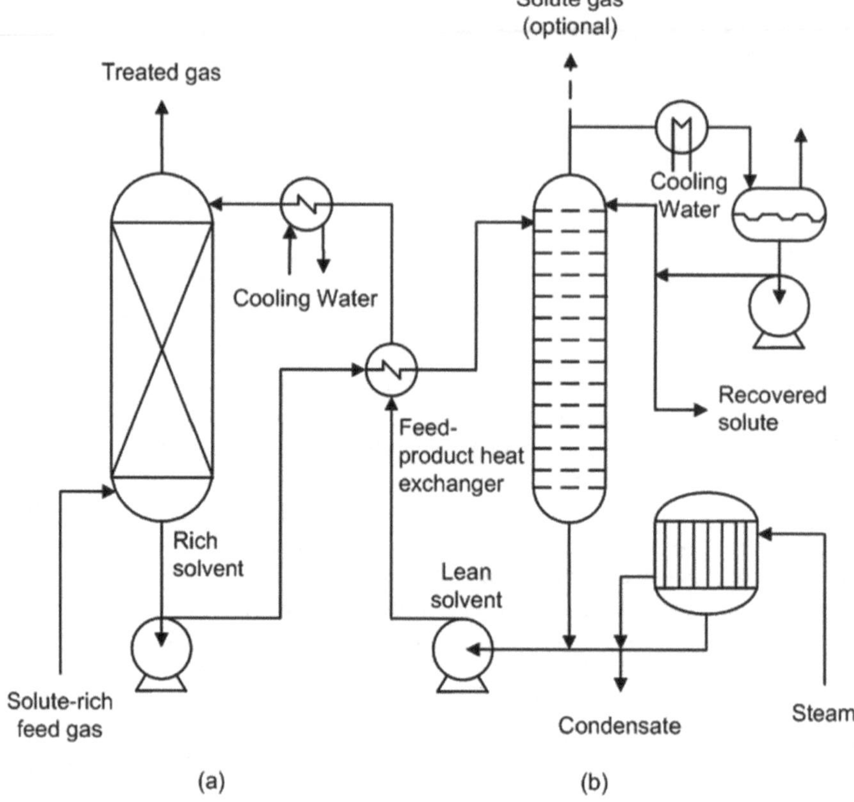

Fig. 3.1 Gas absorber set-up with solvent stripping for regeneration: **a** absorber and **b** stripper

design are discussed, and include absorber height and width. In particular, the number of stages relates directly to the height of the column, which is a primary factor in determining the capital cost of the system. Recall from Chap. 1, that increasing the percent capture of CO_2 increases the minimum thermodynamic work required for separation. Therefore, this factor should influence the real work associated with the process as well. In other words, it will require additional work to capture a higher fraction of the CO_2 from the feed gas. Increasing the number of stages will increase the fraction of CO_2 separated from the feed gas, which results in an increase in the column height. While the height of the column dictates the extent of separation, the diameter of the column is used to control the flows based upon the fluid properties. Increasing the cross-sectional area of the column can aid in minimizing the pressure drop within the column. Gas pressure drop in a system is often overcome by fan or blower power, while liquid pressure drop is often overcome by pumping. These power requirements in addition to the heat required for regeneration comprise the energy requirements of the process subsequently dictating the operating and maintenance costs. Recalling again from Chap. 1, in addition to the percent capture of

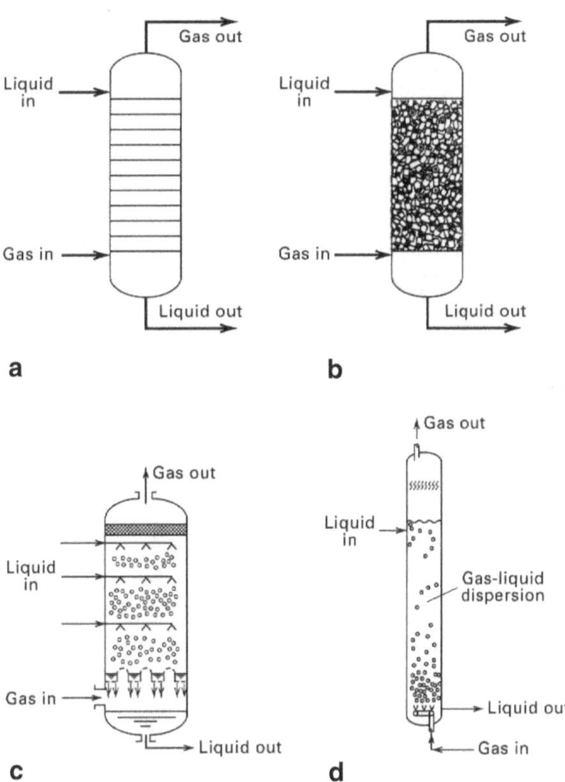

Fig. 3.2 Industrial equipment for absorption and stripping, **a** trayed column; **b** packed column; **c** spray tower; **d** bubble column. (Reprinted with permission of [3])

CO_2, an increase in purity of the CO_2 product stream increases the thermodynamic minimum work of separation as well. As an example, achieving a desired CO_2 purity may require additional heat for effectively separating water from CO_2 if steam is used for regeneration.

The transfer of CO_2 takes place through the gas phase, across the gas-liquid boundary, and then into the liquid-phase as pictured in Steps 1–4 of Fig. 3.4. In step 1, CO_2 diffusion takes place in the gas phase of a given mixture, followed by Step 2, which involves dissolution of CO_2 into a liquid film that separates the gas and liquid phases. The concentration of CO_2 at the gas-liquid interface is typically determined by Henry's law and will be described in greater detail in the next section. In the case of pure physical absorption, this would be the final step in the transfer pathway. However, many absorption processes of CO_2 capture are dependent upon chemical absorption, which involves two additional steps, *i.e.*, diffusion of CO_2 to a binding agent in the liquid phase (Step 3 in Fig. 3.4) and subsequent chemical reaction of CO_2 with the binding agent in the liquid phase (Step 4 in Fig. 3.4). Therefore, it becomes clear that determination of the rate of absorption of CO_2 requires knowledge of diffusion in gas and liquid phases, solubility of CO_2 within the liquid phase, and finally the chemical kinetics between CO_2 and the binding component in solution. These

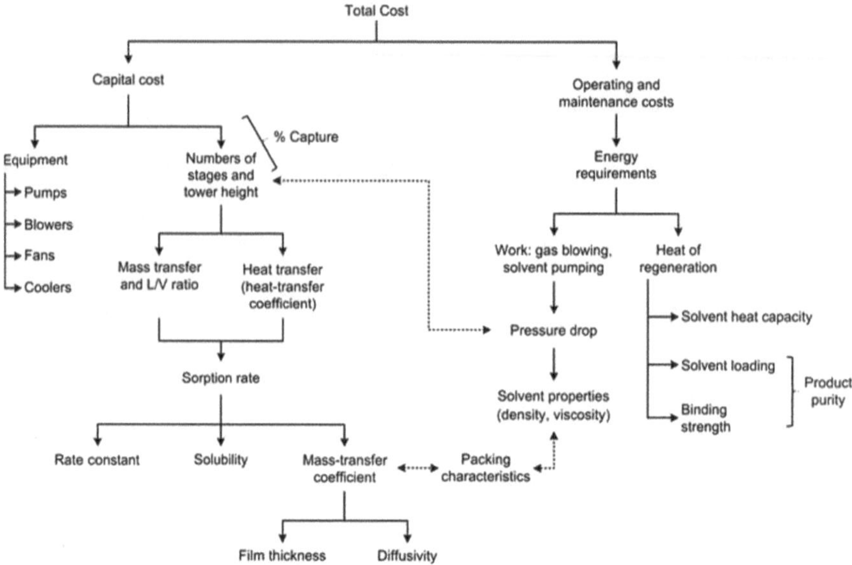

Fig. 3.3 Schematic outlining the components that comprise the cost of the CO_2 absorption separation process

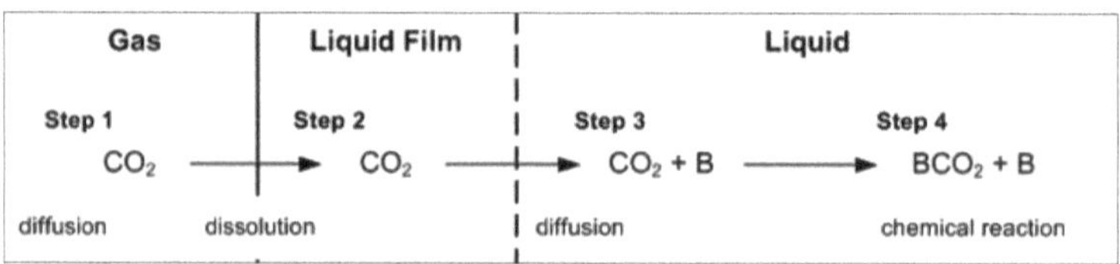

Fig. 3.4 Mass-transfer steps from CO_2 in the gas phase to the solution containing the binding material

aspects are discussed in detail as they ultimately determine the rate of absorption of CO_2 from gas to liquid phases.

3.1 Solubility of CO_2 in Aqueous Solutions

Solubility plays a major role in determining the rate of absorption as it helps to determine the concentration of CO_2 at the interface, according to Henry's Law. It is important to note, however, that Henry's Law is only valid for sufficiently dilute systems. For instance, for flue gas and direct air capture (DAC) applications, Henry's Law is appropriate, but for gasification applications, in which the CO_2 concentration in the gas phase and subsequently in solution, are too high, Raoult's Law (or modified version) may be more appropriate. This will be described in greater detail shortly. At equilibrium, a gas in contact with a liquid has equal *fugacity* in both the gas and liquid phases. The fugacity of an ideal gas is equal to its pressure, but for nonideal gases, the fugacity can be thought of as a corrected pressure due to the nonideal behavior of the gas. For ideal solutions, Raoult's law can be used to determine the relationship between the gas-phase mole fraction of CO_2, y_{CO_2}, and the liquid-phase mole fraction of CO_2, x_{CO_2} as follows:

$$y_{CO_2} = \frac{p_0}{p} x_{CO_2} \tag{3.1}$$

such that p is the total pressure and p_0 is the vapor pressure (saturation pressure) of pure CO_2.

However, if the concentration of dissolved CO_2 in solution is small and the temperature and pressure are far from the critical temperature and pressure of gaseous CO_2, which is often the case for traditional gas mixtures of CO_2, then Henry's law applies and the concentration, x_{CO_2}, of dissolved CO_2 in equilibrium with a partial pressure, p_{CO_2}, is given by:

$$p_{CO_2} = H x_{CO_2} \tag{3.2}$$

such that H is the Henry's Law constant. From Eq. (3.2), it is clear that typical units of H are those of pressure since the mole fraction is dimensionless. However, H is also often expressed in units of pressure multiplied by the inverse of concentration, *i.e.*, atm/M (*i.e.*, atm L/mol). It is important to be comfortable with both types of Henry's law units. If CO_2 reacts in solution, as it does in the case of a chemical absorption process to ensure a driving force for mass transfer and effective separation, Henry's Law does not apply to all of the dissolved gas in solution, but only the unreacted gas in solution. Usually H is temperature-dependent and relatively pressure-independent at moderate conditions. Also, H is a function of the ionic strength of a given solution, as described next.

The solubility of CO_2 in solution can change dramatically based on the type and concentration of electrolyte components contained in a given solution. The solubility of CO_2 in an electrolyte solution may be estimated based upon a method developed by Dirk Willem van Krevelen and his research assistant Jan Hoftijzer [4], which was originally proposed by Setchenow in 1892 [5]. The empirical method relates the Henry's Law constant of CO_2 in pure water, H^0, to that of the electrolyte solution H, by:

$$\log_{10}\left(\frac{H}{H^0}\right) = hI \tag{3.3}$$

such that I (in M) is the ionic strength of the solution defined by:

$$I = \frac{1}{2}\sum c_i z_i^2 \tag{3.4}$$

in which c_i is the concentration of ions with charge z_i. The quantity h (in L/g of ion) in Eq. (3.3) is the sum of the contributions of positive and negative ions associated with the dissociated ionic species in solution, such that:

$$h = h_+ + h_- + h_G \tag{3.5}$$

Table 3.1 Values of h_+, h_-, and h_G (L/g ion) for CO_2 in electrolyte solutions [6b, 7]

Values of h_+ and h_- (L/g ion)			
(L/g h_+ ion)		(L/g h_- ion)	
H^+	0.000	NO_3^-	−0.001
NH_4^+	0.028	Br^-	0.012
Fe^{2+}	0.049	Cl^-	0.021
Mg^{2+}	0.051	CO_3^{2-}	0.021
Ca^{2+}	0.053	SO_4^{2-}	0.022
Co^{2+}	0.058	OH^-	0.066
Ni^{2+}	0.059		
K^+	0.074		
Na^+	0.091		

Values of h_G (L/g ion) for CO_2

°C	0	15	25	40	50
h_G	−0.007	−0.010	−0.019	−0.026	−0.029

where h_G is the ionic species associated with the gas. Table 3.1 provides a list of the solubility data required for determining h for CO_2 in various electrolyte solutions [6]. Note that the ratio within the logarithm of Eq. (3.3) is equivalent to the ratio of the CO_2 solubility in water to that of a given ionic solution. If h is negative, the ratio of solubilities is less than one, implying that CO_2 will be more soluble in the ionic solution than in water. Conversely, if h is positive, the solubility ratio is greater than one implying that CO_2 is less soluble in the ionic solution than in water. As an example, consider the solubility of CO_2 in a 1.5 M aqueous solution of sodium hydroxide (NaOH) at 20°C. The solubility ($1/H^0$) of CO_2 in water is 0.039 mol/L atm. Using Eqs. (3.3–3.5) along with the data of Table 3.1, leads to a solubility of CO_2 in the 1.5 M aqueous NaOH solution of 0.028 mol/L atm. In general, at 20°C, the only cation-anion pairs in Table 3.1 that result in a negative h (*i.e.*, CO_2 is more soluble in the ionic solution) are HNO_3 or HBr.

Example 3.1 Estimate the solubility of CO_2 in 1.5 M NaOH solution at 20°C. The solubility (reciprocal of H^0) of CO_2 in water at 20°C is 0.039 mol/L atm.

Solution: From Table 3.1, we have $h_+ = 0.091$ L/g ion, $h_- = 0.066$ L/g ion for the NaOH solution and $h_G = -0.015$ L/g ion (by interpolation). Using Eq. (3.5), $h = 0.142$ L/g ion. The ionic strength of the solution is determined by Eq. (3.4) as

$$I = \frac{1}{2}[(1.5(+1)^2) + (1.5(-1)^2)] = 1.5 \text{ mol/L}$$

Substituting the value of h and I in Eq. (3.3) yields

$$\log_{10}(0.039H) = (0.142)(1.5)$$
$$1/H = 0.024 \text{ mol/L atm}$$

Therefore, the solubility ($1/H$) of CO_2 in 1.5 M NaOH solution is 0.024 mol/L atm.

Example 3.2 Estimate the solubility of CO_2 in a solution containing 1 M Na_2CO_3 and 1 M NaOH at 20°C.

Solution: In case of mixed electrolytes, assume that the value of H is given by

$$\log_{10}\left(\frac{H}{H^0}\right) = h_1 I_1 + h_2 I_2 + \ldots$$

where I_i is the ionic strength attributable to species i of mixed electrolytes and h_i has the value characteristic of that species. In this example, define Na_2CO_3 and NaOH as species 1 and 2, respectively.

$$h_1 = 0.091 + 0.021 - 0.015 = 0.097$$
$$h_2 = 0.091 + 0.066 - 0.015 = 0.142$$
$$I_1 = \frac{1}{2}\left[\left(1(+1)^2\right) + \left(1(+1)^2\right) + \left(1(-2)^2\right)\right] = 3$$
$$I_2 = \frac{1}{2}\left[\left(1(+1)^2\right) + \left(1(-1)^2\right)\right] = 1$$

Consequently, the solubility ($1/H$) of CO_2 in the mixed electrolytes is

$$\log_{10}(0.039H) = (0.142)(1) + (0.097)(3) = 0.433$$
$$1/H = 0.014 \text{ mol/L atm}$$

A more general form to express CO_2 solubility is through the vapor-liquid equilibrium constant, K_{eq}, defined by:

$$y_{CO_2} = K_{eq}\, x_{CO_2} \tag{3.6}$$

The y–x plot is often referred to as the equilibrium line for absorption separation processes. Solubility data [8] associated with CO_2 in aqueous solution for temperatures and CO_2 partial pressures ranging from 0 to 100°C and 1–36 atm are available in Appendix E. The data show that solubility decreases with increasing temperature and increases with pressure. The largest mole fraction of CO_2 in solution is 0.02623 at 0°C and 32 atm and can be compared to the lowest of 0.000608 at 25°C and 1 atm. Spanning the full temperature and pressure range in Appendix E, the solubility of CO_2 in aqueous solution varies by more than two orders of magnitude.

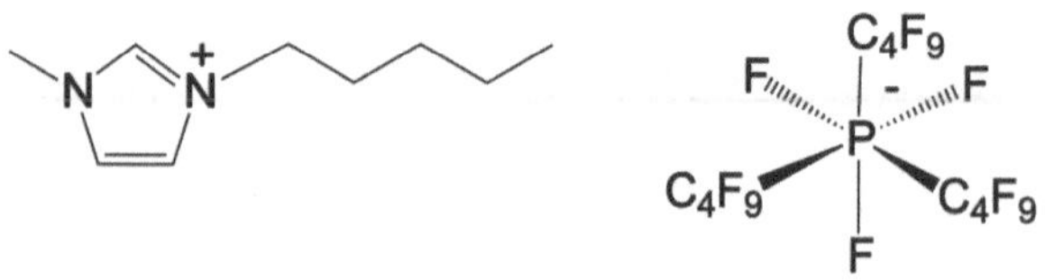

Fig. 3.5 Example structure of an ionic liquid with abbreviation [p$_5$mim][bFAP] and name 1-pentyl-3-methylimidazolium tris(nonafluorobutyl)trifluorophosphate

Ionic Liquids There are certain solvents that have a substantially enhanced CO_2 solubility. An example of such a solvent is an *ionic liquid*. An ionic liquid is a salt in the liquid phase, with the cation-anion pair tunable for enhancing CO_2 solubility. For instance, the ionic liquid termed [p$_5$mim][bFAP], which is pictured in Fig. 3.5, has a solubility of 0.049 at 25°C and 1 atm, which is a significant enhancement from the ambient conditions of a pure aqueous solution. Consider a gas with a CO_2 partial pressure of 10 atm in equilibrium with this ionic solution at 25°C. At these conditions, H_{CO_2} of [p$_5$mim][bFAP] is approximately 20 atm, which yields an approximate CO_2 liquid-phase mole fraction of 0.5 [9]. However, it is important to recognize that the approximate partial pressure of CO_2 in a typical flue gas mixture is only 0.12 atm and in air a mere 0.000391 atm, which correspond to CO_2 liquid-phase mole fractions of 0.006 and 0.0000195, respectively in this particular ionic solution.

The mechanism by which CO_2 absorbs into an ionic liquid is based upon the free volume between the ionic framework of the solution. The anions and cations participate in Coulombic interactions, which lead to a more rigid packing than found in traditional molecular solvents. This is illustrated by the lower thermal expansion coefficients of ionic liquids compared to typical organic solvents, *i.e.*, approximately $0.4–0.7 \times 10^{-3}K^{-1}$ (ionic liquids) versus $1.1–1.8 \times 10^{-3}K^{-1}$ (organic solvents) [10]. Molecular dynamics simulations have provided insight into the mechanism by which CO_2 absorbs into an ionic liquid, and it was found that the anions of the solution play the major role, in that upon introduction of CO_2, the anions will rearrange themselves to create space to accommodate the CO_2 into the void spaces of the ionic framework [11]. The cations are thought to play a secondary role in the CO_2 absorption mechanism. The cation pictured in Fig. 3.5 contains an imidazolium ring bonded to an alkyl chain. Cations with longer alkyl chains have exhibited slightly higher solubilities. It has been suggested that the longer alkyl chain leads a larger free volume in solution, which increases the CO_2 capacity and related CO_2 solubility [12].

3.2 Diffusion of CO_2 in Aqueous Solutions

Rates of diffusion are most commonly presented in terms of a molar flux, *i.e.*, mol/area time, where the area is the cross-sectional area perpendicular to the direction of flow. The simplest form of transport is that in which diffusion occurs within a single phase, in which the diffusivity and concentration of diffusing species are the quantities of interest. The diffusivity and species' concentrations are functions of time,

Table 3.2 Experimental-based diffusivities of common gas-gas pairs at 1 atm [13]

Gas pair	Temperature (K)	Diffusivity (cm^2/sec)
CO_2–H_2	298.2	0.646
CO_2–H_2O	307.4	0.202
CO_2–N_2	298.2	0.165
CO_2–CO	325.4	0.185
CO_2–Air	282.0	0.148
CO_2–O_2	296.0	0.156
$^{12}CO_2$–$^{14}CO_2$	312.8	0.125
CO_2–SO_2	263.0	0.064
CO_2–propane	298.1	0.087
H_2O–O_2	308.1	0.282
H_2O–Air	298.2	0.260
H_2O–Air	333.2	0.305
N_2–O_2	293.2	0.220

system geometry, velocity field, temperature, pressure, initial and boundary conditions, and chemical nature of the diffusing species. The diffusivity, D_A of species A in a given solution is defined as the ratio of the species *flux*, J_A to its concentration gradient. This relationship is known as Fick's first law and is expressed by:

$$J_A = -D_A \frac{dc_A}{dx} \tag{3.7}$$

such that J_A is the flux of a species relative to the average molar velocity of all of the components. The negative sign indicates that the diffusion occurs in the direction of decreasing concentration.

Molecular diffusion, represented by D, is the result of the thermal motion of molecules and can occur as a result of concentration, temperature, pressure or electrical gradients applied to a given mixture. The diffusion coefficient, D_A, for the diffusion of gas A in a given liquid is much smaller than the diffusion of a gas within a gas. Typical gas diffusivities in nonviscous liquids at 25°C range from 0.5×10^{-5} to 2×10^{-5} cm^2/s compared to values of 0.1–1.0 cm^2/s for common gas pairs at atmospheric pressure as shown in Table 3.2 [13]. Table 3.3 lists values of D for CO_2 in various solvents at 25°C [14]. Although the temperature conditions of the various experimental diffusivities reported in Table 3.2 are not the same, they are close enough such that the inverse relationship between molecular weight and diffusivity is clear.

Theoretical estimation of gaseous diffusion may be determined from the method developed by Chapman and Enskog [15] and can be estimated by:

$$D = \frac{1.86 \times 10^{-3} T^{\frac{3}{2}} \left(\frac{1}{M_1} + \frac{1}{M_2}\right)^{\frac{1}{2}}}{p\sigma_{12}^2 \Omega} \tag{3.8}$$

such that D is the gas-phase diffusivity, in units of cm^2/sec, T is the absolute temperature in K, p is the pressure in atm, M_i are the molecular weights of the different species, σ_{12} is the Lennard-Jones collision diameter, expressed in Å, Ω is a function of ε_{12}, which is the Lennard-Jones well depth, expressed in K.

Table 3.3 Diffusivities (STP) of CO_2 in various aqueous-based solvents with their corresponding viscosities

Solvent	Electrolyte concentration (mol/L)	$D_{AB} \times 10^5$ (cm^2/s)	Viscosity (cP)
Water [16]	–	1.92, 1.5[a], 1.15[b]	0.89
Amine-based solvents (30 wt%) [24, 14a, 14b] :			
MEA[c]	–	1.60	2.20
DEA[d]	–	1.18	3.26
MDEA[e]	–	0.98	3.02
AMP[f]	–	0.88	3.79
Electrolyte solvents [25, 26]:			
NaCl	1.041, 3.776	1.73, 1.30	0.972, 1.350
$NaNO_3$	1.076, 3.602	1.76, 1.34	0.952
Na_2SO_4	0.318, 0.898	1.74, 1.50	1.098
$MgCl_2$	0.377, 1.262	1.80, 1.43	1.074
$MgSO_4$	0.195, 0969	1.89, 1.28	1.222
$Mg(NO_3)_2$	0.215, 1.219	1.85, 1.64	1.046
Ionic liquids[g]:			
[bmim][PF_6]	–	0.06 [27] 0.27 [28], 0.40 [11b] (30°C)	182.0 [29]
[bmim][BF_4]	–	0.08 [27]	75.0 [29–30]
[emim][TfO]	–	0.52 [28]	45.0 [31]
[emim][Tf_2N]	–	0.66 [28], 0.79 [32]	27.0 [29]
Other solvents [33, 34] :			
Amyl alcohol	–	1.91	4.00
i-Butanol	–	2.20	3.96
Ethanol	–	3.42, 3.20 (17°C)	1.074
Heptane	–	6.03	0.41

[a]Diffusivity of H_2CO_3 in water [35]
[b]Diffusivity of HCO_3^- in water [36]
[c]MEA = monoethanolamine (primary)
[d]DEA = diethanolamine (secondary)
[e]MDEA = methyldiethanolamine (tertiary)
[f]AMP = isobutanolamine (hindered)
[g][bmim][Tf_2N], [pmmim][Tf_2N], [bmpy][Tf_2N], and [bmim][BF_4] in the range of $0.3–1.0 \times 10^{-5}$ cm^2/s [37]

These parameters are discussed in detail in Chap. 4 since they play a key role in the understanding of physical adsorption processes. The Lennard-Jones parameters are available for several common gas pairs in Appendix F. The collision diameter can be expressed as the arithmetic mean of the two species present as:

$$\sigma_{12} = \frac{1}{2}(\sigma_1 + \sigma_2) \tag{3.9}$$

and the well-depth can be described as the geometric mean of the two species as:

$$\varepsilon_{12} = \sqrt{\varepsilon_1 \varepsilon_2} \tag{3.10}$$

Example 3.3 An Integrated Gasification-Combined Cycle (IGCC) plant generates a fuel gas mixture of H_2 and CO_2 with plans to carry out separation using a Selexol-based physical solvent. In order to maintain the high efficiency that is characteristic of an IGCC plant, the separation needs to be done at high temperatures, greater than 200°C and pressures up to 10 atm. To determine whether mass-transfer resistance will take place in the gas or liquid on either side of the interface that separates the fuel gas from the solvent, determine the gas–gas diffusivity, D, for H_2 and CO_2 at these conditions. Assume that the liquid-phase diffusivity is 1.0×10^{-5} cm^2/s. In addition, compare the diffusivity to the value in Table 3.2 at 25°C and 1 atm.

Solution: The diffusivity can be calculated using Eq. (3.8) and values from Appendix F.

$$D = \frac{1.86 \times 10^{-3} T^{\frac{3}{2}} \left(\frac{1}{M_1} + \frac{1}{M_2}\right)^{\frac{1}{2}}}{p\sigma_{12}^2 \Omega}$$

where $M_1 = 2.02$, $M_2 = 44.01$, $T = 473.15$ K and $p = 10$ atm. σ_{12} is defined by Eq. (3.9) as:

$$\sigma_{12} = \frac{1}{2}(\sigma_1 + \sigma_2) = \frac{1}{2}(2.826 + 3.941) = 3.383\,\text{Å}$$

Ω can be determined using Table F.2 and the geometric mean of ε as:

$$\varepsilon_{12} = k_b\sqrt{(59.7)(195.2)} = 107.9k_b$$

$$\frac{k_bT}{\varepsilon_{12}} = \frac{473.15k_b}{107.9k_b} = 4.4$$

At a value of 4.4, $\Omega = 0.8652$.

$$D = \frac{(1.86 \times 10^{-3})(473.15^{\frac{3}{2}})\left(\frac{1}{2.02} + \frac{1}{44.01}\right)^{\frac{1}{2}}}{(10)(3.383^2)(0.8652)} = 0.139\ \text{cm}^2/\text{s}$$

The gas–gas diffusivity for H_2 and CO_2 is a factor of six less than that at 25°C and 1 atm, 0.646 cm^2/s found in Table 3.2. The liquid-phase diffusivity is 4 orders of magnitude lower implying gas-phase resistance is likely minimal.

In 1955, Charles Wilke and Pin Chang [16] developed the following correlation to estimate the diffusion of a gas A (solute) at low concentrations in a given solvent:

$$D_A = 7.4 \times 10^{-8} \left[(\phi M)^{\frac{1}{2}} \frac{T}{\mu \tilde{V}_A^{0.6}} \right] \tag{3.11}$$

such that D_A is the liquid-phase diffusivity of a given solute A at very low concentrations in a given solvent, in units of cm^2/s, ϕ represents the association parameter of the solvent, M is the molecular weight of the solvent, T is the temperature, in K, μ is the solvent viscosity, in units of cP, $\tilde{V}_A$ is the molar volume of the solute at its normal boiling point, in units of cm^3/mol.

Viscosity, μ, is the measure of the fluid's ability to resist stress. A highly viscous fluid will not flow as well as a moderately viscous fluid, *e.g.*, honey (high μ) vs. water (moderate μ). Viscosity can be measured for both gases and liquids, with gases roughly 2 orders of magnitude less viscous than liquid. For example, at 20°C, the viscosity of water is 1 cP while the viscosity of air is 0.02 cP. The unit cP is centipoise, which is equivalent to 1 mPa s, and is the common unit of viscosity. In the gas phase, the molar volume of CO_2 is the volume occupied by one mole of CO_2 at a given temperature and pressure. In a solution, the CO_2 also has a molar volume, but it is much more difficult to determine as it is highly dependent upon the solvent, temperature, and pressure conditions. The approximate molar volume of dissolved CO_2 in aqueous solution at its normal boiling point is 34.0 cm^3/mol solvent [17]. The association parameter of the solvent describes the solvent interactions, with the solvent molecules behaving as large molecules and likely diffusing more slowly. The following association parameters have been recommended by Wilke and Chang [16] for some common solvents: water, 2.26 [18]–2.6; methanol, 1.9; ethanol, 1.5; benzene, ether, and heptanes, 1.0. In general, an association parameter of 1.0 for unassociated solvents is a reasonable approximation. Further discussion of association parameters can be found in the work of Lusis and Ratcliff [19], Akgerman and Gainer [20], Gainer and Metzner [21], and Hiss and Cussler [22]. Average errors using this correlation for diffusivity predictions are expected to be between 10 to 15% [23].

3.3 Chemical Kinetics of CO_2 in Solution

The separation of CO_2 from a gas mixture may occur via a physical or chemical separation process. A chemical separation involves a reactive solvent species that has some degree of selectivity toward CO_2. Carbon dioxide is a linear molecule in the gas phase and within a physical solvent. However, when CO_2 interacts and bonds directly with an oxygen atom of OH^- or a nitrogen atom of an amine group, it will have to bend from its linear structure to accommodate the additional atom in each case. This accommodation requires a significant activation barrier compared to molecules that already possess a bent structure, such as SO_2, water, and NO_2. The activation barrier for $CO_2 + OH^-$ in solution has been estimated from both experiments [38] and theoretical calculations [39] to range between 50 and 80 kJ/mol, compared to an activation barrier of approximately [40] 26 kJ/mol for the same reaction, but with SO_2 in place of CO_2. Due to the exponential relationship between the rate constant and the activation energy, a change in the activation barrier of just 10 kJ/mol can change the rate constant by up to several orders of magnitude. Specifically, the rate constant for the reaction of OH^- with SO_2 may be up to 5 orders of magnitude higher with a barrier of 26 kJ/mol than OH^- with CO_2 having a barrier of 50 kJ/mol. Whether the rate of reaction with SO_2 will be faster than that of CO_2 will depend upon the relative concentrations of each of these species.

A reaction occurs within the liquid phase between the reactive component of the solvent and solution-phase CO_2. Provided the chemical reaction occurs close enough to the gas-liquid boundary where CO_2 dissolves, the reaction will actively assist mass transfer and enhance the CO_2 uptake in the solvent. Often diffusion and reaction phenomena are coupled closely, occurring in the same region of the solvent. As previously described, gas-phase CO_2 diffuses through the gas film to the gas-liquid boundary, diffuses through the boundary at a rate dependent on its solubility in the given solution, then diffuses through the solution phase upon which it approaches a reactive molecule, reacts to form a product species, which is a bound species of CO_2 either in the form of a carbamate (C–N) or carbonate (C–O) bond.

The most simple type of reaction is the irreversible first-order reaction where the rate of reaction of CO_2 is proportional to its local concentration and a first-order rate constant k_1, *e.g.*, $r = k_1 c_{CO_2}$. In a gas-phase process transition-state theory [41] may be used to estimate the rate constant, k of a given reaction. A similar approach to reactions in solution was first applied by Wynne-Jones and Eyring [42] and later by Evans and Polanyi [43] and by Bell [44]. The approach to determine estimates of k for reactions in solution account for the nonideality of the reactive species due to the solution through the inclusion of the *activity coefficient* of each species into the general rate constant expression as follows:

$$k = k_0 \frac{\gamma_{CO_2} \gamma_B}{\gamma_{AC}} \tag{3.12}$$

such that k_0 is the gas-phase rate constant and γ_{CO_2}, γ_B, and γ_{AC} are the solution-phase activity coefficients of CO_2, reactive binding species, B, and activated complex, respectively. The activity coefficient provides an indication of nonideality of the reactive species due to its presence in a given solution, with an activity coefficient of 1.0 representative of ideal behavior, *i.e.*, the species acts as it would in an ideal pure phase.

Often times in solution reactions are reversible, proceeding in both directions at the same time but at unequal rates until *chemical equilibrium* is reached, at which point the forward and reverse rates become equal and the net rate is zero. If the reversible reaction were first-order in both directions the rate would be equal to:

$$r = k_1 c_{CO_2} - k_{-1} c_p, \tag{3.13}$$

such that k_{-1} is the first-order rate constant of the reverse reaction and c_p is the concentration of the product species. At equilibrium, $r = 0$ with

$$k_1 c_{CO_2} = k_{-1} c_p \tag{3.14}$$

and

$$\left(\frac{c_p}{c_{CO_2}}\right)_{eq} = \frac{k_1}{k_{-1}} = K_{eq} \tag{3.15}$$

such that K_{eq} is the equilibrium constant, or the ratio of the product concentration to the reactant concentration. As previously discussed, a second-order reaction between CO_2 and a reactant B to form products, P, can be considered as:

$$CO_2 + zB \overset{k_2}{\rightarrow} P \tag{3.16}$$

with the rate expression,

$$r = k_2 c_{CO_2} c_B^Z \tag{3.17}$$

in which z is the stoichiometric coefficient, and c_B and c_{CO_2} are the concentrations of CO_2 and the reactive binding species in aqueous solution, respectively. The reaction order depends upon whether the reaction proceeds in an elementary fashion (*i.e.*, as written) or non-elementary, in which the reaction is expressed in an overall fashion consisting of individual steps that each proceed elementarily.

CO_2-Solvent Chemistry for Pure Water Since most solvents are in aqueous solution, the chemical reactions that occur between CO_2 and water play some role in the overall thermodynamic and kinetic mechanisms associated with CO_2 absorption using solvent-based approaches. The chemical reactions occurring between CO_2 and pure water have been extensively investigated, and can be fully described by the following reactions:

$$CO_2 \overset{H_{CO_2}}{\Longleftrightarrow} CO_{2(g)} \tag{3.18}$$

$$CO_2 + H_2O \overset{K_1}{\Longleftrightarrow} H_2CO_3 \tag{3.19}$$

$$CO_2 + OH^- \overset{K_2}{\Longleftrightarrow} HCO_3^- \tag{3.20}$$

$$H^+ + CO_3^{2-} \overset{K_3}{\Longleftrightarrow} HCO_3^- \tag{3.21}$$

$$H^+ + HCO_3^- \overset{K_4}{\Longleftrightarrow} H_2CO_3 \tag{3.22}$$

$$H^+ + OH^- \overset{K_5}{\Longleftrightarrow} H_2O \tag{3.23}$$

such that H_{CO_2} is the Henry's law constant, and K_i is the equilibrium constant of a given reaction, as previously defined in Eq. (3.15). The rate and equilibrium constant for each reaction have been determined previously and are available in Appendix G1. Using the available equilibrium data from Appendix G1, a comparison of the normalized concentrations of the three carbonate species, *i.e.*, HCO_3^-, H_2CO_3, and CO_3^{2-}, formed in aqueous solution are plotted in Fig. 3.6 as a function of pH. As expected, at low pH, carbonic acid (H_2CO_3) is the dominant species. As pH increases, the concentration of H_2CO_3 decreases with a subsequent rise in bicarbonate anion (HCO_3^-) concentration. As the pH increases further, resulting in an increasingly basic solution, the HCO_3^- concentration plateaus through the pH range of 5–9, and

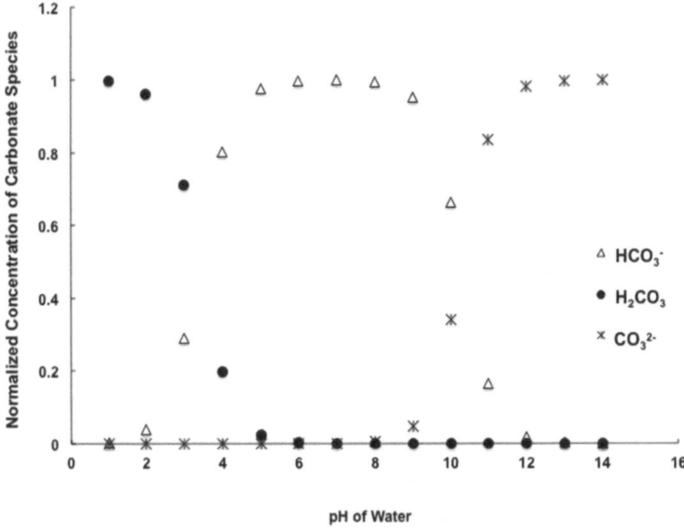

Fig. 3.6 Comparison of the normalized concentration of the three carbonate species in aqueous solution as a function of pH

then decreases, while the carbonate anion (CO_3^{2-}) increases, thereby dominating the solution at pH values greater than 12.

Using the equlibruim data associated with Reaction (3.22) combined with Henry's Law, the pH of "pristine rainwater" may be predicted. To estimate the pH of a drop of rain, one must consider the transport of the acid gas (*i.e.*, CO_2) across the gas-liquid boundary of the rain drop. Assuming the CO_2 concentration in air is approximately 0.039 mol% in addition to the Henry's law constant of CO_2 in water, leads to a concentration of CO_2 at the gas-liquid interface of 1.5×10^{-5} M. Recalling the relationship between the equilibrium constant, K_{eq} and the concentrations of reactants and products of Reaction (3.22),

$$K_{\mathrm{eq}} = \frac{[\mathrm{H_2CO_3}]}{\left[\mathrm{H^+}\right]\left[\mathrm{HCO_3^-}\right]} \tag{3.24}$$

Using the equilibrium data from Appendix G1 for Eq. (3.24) and the assumption that the dominantly negatively charged ion is HCO_3^- and that $[H^+] \approx [HCO_3^-]$, the concentration of H^+ can be determined, which relates directly to the pH by taking the negative log of the concentration, thereby arriving at a pH of rainwater of 5.6. Due to the presence of CO_2 in the air, 5.6 represents the upper bound pH of acid rain, rather than 7.0.

CO_2-Solvent Chemistry for Aqueous Solutions Sodium hydroxide is a strong base and dissociates completely into solvated sodium and hydroxide ions in water. One mole of NaOH dissolved in a liter of water leads to a solution that is 1 M in OH^- anion. The OH^- will participate in the equilibrium toward water dissociation. Water is *amphoteric* (*i.e.*, it can act as an acid or a base) and in solution, water can dissociate into ions, H^+ and OH^-. This dissociation reaction has an equilibrium constant, K_{eq} of 1.79×10^{-16} M at room temperature. The concentration of H^+ and OH^- can be calculated as:

$$K_{eq} = \frac{[H^+][OH^-]}{[H_2O]} \tag{3.25}$$

In pure water, the concentrations of H^+ and OH^- are equivalent. The molar concentration of H_2O is the density of water (1000 g/L) divided by its molecular weight (18.01 g/mol), or 55.6 M.

$$\begin{aligned}[H^+] &= \sqrt{K_{eq}[H_2O]} = \sqrt{(1.79 \times 10^{-16})(55.6)} \\ &= \sqrt{1.00 \times 10^{-14}} = 1 \times 10^{-7}\text{M}\end{aligned} \tag{3.26}$$

Since the pH of a solution is defined as the negative log of the H^+ concentration in solution, the pH of pure water is 7. Notice that in Eq. (3.25), the concentration of water is much greater than that of H^+ or OH^-. When the molar concentration of water is taken as constant (55.6 M), the product of the concentration and K_{eq} is termed the acid dissociation constant, K_w, which is equal to 1.00×10^{-14} M^2. By introducing 1 M OH^- into the solution via NaOH, the H^+ concentration in solution drops to approximately 1×10^{-14} M, or a pH of 14 to maintain equilibrium. The mechanism of CO_2 reactivity in an aqueous solution of NaOH is the formation of bicarbonate anion, HCO_3^-, and it is this reactivity that assists in driving the mass transfer of CO_2 across the gas-liquid phase boundary. Depending upon the pH of the solution, bicarbonate can also react with water or protons in solution to form carbonic acid, H_2CO_3, or water and hydroxyl anions to form carbonate anion, $CO_3{}^{2-}$. Pinsent et al. [45] have experimentally determined the rate constant of the reaction of NaOH with CO_2 and found that over varying ionic strengths, it ranges between 6.28×10^3 and 13.7×10^3 L/mol s. In addition, they determined the exothermic heat of absorption to range between −85.6 to −103.5 kJ/mol CO_2. Recall that the forward rate of a chemical reaction is equal to the product of the rate constant and the reactant concentrations raised to their respective stoichiometric coefficients. Assuming the reaction of CO_2 with NaOH proceeds as:

$$CO_2 + 2NaOH \rightarrow Na_2CO_3 + H_2O \tag{3.27}$$

and assuming the rate constant is on the order of 10^3L/mol s and the CO_2 and NaOH concentrations are 0.024 and 1 M, respectively, the forward reaction rate with respect to CO_2, *i.e.*, defined as the change in the concentration of CO_2 per time, is equal to:

$$\frac{dc_{CO_2}}{dt} = \sim \left(\frac{10^3}{\text{Ms}}\right)(0.024\ \text{M})(1\ \text{M})^2 = 24\ \text{M}^2/\text{s} \tag{3.28}$$

This rate can be thought of as the consumption of CO_2 over time. When the base monoethanolamine (abbreviated as MEA, with molecular formula, $NH_2CH_2CH_2OH$)

is introduced into water, it reacts with the water as a weak base to generate OH^- in solution, which increases the pH. The nitrogen in the MEA extracts a proton from water to form a four-coordinate nitrogen species with a positive charge, *i.e.*, $NH_2CH_2CH_2OH + H_2O \leftrightarrow {}^+NH_3CH_2CH_2OH + OH^-$; however, MEA is a weak base relative to NaOH, and the reaction does not proceed completely to the right, as it does with NaOH. The pH of a 1-M MEA solution is approximately 11.7, which implies an OH^- concentration in solution of approximately 0.005 M; therefore, when one mole of MEA is added to water, about 99.5% of the MEA will be present in an unionized (*i.e.*, molecular) form, with only approximately 0.5% present as the protonated ammonium ion form. The reactivity of CO_2 in solution with amines is complex in that it involves a combination of carbonate chemistry (OCO_2) and carbamate chemistry (NCO_2).

The reaction of CO_2 with *primary* and *secondary amines* involves a direct reaction between the carbon atom of CO_2 and the nitrogen atom of the amine, forming a *carbamate bond*. There are two proposed mechanisms in the literature by which this carbamate complex is thought to form: a two-step process involving the formation of a *zwitterion* complex [46] or a direct *termolecular reaction* involving the amine, CO_2, and water in a single step [47].

Mechanism 1

$$CO_2 + R_2NH \leftrightarrow R_2NH^+CO_2^- \tag{3.29}$$

$$R_2NH^+CO_2^- + H_2O \rightarrow R_2NCO_2^- + H_3O^+ \tag{3.30}$$

In this case CO_2 is reacting with a secondary amine, *i.e.*, the nitrogen atom is bound to two R groups (*i.e.*, carbon-based side chains), or arbitrarily long hydrocarbon chains. In Reaction (3.29), the product zwitterion is a short-lived complex in which both a positive and negative charge are momentarily stabilized simultaneously within the complex. The formation of the zwitterion is thought to be the rate-limiting step, where Reaction (3.30) is thought to take place instantaneously.

Mechanism 2

$$CO_2 + R_2NH + H_2O \leftrightarrow R_2NCO_2^- + H_3O^+ \tag{3.31}$$

Although termolecular reactions are rare due to the order required in assembling three molecules in the correct orientation for the reaction to proceed, this mechanism has also been proposed in the literature. The reaction is second-order with the rate constant dependent on the concentrations of amine and CO_2. In addition to amine reactivity with CO_2, amine protonation also takes place in solution due to the newly formed hydronium ion. Therefore, the following reaction is likely coupled to carbamate formation:

$$R_2NH + H_3O^+ \leftrightarrow R_2NH_2^+ + H_2O \tag{3.32}$$

This leads to a theoretical CO_2 loading limit of 1 mol of CO_2 per two moles of amine. There have been many investigations [48] on the kinetics of CO_2 with primary amines

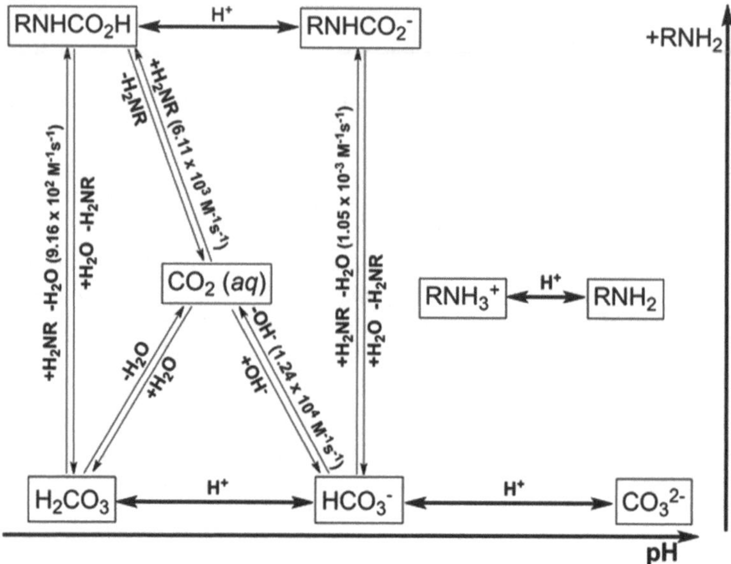

Fig. 3.7 General reaction scheme including all reactions between primary amine, the CO_2/carbonate group and protons. ↔ represents instantaneous protonation equilibria, and ⇄ represents the kinetically observable reactions determined from previous studies. (Reprinted with permission from [2])

(*e.g.*, MEA) and estimates of rate constants range from 5.0×10^3 to 7.0×10^3 L/mol s with heats of absorption averaging approximately −80 kJ/mol CO_2, somewhat tunable by the chemical structure of the amine [49]. Both the rate constant and the heat of absorption are similar to the reaction of CO_2 with NaOH.

The reactivity of CO_2 via tertiary amines is different, in that a bicarbonate product forms since the direct bond with nitrogen is not possible due to the steric hinderance of the three R groups. The reaction is slow and thought to be termolecular via the pathway:

$$R_3N + CO_2 + H_2O \leftrightarrow R_3NH^+ + HCO_3^- \tag{3.33}$$

The reaction is typically represented as second-order with respect to the amine and CO_2 concentrations. The theoretical capacity in the case of CO_2 reaction with tertiary amines is 1 mol of CO_2 captured per 1 mol of amine in solution. Rate constant estimates range between 0.5 to 4 L/mol s, which are 3 to 4 orders of magnitude slower than NaOH and primary and secondary amines [47a, 50]. Heats of absorption average approximately −55 kJ/mol CO_2, which is on average 25–30 kJ/mol lower than the heats of absorption for NaOH and primary and secondary amines [14c, 51].

The CO_2-amine chemistry previously described is a simplification of the chemical complexities of a real system, in which both carbamate and carbonate chemistries occur simultaneously. A more accurate picture includes the chemistry of both carbamate and carbonate pathways as shown in the equilibrium diagram of Fig. 3.7. For

Table 3.4 Second-order rate constants and pK_a's for CO_2 reactions with bases in water at 25°C

Base	pK_a	k(L/mol s)
OH^- [52]	NA	12100
H_2O	14	6.7×10^{-4}
Carbonic anhydrase [54]	~7.0	2.0×10^5; 1.4×10^6
NH_3 [45, 55]	9.2	417; 5070
MEA [55b, 56]	9.5	7570
Piperazine [53]	9.8	53700
CH_3NH_2 [56–57]	10.6	24000
Piperidine [57–58]	11.1	60300

instance, when CO_2 interacts with a primary or secondary amine, a carbamate may be formed as an intermediate in the pathway of carbonate formation. The three types of reaction chemistries that take place in CO_2 absorption processes via amine-based solvents are: 1) CO_2 and carbonate reactions with water, which include carbonic acid, bicarbonate, and carbonate ions, 2) amines and their deprotonated and protonated forms, and 3) carbamate species, including protonated and deprotonated states. The horizontal axis in Fig. 3.7 represents the pH, while the vertical axis represents the CO_2-H_2O-amine interactions. The rate constant for the CO_2 reaction with OH^- to form the bicarbonate anion is 1.24×10^4 mol/L s, and with MEA is 6.11×10^3 mol/L s. Although the rate constant of the CO_2-MEA reaction is slower, CO_2 is more likely to take the carbamate path since the concentration of amine is greater than that of OH^-. In addition to the direct CO_2-amine interaction as outlined in Eqs. (3.29–3.31), CO_2 may interact with amines in the form of carbonic acid (H_2CO_3) or bicarbonate anion (HCO_3^-). The forward rate constant for the reaction of MEA with carbonic acid is 9.16×10^2 mol/L s and with the bicarbonate anion, 1.05×10^{-3} mol/L s, compared with the direct CO_2 interaction, 6.11×10^3 mol/L s for similar conditions [52]. A complete list of the rate and equilibrium constants of the CO_2-amine interactions is available in Appendix G2.

3.3.1 *Role of Catalysis in Absorption*

Within a chemical reaction, a catalyst is a substance that increases the rate of reaction without being modified or consumed in the reaction. The acid-base chemistry mechanisms previously discussed represent examples of homogeneous catalysis *if* the amine plays the role of the catalyst through the intermediate formation of a carbamate complex on the path to a carbonate species. In solution, acids and bases are capable of catalyzing reactions provided they are not themselves consumed in the reaction. Table 3.4 displays forward rate data for the reaction of CO_2 in aqueous solution with various bases at 25°C [53]. While the products may vary, the rate-limiting step is the attack of the slightly positive (*i.e.*, acidic) carbon atom of CO_2 by a base, either a partially negative N or O atom to form a carbamate or carbonate interaction, respectively. It is clear from the *pKa* values that the basicity is important in determining the extent of reactivity, but it is not the only factor; for instance, secondary amines are less basic than OH^-, but react more rapidly.

However, for amines with enhanced kinetics such as piperazine (PZ) and others shown in Table 3.4, the reactions occur more rapidly than with OH^-. In particular, model investigations of amine mixtures of PZ-AMP (2-amino-2-methyl-1-propanol) and PZ-MDEA (methyldiethanolamine) carried out by Puxty and Rowland [55b] provide insight into the role that PZ plays in enhancing CO_2 absorption. They examined the concentration profiles of PZ and each amine through the liquid film region beyond the gas-liquid interface. In the case of the PZ-AMP mixture with 3.14 M AMP and 0.23 M PZ, although both amines have similar pK_a values, AMP ultimately accepts more protons due to its higher concentration, resulting in a higher concentration of PZ in solution to enhance the mass transfer of CO_2 across the interface. In the presence of AMP, the concentration of PZ at the interface after 0.44 s of exposure is 0.11 M, compared to 0.045 M without AMP. In the case of the PZ-MDEA mixture, a similar mechanism was determined, but with less effect due to MDEA being a poorer base than AMP, with a pKa of 8.22 compared to 9.29, respectively at 313 K. In general, AMP and MDEA are acting as buffers to maintain a high pH while CO_2 is continuously absorbed in solution via the reactive PZ amine species. A complete list of the rate and equilibrium constants of the CO_2-PZ and CO_2-AMP interactions is available in Appendix G3. A potential limitation of such fast kinetics is the possibility of diffusion limitations that inhibit effective mass transfer across the interface [59]. In other words, if the rate of reaction at the gas-liquid interface is taking place faster than the rate of absorption into the solution, the mass-transfer is diffusion-limited (or diffusion-controlled). This is also the case with the catalyst, carbonic anhydrase, which is discussed next.

Carbonic Anhydrase Aside from amine-enhanced absorption of CO_2 into solution, there exists *biomimetic* catalysts such as carbonic anhydrase that can hydrate and dehydrate CO_2 orders of magnitude faster than amines or water, as illustrated by the rate constant comparison of Table 3.4. Carbonic anhydrase is a natural zinc enzyme found in the red blood cells of mammals, and is used to capture CO_2 and transform it into a bicarbonate ion and a proton, which can easily dissolve in blood and can be subsequently removed and transported to the lungs. In the lungs, the enzyme is "regenerated" and releases CO_2, which is then exhaled. The kinetics of bicarbonate formation via carbonic anhydrase are up to 10 orders of magnitude faster than its formation in an aqueous solution and approximately 3 orders of magnitude faster than CO_2 binding via the fastest amine-based solvents. Carbonic anhydrase is a natural enzyme with a zinc center coordinated with three histidine groups and catalyzes the conversion of CO_2 to H_2CO_3 and HCO_3^- in near neutral water at high rates, moderate binding energies and high selectivity [60]. The mechanism for the conversion from CO_2 to HCO_3^- using carbonic anhydrase, and related catalysts, is shown in Fig. 3.8. Within this mechanism, the zinc atom of the enzyme binds water, which dissociates into H^+ and OH^-, which is left bound directly to zinc (Step 1 in Fig. 3.8). The bound OH^- ion binds to CO_2 in the same way that it would in an aqueous solution, and creates HCO_3^- (Steps 2 and 3). The HCO_3^- is released to the solution and is replaced by a H_2O to complete the catalytic cycle (Step 4). The mechanism is carried out in reverse to regenerate CO_2. Because carbonic anhydrase converts H_2O to OH^-

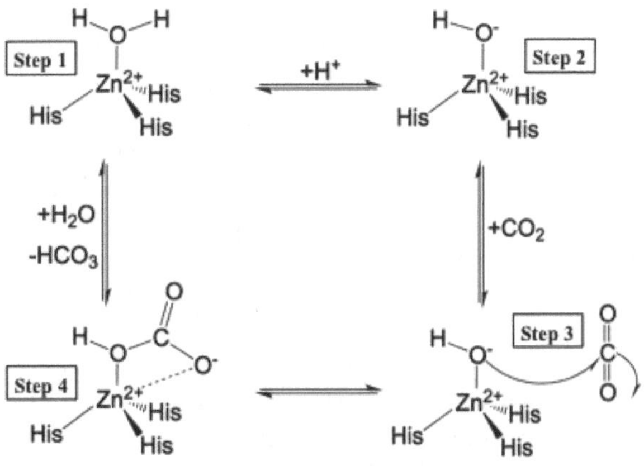

Fig. 3.8 The mechanism for the conversion of CO_2 to HCO_3^- by carbonic anhydrase and related metalloenzymes

in the first step of this mechanism, there is a significant amount of H^+ formed and released to the solution. A buffer, such as a phosphate buffer commonly found in biological systems, is used to maintain the pH of the solution so that excess protons react with the phosphate leaving the solution equilibrium unchanged. The rate of this reaction is 1.2×10^8 mol/L s. However, the reaction enthalpy depends on the ultimate speciation of the adsorbed CO_2 and will depend on the buffer or amine that is used to provide capacity for the CO_2. Additionally, the final bound form of CO_2 is the same since carbonic anhydrase is hydrating CO_2 to form bicarbonate, similar to the case of water.

3.4 Determining the Rate of Absorption

The molar flux of CO_2 at the gas-liquid boundary, J_{G,CO_2} is dependent upon the *advection* of CO_2 from the bulk transport of the gas at velocity, G and the molecular diffusivity of CO_2 in the gas phase, D_{G,CO_2} such that:

$$\begin{aligned} J_{G,CO_2} &= c_{CO_2}G - D_{G,CO_2}\left.\frac{dc_{CO_2}}{dz}\right|_i = c_{CO_2}G - \frac{D_{G,CO_2}}{\delta}(c_{\infty,CO_2} - c_{i,CO_2}) \\ &= c_{CO_2}G - k_{G,CO_2}(c_{\infty,CO_2} - c_{i,CO_2}) \end{aligned} \tag{3.34}$$

where, $D_{G,\,CO_2}$ is typically in units of cm^2/s, $k_{G,\,CO_2}$ is the *mass-transfer coefficient*, also commonly referred to as the "mass-transfer velocity," in units of cm/s, δ is the thickness of the gas-film at the gas-liquid interface, $c_{\infty,\,CO_2}$ is the concentration of CO_2 in the bulk gas phase, $c_{i,\,CO_2}$ is the concentration of CO_2 at the gas-liquid interface.

The mass-transfer coefficient is the rate at which a component is transferred from one phase to another. Mass-transfer coefficients are a central component to separation in that they dictate the rate at which equilibrium is reached and control the time required to achieve a given level of separation, thereby determining the size and cost of the equipment used. The gas-phase mass-transfer coefficient as stated in Eq. (3.34), is equal to the quotient of the diffusivity and the film thickness on the gas-side of the gas-liquid interface. The concentration difference represents the driving force that is responsible for the diffusion from the bulk of the gas across the interface. The diffusion resistance in the gas phase is significantly less than that of the liquid phase due to the greater density and subsequent increase in molecular collisions taking place in the liquid phase. As will be discussed shortly, the gas-phase diffusivities can be up to five orders of magnitude greater than liquid-phase diffusivities. However, the mass-transfer resistance is also dependent upon the film thickness, δ, which is somewhat wider in the gas film than the liquid film. The film thickness is proportional to the ratio of the viscosity of the fluid to the fluid density [61], which may be up to five orders of magnitude greater for gases than liquids[1]. Despite the fact that the gas film is larger than the liquid film, the lower diffusion resistance present in the gas will lead to an overall lower mass-transfer resistance in the gas phase compared to the liquid.

Another important aspect of the mass-transfer coefficient is that its reciprocal represents the resistance of CO_2 transfer across the interface. In general, $k_{G,\,CO_2}$ is directly proportional to the average velocity of the gas through the mass-transfer system; therefore, the resistance to CO_2 transfer by diffusion general decreases with increasing fluid velocity, L. However, increasing the fluid velocity may increase cost since additional energy is required to move the liquid through the system.

The molar flux of CO_2 in solution, $J_{L,\,CO_2}$ is dependent upon the advection of CO_2 from the bulk transport of the liquid at velocity, L and the molecular diffusivity, $D_{L,\,CO_2}$ of CO_2 in solution:

$$\begin{aligned} J_{L,\,CO_2} &= c_{CO_2}L - D_{L,\,CO_2}\left.\frac{dc_{CO_2}}{dz}\right|_i = c_{CO_2}L - \frac{D_{L,CO_2}}{\delta}(c_{i,\,CO_2} - c_{\infty,\,CO_2}) \\ &= c_{CO_2}L - k_{L,\,CO_2}(c_{i,\,CO_2} - c_{\infty,\,CO_2}) \end{aligned} \tag{3.35}$$

The first term on the right-hand side of Eq. (3.35) represents the advective transport of CO_2 and the second term the molecular diffusion of CO_2 in the liquid phase. In particular, δ is the liquid-film thickness at the gas-liquid interface, $k_{L,\,CO_2}$ is the liquid-phase mass-transfer coefficient, and $c_{i,\,CO_2}$ and $c_{\infty,\,CO_2}$ are the concentrations of CO_2 at the gas-liquid interface and within the bulk fluid, respectively.

Understanding the physical limits placed on the parameters that dictate the rate of mass transfer in an absorption process allows scientists to design new solvents that can approach these limits. As previously demonstrated, absorption involves simultaneous diffusion, advection, and reaction processes. Absorption processes may involve (a) a liquid flowing over an inclined or vertical surface, potentially in turbulent flow

[1] In the example the air-water interface, the viscosities of air and water are 18.6 and 0.89 Pa s, respectively and the densities of air and water are 1.22 and 1000 kg/m^3 respectively; the air-to-film ratio is ~17000.

Table 3.5 Tunable mass transfer parameters

Parameter	Name	Typical units	Likely range
D	Gas diffusivity	cm^2/s	0.1–1.0
H_{CO_2}	Henry's law constant	atm	20–1700
D_{L,CO_2}	Liquid diffusivity	cm^2/s	$(0.5–2.0) \times 10^{-5}$
D_B	Liquid diffusivity of absorbent		
k_2	Reaction rate constant	L/mol s	6.7×10^{-4}–1.2×10^8
c_B	Bulk concentration of absorbent	mol/L	0.1–8[a]
k_{L,CO_2}	Liquid-phase mass-transfer coefficient	cm/s	0.01–0.1
c_{i,CO_2}	Concentration of CO_2	mol/L	Set by p_{CO_2} and H_{CO_2}
$D_B/D_{L,CO_2}$	Diffusivity ratio		0.2–2.0

[a] Depending on the corrosive nature, typically less than 1.0 mol/L

similar to that of a wetted-wall column operating at a sufficiently high *Reynolds number*, *Re*, (b) a gas being blown in the form of bubbles through the liquid, (c) an agitated liquid from stirring with entrained gas bubbles in the liquid, or (d) a liquid sprayed through the gas as drops or jets. Examples of absorption processes are shown in Fig. 3.2.

Based upon the previous discussion of CO_2 solubility, diffusion, and reaction kinetics, the potentially "tunable" parameters with ranges of values physically possible today are listed in Table 3.5. Next, the rate of absorption are discussed, which relies on knowledge of many of these parameters.

3.4.1 Liquid-Phase Resistance: Mass Transfer Theories

Physical Absorption Physical absorption of CO_2 involves its separation based purely upon its solubility within a given solvent, which can be determined by the Henry's law relationship as previously described. Common physical solvents for CO_2 capture include methanol (CH_3OH), which is often used for the separation of CO_2 from natural gas, propylene carbonate ($C_4H_6O_3$), and dimethyl ethers of polyethylene glycol ($CH_3(CH_2CH_2O)_nCH_3$). Neglecting advective transport, the physical absorption process is one in which gas dissolves in a liquid without reacting, with the rate of absorption at the interface represented by the second term on the right hand side of Eq. (3.35):

$$J_{L,\,CO_2} = k_{L,\,CO_2}(c_{i,\,CO_2} - c_{\infty,\,CO_2}) \tag{3.36}$$

Both the rate of absorption and the mass-transfer coefficient are represented on the basis of the area of the interface.Three models will be explicitly considered, with the first being the film model. This model proposed in 1923 by Walter Gordon Whitman [62] owes its original foundation to the "diffusion layer" concept of Walther Hermann Nernst [63] in 1904 and focuses on the rate of mass transfer within the liquid film at the gas-liquid interface. Within the film advection is neglected and mass transfer takes place solely from molecular diffusion. The film thickness, δ, depends upon the surface geometry, liquid agitation, physical properties of the fluid, etc. Due to its simplicity this model is often used, but more complex models exist that often better represent the complexities of absorption in agitated liquids, which require a mechanism that takes mass transfer via advection into account.

Mechanisms of absorption that include the time scale on which a given fluid element within the film is exposed to the interface were later developed. The exposure time of the fluid element at the interface is dependent upon the time associated with the replacement of fluid at the interface from fluid in the bulk. In 1935, Ralph Higbie [64] developed the penetration model, which separates the liquid exposed to the gas at the interface into a series of fluid elements, each of which are exposed to the gas phase for some length of time, θ. The change in concentration of gas A with distance and time is governed by Fick's second law:

$$\frac{dc_A}{d\theta} = D_A \frac{d^2 c_A}{dx^2} \tag{3.37}$$

and the average flux of CO_2, $J_{L,\,CO_2}$ at time θ is given by:

$$J_{L,\,CO_2} = 2\sqrt{\frac{D_{L,\,CO_2}}{\pi\theta}}(c_{i,\,CO_2} - c_{\infty,\,CO_2}) \tag{3.38}$$

and the liquid-phase mass-transfer coefficient is expressed by:

$$k_{L,\,CO_2} = \frac{J_{L,\,CO_2}}{(c_{i,\,CO_2} - c_{\infty,\,CO_2})} = 2\sqrt{\frac{D_{L,\,CO_2}}{\pi\theta}} \tag{3.39}$$

In 1951, Peter Victor Danckwerts [65] extended the penetration model to the surface-renewal model, which assumes that there is a distribution of exposure times in which the gas is in contact with the liquid. In this model, the rate of absorption is expressed as:

$$J_{L,\,CO_2} = s(c_{i,\,CO_2} - c_{\infty,\,CO_2})\sqrt{\frac{D_{L,\,CO_2}}{\pi}} \int_0^\infty \frac{e^{-s\theta}}{\sqrt{\theta}} d\theta \tag{3.40}$$

such that s is the fraction of surface area replaced with fresh liquid per unit time. These models can be used to determine the approximate rate of CO_2 absorption into an agitated liquid, with the Higbie and Danckwerts models showing promise for systems involving CO_2 specifically [66]. These theoretical models are focused on mass transfer in the liquid phase, yet CO_2 separation via absorption involves a two-phase system, *i.e.*, a gas phase from which the CO_2 is transferred into a liquid phase. Two-film theory includes both gas- and liquid-phase resistances and will be revisited when mass-transfer correlations are discussed, which also taken into account both gas- and liquid-phase mass transfer.

In general, the presence of a chemical reaction may significantly influence the mass transfer of CO_2 into the liquid phase. The enhancement factor, *E*, is the ratio of the average rate of absorption into an agitated liquid in the presence of reaction to the average rate of absorption without reaction.

Absorption with Reaction If dissolved CO_2 gas reacts with a dissolved binding material, B by an irreversible second-order reaction, $CO_2 + zB \xrightarrow{k_2} P$, the rate of CO_2 absorption into solution is influenced by a number of parameters, including CO_2 solubility, mass transfer, diffusivity and reaction kinetics. The stoichiometric coefficient, z of the reactive binding species, B, is also an important parameter that is included in the absorption calculation. If the molar ratio between CO_2 and the binding species in a given reaction is 1:2 (*i.e.*, NaOH), then $z = 2$. The balanced chemical reaction must be written out in order to determine z accurately. In general, regardless of the mass-transfer model assumed, the rate of absorption with reaction is related to the liquid-phase (physical) mass-transfer coefficient, $k_{L,\,CO_2}$ as:

$$J_{L,CO_2} = c_{i,\,CO_2} k_{L,\,CO_2} E \tag{3.41}$$

with the functionality of the enhancement factor, E having varying dependence on the potentially tunable parameters previously listed in Table 3.5. Additionally, if CO_2 reacts fairly close to the gas-liquid interface and it can be assumed that it reacts fairly fast then the concentration of CO_2 distant from the interface (*i.e.*, c_{∞,CO_2}) is approximately zero. In practical application, this scenario is not necessarily favored since an energy-efficient reversible process requires near-equilibrium conditions.

In the case of the film model, van Krevelen and Hoftijzer [4] computed the approximate solution to the enhancement factor as:

$$E = \frac{\sqrt{M\left(\dfrac{E_i - E}{E_i - 1}\right)}}{\tanh\sqrt{M\left(\dfrac{E_i - E}{E_i - 1}\right)}} \tag{3.42}$$

such that

$$\sqrt{M} = \frac{\sqrt{D_{L,\,CO_2} k_2 c_B}}{k_{L,\,CO_2}} \tag{3.43}$$

and

$$E_i = \left(1 + \frac{D_B c_B}{z D_{L,CO_2} c_{i,CO_2}}\right) \tag{3.44}$$

where E_i is the enhancement factor corresponding to an instantaneous reaction, $D_{L,\,CO_2}$ is the CO_2 diffusivity in the liquid phase, k_2, is the forward rate constant of CO_2 with the binding species, $k_{L,\,CO_2}$ is the liquid-phase mass-transfer coefficient, $c_{i,\,CO_2}$ is the concentration of CO_2 at the gas-liquid interface, and c_B is the bulk concentration of the binding species in solution. For a fixed value of E_i, an increase in $\sqrt{M}$ leads to an increase in E as a limiting value is approached, in which $E = E_i$. Several criteria associated with $\sqrt{M}$ that can dominate the calculation of E are discussed and are outlined in Table 3.6. For instance, when $\sqrt{M} > 10E_i$, the enhancement factor

Table 3.6 Limiting cases of the film model

Limiting case	Enhancement factor	Physical explanation
1. $\sqrt{M} > 10E_i$	$E = E_i$	Instantaneous reaction
2. $\sqrt{M} < 1/2E_i$	$E = \left(\frac{\sqrt{M}}{\tanh\sqrt{M}}\right)$	Pseudo-first-order reaction
3. $\neq 2$ satisfied and $\sqrt{M} > 3$	$E = \sqrt{D_{L,CO_2} k_2 c_B}$	Fast pseudo-first-order reaction
4. None of the above	Determine E from Fig. 3.9	

is approximately equal to that of an instantaneous reaction. Physically, this is the case when the reaction rate constant is high and the rate of mass transfer is limited by the rate of the diffusion of reaction products away from the gas-liquid interface.

If $\sqrt{M} < 1/2E_i$, the reaction is pseudo-first-order and the enhancement factor is calculated from $\sqrt{M}/\tanh\sqrt{M}$ as shown in Table 3.6. Physically, this is the case in which the reaction is sufficiently slow, or the physical mass-transfer coefficient is sufficiently large such that the concentration of the CO_2 binding material is nearly undepleted with its bulk concentration, c_B, consistent throughout the liquid-phase film depth. If this condition, in addition to $\sqrt{M} > 3$ holds, then to a reasonable approximation,

$$E = \sqrt{D_{L,CO_2}\, k_2 c_B} \tag{3.45}$$

and the rate of absorption with reaction may be simply approximated by:

$$J_{L,CO_2} = c_{i,CO_2}\sqrt{D_{L,CO_2}\, k_2 c_B} \tag{3.46}$$

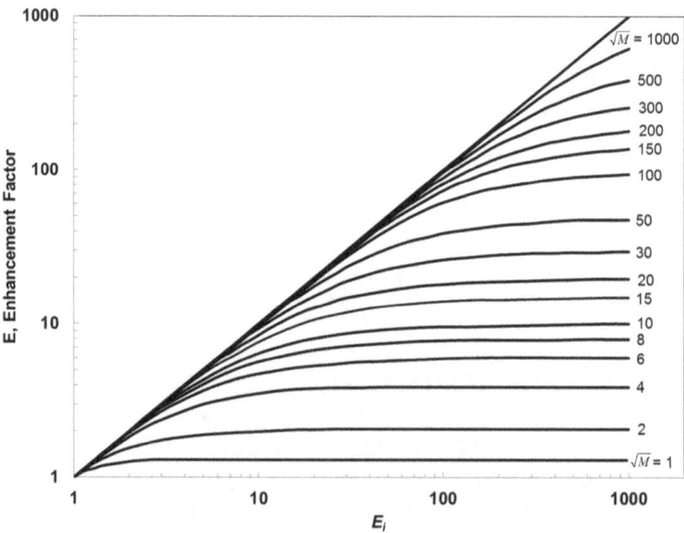

Fig. 3.9 Enhancement factors for second-order reaction; for use with the Film and Higbie models, based on Eq. (3.42). (Reprinted from [1])

This final definition of the rate of absorption of CO_2 into solution in terms of the film model corresponds to a fast pseudo-first-order reaction. In the case in which none of these scenarios hold, the enhancement factor must be obtained from Eq. (3.42) or Fig. 3.9, after calculation of $\sqrt{M}$ and E_i from Eqs. (3.43) and (3.44), respectively.

In the case in which the diffusivity ratio between the binding material and dissolved CO_2, *i.e.*, $D_B/D_{L,CO_2}$, significantly deviates from unity, the Higbie model is preferred over the film model.

Example 3.4 Using the Film Model, estimate the enhancement factor, E, and the resultant flux, J_{L,CO_2}, for the following cases: (a) a postcombustion capture system using a 5 M MEA solution (approximately 30 wt% MEA); (b) a direct air capture system using a 1 M NaOH solution; and (c) a postcombustion capture system using a 10 M unobtainium solution, which has the reaction rate of carbonic anhydrase, $z=1$, and $H=20$ atm. For all, assume $k_L=10^{-2}$ cm/s, $D_{L,CO_2}=D_B=1\times10^{-5}$ cm^2/s.

Solution:

(a) Using Eq. (3.43) and Eq. (3.44) calculate $\sqrt{M}$ and E_i, respectively, then use those to determine E from Fig. 3.9. Using a rate constant for MEA of 7570 L/mol s with the given data yields,

$$\sqrt{M}=\frac{\sqrt{\left(1\times10^{-5}\,\frac{\text{cm}^2}{\text{s}}\right)\left(7570\frac{\text{L}}{\text{mol s}}\right)(5\text{ mol/L})}}{10^{-2}\text{ cm/s}}=61.5$$

If H for CO_2 in a 30% MEA/H_2O solution is assumed to be approximately the same as H for CO_2 in pure water, then $H=1630$ atm. The mole fraction of CO_2 at the interface of the liquid with ~12% CO_2 in flue gas is

$$x_{CO_2}=\frac{p_{CO_2}}{H}=\frac{0.12}{1630}=7.4\times10^{-5}$$

which corresponds to a concentration of 0.004 M CO_2 if the solution were pure water. Given that $z=1$ for MEA, we have

$$E_i=1+\frac{\left(1\times10^{-5}\,\frac{\text{cm}^2}{\text{s}}\right)(5\text{ M})}{(1)\left(1\times10^{-5}\,\frac{\text{cm}^2}{\text{s}}\right)(0.004\text{ M})}=1251$$

Using Table 3.6, $\sqrt{M}<1/2E_i$ and $\sqrt{M}>3$; therefore, this is a fast pseudo-first-order reaction, with

$$E=\sqrt{D_{L,CO_2}k_2c_B}=\sqrt{(1\times10^{-5})(7570)(5)}=0.378\,\frac{\text{cm}}{\text{s}}$$

$$J_{L,\,CO_2} = c_{i,\,CO_2} E = \left(4.0 \times 10^{-6}\ \frac{\text{mol}}{\text{cm}^3}\right)\left(0.378\ \frac{\text{cm}}{\text{s}}\right) = 1.5 \times 10^{-6}\ \frac{\text{mol}}{\text{cm}^2\ \text{s}}$$

(b) Using Eq. (3.41), and assuming an average rate constant of NaOH of 12100 L/mol s NaOH yields,

$$\sqrt{M} = \frac{\sqrt{\left(1 \times 10^{-5}\ \frac{\text{cm}^2}{\text{s}}\right)\left(12100\ \frac{\text{L}}{\text{mol s}}\right)(1\ \text{mol/L})}}{10^{-2}\ \text{cm/s}} = 34.8$$

The solubility of CO_2 in 1 M NaOH solution can be determined with Eqs. (3.3–3.5) following the steps of Example 3.1. From this procedure, $H = 35.5$ atm/M; therefore, for a system encountering 400 ppm CO_2 in the gas phase,

$$c_{i,\,CO_2} = \frac{p_{CO_2}}{H} = \frac{0.0004}{35.5} = 1.1 \times 10^{-5}\ \text{M}$$

Given that $z = 2$ for NaOH,

$$E_i = 1 + \frac{\left(1 \times 10^{-5}\ \frac{\text{cm}^2}{\text{s}}\right)(1\ \text{M})}{(2)\left(1 \times 10^{-5}\ \frac{\text{cm}^2}{\text{s}}\right)(1.1 \times 10^{-5}\ \text{M})} = 45{,}455$$

Using Table 3.6, $\sqrt{M} < 1/2E_i$ and $\sqrt{M} > 3$; therefore, this is a fast pseudo-first-order reaction, with

$$E = \sqrt{D_{L,\,CO_2} k_2 c_B} = \sqrt{(1 \times 10^{-5})(12100)(1)} = 0.348\ \frac{\text{cm}}{\text{s}}$$

$$J_{L,\,CO_2} = c_{i,\,CO_2} E = \left(1.1 \times 10^{-8}\ \frac{\text{mol}}{\text{cm}^3}\right)\left(0.348\ \frac{\text{cm}}{\text{s}}\right) = 3.8 \times 10^{-9}\ \frac{\text{mol}}{\text{cm}^2\ \text{s}}$$

The flux of CO_2 in DAC is approximately 3 orders of magnitude lower than the PCC case.

(c) Using Eq. (3.43), and assuming an average rate constant for carbonic anhydrase of 1.4×10^6 L/mol s from Table 3.4,

$$\sqrt{M} = \frac{\sqrt{\left(1 \times 10^{-5}\ \frac{\text{cm}^2}{\text{s}}\right)\left(1.4 \times 10^{6}\ \frac{\text{L}}{\text{mol s}}\right)(10\ \text{M})}}{10^{-2}\ \frac{\text{cm}}{\text{s}}} = 1183$$

Since $H = 20$ atm, the mole fraction of CO_2 at the interface of the liquid with ~12 mol% CO_2 in flue gas is

$$x_{CO_2} = \frac{p_{CO_2}}{H} = \frac{0.12}{20} = 0.006$$

which corresponds to a concentration of 0.33 M CO_2 if the solution were pure water. Since $z = 1$ (CO_2 reacts in a 1:1 molar ratio with OH^- of carbonic anhydrase),

$$E_i = 1 + \frac{\left(1 \times 10^{-5}\ \frac{\text{cm}^2}{\text{s}}\right)(10\ \text{M})}{(1)\left(1 \times 10^{-5}\ \frac{\text{cm}^2}{\text{s}}\right)(0.33\ \text{M})} = 31$$

Using Table 3.6, $\sqrt{M} > 10E_i$; therefore, this is an instantaneous reaction, so $E = E_i$ and,

$$J_{L,CO_2} = c_{i,CO_2} k_L E = \left(3.3 \times 10^{-4}\ \frac{\text{mol}}{\text{cm}^3}\right)\left(0.01\ \frac{\text{cm}}{\text{s}}\right)(31)$$
$$= 1.0 \times 10^{-4}\ \frac{\text{mol}}{\text{cm}^2\ \text{s}}$$

In this case, the rate of absorption is approximately 2 orders of magnitude faster than with MEA.

Several gas mixture scenarios are considered with the results presented in Tables 3.7 and 3.8. The enhancement factor estimates and corresponding rates of CO_2 absorption have been calculated for the following two CO_2 capture applications: 1) from air at 1 atm and 0.0390 mol% CO_2 and 2) from flue gas at 1 atm and 12 mol% CO_2. The following limiting cases have been considered in terms of the absorption rate parameters:

Table 3.7 Comparison of rates of absorption predictions from various models for the case of high CO_2 solubility (low H), fast reaction, and a diffusion ratio of unity

	$J\left(\frac{\text{mol}}{\text{cm}^2\text{s}}\right)$	
	Air	Flue
Instantaneous	1.001 (−4)	1.332 (−4)
Pseudo-first-order	3.741 (−5)	1.150 (−2)
Fast pseudo-first-order	3.741 (−5)	1.150 (−2)
Film model	3.108 (−5)	1.332 (−4)
Higbie model	3.108 (−5)	1.332 (−4)

Numbers in parentheses represent exponents of 10

Table 3.8 Enhancement factor predictions from various models for scenarios involving kinetics, solubility, and diffusivity

Scenario	Cases satisfied from Table 3.6	$E = \frac{\sqrt{M\left(\frac{E_i-E}{E_i-1}\right)}}{\tanh\left(\sqrt{M\left(\frac{E_i-E}{E_i-1}\right)}\right)}$		$E = \sqrt{D_{L,CO_2} k_2 c_B}$	$E = \left(\frac{\sqrt{M}}{\tanh\sqrt{M}}\right)$	$E = \left(1 + \frac{D_B c_B}{z D_{L,CO_2} c_{i,CO_2}}\right)$
		Higbie	Film	Fast PFO	PFO	Instantaneous
Direct air capture						
Slow, low H, DR=1	Fast PFO	3.17	3.17	0.316	3.17	927
Fast, low H, DR=1	Fast PFO	288	288	34.6	346	927
Slow, low H, DR=2	Fast PFO	3.17	3.17	0.316	3.17	1.85 (3)
Fast, low H, DR=2	Fast PFO	304	316	34.6	346	1.85 (3)
Slow, high H, DR=1	Fast PFO	3.17	3.17	0.316	3.17	7.87 (4)
Fast, high H, DR=1	Fast PFO	346	346	34.6	346	7.87 (4)
Slow, high H, DR=2	Fast PFO	3.17	3.17	0.316	3.17	1.57 (5)
Fast, high H, DR=2	Fast PFO	346	346	34.6	346	1.57 (5)
Flue gas capture						
Slow, low H, DR=1	none	2.38	2.38	0.316	3.17	4.01
Fast, low H, DR=1	IR	4.01	4.01	34.6	346	4.01
Slow, low H, DR=2	Fast PFO	2.52	2.70	0.316	3.17	7.02
Fast, low H, DR=2	IR	4.97	7.02	34.6	346	7.02
Slow, high H, DR=1	Fast PFO	3.16	3.16	0.316	3.17	257
Fast, high H, DR=1	none	257	257	34.6	346	257
Slow, high H, DR=2	Fast PFO	3.16	3.17	0.316	3.17	513
Fast, high H, DR=2	none	219	249	34.6	346	513

Numbers in parentheses represent exponents of 10; pseudo-first-order (PFO); instantaneous reaction (IR)

- High and low CO_2 solubilities, indicated by ↓HL (20 atm) and ↑HL (1700), respectively
- Base-to-CO_2 solution-phase diffusivity ratios of 1 and 2, indicated by DR = 1, and DR = 2, respectively
- Fast (1.2×10^8 L/mol s) and relatively slow (1×10^4 L/mol s) rate constants for the reaction of CO_2 with the binding species
- The CO_2 diffusivity was selected as 1×10^{-5} cm^2/s, base concentration as 1 M, and liquid-phase CO_2 mass-transfer coefficient as 0.1 cm/s

The results are presented in Tables 3.7, 3.8 and 3.9. Using Eq. (3.43), $\sqrt{M} = 346$ in the case of a "fast" reaction (high rate constant), and 3.16 in the case of a "slow" reaction (low rate constant). The parameters E_i and E have been calculated for each of the scenarios listed above for model estimates including instantaneous, pseudo-first-order (PFO), and fast PFO reactions, in addition to the Film and Higbie approximations with the calculated values of E presented in Table 3.8. In the case of the Higbie model, rather than calculating the parameter E_i in the same manner as the Film-model approach using Eq. (3.44), this term is calculated by,

$$E_i = \sqrt{\frac{D_{L,CO_2}}{D_B}} + \frac{c_B}{z c_{i,CO_2}} \sqrt{\frac{D_B}{D_{L,CO_2}}} \qquad (3.47)$$

Although E_i is calculated differently for the Higbie and Film models, E is calculated using Eq. (3.42) for both approaches.

Notice that when the ratio of the CO_2 and binding species diffusivities equals one, Eq. (3.47) collapses to the equation of E_i for the Film model. Once E_i and $\sqrt{M}$ are known, the criteria from Table 3.6 may be considered to determine E. Many of the scenarios result in the application of the estimation of E using Eq. (3.45), thereby justifying a fast pseudo-first-order reaction mechanism. Comparing the rates of CO_2 absorption for air capture versus flue gas capture from Table 3.7, it is clear that the flue gas case is more sensitive to the model assumptions used. Table 3.7 examines the scenario of a high CO_2 solubility (low H), fast reaction (high k), and a base-to-CO_2 diffusion ratio of unity. For this case, if one assumes a fast PFO reaction, the rate of absorption would be overestimated by approximately two orders of magnitude. The air capture case is much less sensitive, with an approximate 20% overestimation. Using the enhancement factors calculated according to Table 3.6 and 3.8, the rates of absorption for capture from air and flue gas based upon the various model approaches are listed in Table 3.9. In the case of air capture, the fast PFO assumption is valid for all cases. In all cases, the fast PFO estimate agrees with the more rigorous Film estimate to within 20%, with the Film estimate in agreement with the Higbie estimate to within 6%. In the case of CO_2 capture from flue gas, the analysis is not as straightforward since the flue gas stream is significantly more concentrated in CO_2 than ambient air. There are several cases in which none of the criteria of Table 3.6 are satisfied. In these cases, as noted in Table 3.9, assuming a PFO or fast PFO reaction can lead to an order of magnitude overestimation in the rate of absorption. There are two scenarios that satisfy the condition of instantaneous reaction. These are both fast reactions (high k) and high solubilities (low H)

Table 3.9 Rates of absorption predictions from various models for scenarios involving kinetics, solubility, and diffusivity

Scenario	Cases satisfied from Table 3.6	$J_{L,CO_2} = c_{i,CO_2} k_{L,CO_2} E$		$J_{L,CO_2} = c_{i,CO_2} E$	$J_{L,CO_2} = c_{i,CO_2} k_{L,CO_2} E$	
		Higbie	Film	Fast PFO	PFO	Instantaneous
Direct air capture						
Slow, low H, DR=1	FPFO	3.424 (−7)	3.424 (−7)	3.415 (−7)	3.428 (−7)	1.001 (−4)
Fast, low H, DR=1	FPFO	3.108 (−5)	3.108 (−5)	3.741 (−5)	3.741 (−5)	1.001 (−4)
Slow, low H, DR=2	FPFO	3.425 (−7)	3.426 (−7)	3.415 (−7)	3.428 (−7)	2.001 (−4)
Fast, low H, DR=2	FPFO	3.280 (−5)	3.409 (−5)	3.741 (−5)	3.741 (−5)	2.001 (−4)
Slow, high H, DR=1	FPFO	4.030 (−9)	4.030 (−9)	4.016 (−9)	4.031 (−9)	1.000 (−4)
Fast, high H, DR=1	FPFO	4.390 (−7)	4.390 (−7)	4.399 (−7)	4.399 (−7)	1.000 (−4)
Slow, high H, DR=2	FPFO	4.030 (−9)	4.030 (−9)	4.016 (−9)	4.031 (−9)	2.000 (−4)
Fast, high H, DR=2	FPFO	4.393 (−7)	4.395 (−7)	4.399 (−7)	4.399 (−7)	2.000 (−4)
Flue gas capture						
Slow, low H, DR=1	none	7.887 (−5)	7.887 (−5)	1.050 (−4)	1.054 (−4)	1.332 (−4)
Fast, low H, DR=1	IR	1.332 (−4)	1.332 (−4)	1.150 (−2)	1.150 (−2)	1.332 (−4)
Slow, low H, DR=2	FPFO	8.363 (−5)	8.976 (−5)	1.050 (−4)	1.054 (−4)	2.332 (−4)
Fast, low H, DR=2	IR	1.649 (−4)	2.332 (−4)	1.150 (−2)	1.150(−2)	2.332 (−4)
Slow, high H, DR=1	FPFO	1.236 (−6)	1.236 (−6)	1.236 (−6)	1.241 (−6)	1.004 (−4)
Fast, high H, DR=1	none	1.004 (−4)	1.004 (−4)	1.354 (−4)	1.354 (−4)	1.004 (−4)
Slow, high H, DR=2	FPFO	1.237 (−6)	1.238 (−6)	1.236 (−6)	1.241 (−6)	2.004 (−4)
Fast, high H, DR=2	none	8.545 (−5)	9.726 (−5)	1.354 (−4)	1.354 (−4)	2.004 (−4)

Numbers in parentheses represent exponents of 10; fast pseudo-first-order (FPFO)

with either a diffusivity ratio of 1 or 2. In these cases, assuming either a PFO or fast PFO reaction would lead to overestimates in the rate of absorption by up to 2 orders of magnitude. Of the cases in which the conditions for a fast PFO reaction are satisfied, all of the estimates agree with both the more rigorous Film and Higbie estimates, except for the case of the slow reaction with high CO_2 solubility and diffusion ratio of 2. For this case, although the conditions for fast PFO are satisfied, they deviate from the Film and Higbie models by approximately 20%. When comparing the Film to the Higbie model estimates, for diffusion ratios of 2, the Film model overestimates the rate of absorption by at most 40%, which is in the case of an instantaneous reaction. In general, if one follows the approach outlined in Table 3.7 for calculating the enhancement factor, reasonably accurate rates of CO_2 absorption are possible.

Example 3.5 Carbon dioxide is absorbed into a solution of NaOH in a packed column at 20°C. The dissolved CO_2 undergoes an irreversible second-order reaction ($z=2$). Its diffusivity is 1.8×10^{-5} cm²/s and its solubility is 4×10^{-5} mol/cm³. The rate constant for the reaction between CO_2 and OH^- in the solution can be taken as 10^4 L/mol s. The diffusivity ratio between NaOH and CO_2 is 1.7. The liquid-phase mass transfer coefficient is 10^{-2} cm/s. Find the rate of absorption (per unit volume of packed space) at a point where the partial pressure of CO_2 is 1 atm and the concentration of NaOH is 0.5 M. The interfacial area per unit packed volume is 1 cm^{-1}. Consider gas-side resistance to be negligible.

Solution: The reaction is $CO_2 + 2NaOH \rightarrow Na_2CO_3 + H_2O$ and the following properties are given,

k_{L,CO_2} $= 10^{-2}$ cm/s
k_2 $= 10000$ L/mol s
c_{i,CO_2} $= 4 \times 10^{-5}$ mol/cm³
c_B $= 0.5\text{ M} = 0.5\text{ mol/L} \equiv 5 \times 10^{-4}$ mol/cm³
D_{L,CO_2} $= 1.8 \times 10^{-5}$ cm²/s
$D_B/D_{L,CO_2}$ $= 1.7$
a $= 1$ cm^{-1}

Since $D_B/D_{L,CO_2}$ is significantly different from unity, the Higbie model will be used as it is likely more accurate than the Film model. The rate of absorption with reaction can be determined from Eq. (3.41)

$$J_{L,CO_2} = c_{i,CO_2} k_{L,CO_2} E$$

where the enhancement factor, E, may be read from Fig. 3.9. In order to use Fig. 3.9, the values of E_i and $\sqrt{M}$ must be known. For the Higbie model, E_i is calculated from:

$$E_i = \sqrt{\frac{D_{L,CO_2}}{D_B}} + \frac{c_B}{z c_{i,CO_2}} \sqrt{\frac{D_B}{D_{L,CO_2}}}$$

$$= \sqrt{\frac{1}{1.7} + \frac{(5 \times 10^{-4}\ \text{mol/cm}^3)}{(2)\left(4 \times 10^{-5}\ \text{mol/cm}^3\right)}\sqrt{1.7}} = 8.92$$

Generally, this equation may be used when E_i is much greater than one, which it is the case for this example. To determine $\sqrt{M}$, the following equation is used:

$$\sqrt{M} = \frac{\sqrt{k_2 c_B D_{L,\,CO_2}}}{k_{l,\,CO_2}}$$

$$\sqrt{M} = \frac{\sqrt{\left(\frac{10^4\ \text{L}}{\text{mol s}}\right)(1000\ \text{cm}^3/\text{L})\left(5 \times 10^{-4}\ \text{mol/cm}^3\right)\left(1.8 \times 10^{-5}\ \text{cm}^2/\text{s}\right)}}{10^{-2}\ \text{cm/s}}$$

$$\sqrt{M} = 30$$

Using Goal Seek in Excel or Fig. 3.9 and with $E_i = 8.92$, $E \approx 8$ when $\sqrt{M} = 30$. The rate of absorption is:

$$J_{L,\,CO_2} = \left(4 \times 10^{-5}\ \text{mol/cm}^3\right)\left(10^{-2}\ \text{cm/s}\right)(8) = 3.2 \times 10^{-6}\ \frac{\text{mol}}{\text{cm}^2\ \text{s}}$$

To find the rate of absorption per unit volume of packing, multiply $J_{L,\,CO_2}$ by the interfacial area per unit volume (a). Thus, the rate of absorption is 3.2×10^{-6} mol/cm^3 s.

3.4.2 Liquid-Phase Resistance: Mass Transfer Relations

In addition to the mass-transfer theory predictions for absorption rates, mass-transfer relations based upon experiment and dimensional analysis are also used. In absorption, mass transfer is a very complex process involving both turbulent and laminar flow regimes as liquid flows across packing material through the column. Due to these complexities a number of theories have been developed in an attempt to establish correlations that relate the rate of mass transfer to the solute diffusivity. Mass transfer depends upon solute diffusivity and the variables that control the fluid flow, *i.e.*, fluid velocity u, fluid viscosity μ, and fluid density ρ. The shape of the interface (*i.e.*, packing material) will influence the mass-transfer process so that a unique relation exists for each shape, but for any given shape the mass-transfer coefficient, $k = \text{f}(D, l, u, \mu, \rho)$ such that l is some linear dimension and D is diffusivity. Dimensional analysis yields:

$$\frac{kl}{D} = f\left(\frac{\mu}{\rho D}, \frac{\rho u l}{\mu}\right) \tag{3.48}$$

Table 3.10 Common dimensionless numbers used in mass transfer

Name	Dimensionless number	Physical meaning
Sherwood (Sh)	$\frac{kl}{D}$	$\frac{\text{mass transfer velocity}}{\text{diffusion velocity}}$
Schmidt (Sc)	$\frac{\mu}{\rho D} = \frac{\upsilon}{D}$	$\frac{\text{diffusivity of momentum}}{\text{diffusivity of mass}}$
Reynolds (Re)	$\frac{\rho ul}{\mu} = \frac{lu}{\upsilon}$	$\frac{\text{inertial forces}}{\text{viscous forces}}$

such that

kl/D is the Sherwood number, denoted Sh,
$lu\rho/\mu$ is the Reynolds number, denoted Re, and
$\mu/\rho D$ is the Schmidt number, denoted Sc

These dimensionless groups are outlined in further detail in Table 3.10. Often the kinematic viscosity, ν is used to replace the quotient of the fluid viscosity and the fluid density.

A number of correlations associated with the mass-transfer coefficients at fluid-fluid interfaces for various scenarios are listed in Table 3.11. Knowledge of fluid properties such as the viscosity, density, and diffusivity, in addition to system properties such as the characteristic length and *superficial velocity* are required to determine the mass-transfer coefficient, k, as listed in Table 3.11.

Solvent Selection The desired properties of a solvent for CO_2 separation include:

- high CO_2 capacity
- fast kinetics with CO_2

Table 3.11 Mass-transfer correlations for various fluid-fluid interfaces

Scenario	Correlation	Remarks
Liquid in packed-bed	$k\left(\frac{1}{\upsilon g}\right)^{\frac{1}{3}} = 0.0051\left(\frac{u}{\upsilon a}\right)^{0.67}\left(\frac{D}{\upsilon}\right)^{0.5}(ad_p)^{0.4}$	Known as best available correlation for liquids
	$\frac{kl}{D} = 25\left(\frac{lu}{\upsilon}\right)^{0.45}\left(\frac{\upsilon}{D}\right)^{0.5}$	Classical result and widely referenced
Gas bubbles in stirred tank	$\frac{kl}{D} = 0.13\left(\frac{(P/V)l^4}{\rho\upsilon^3}\right)^{\frac{1}{4}}\left(\frac{\upsilon}{D}\right)^{\frac{1}{3}}$	k is not dependent on bubble size
Gas bubbles in unstirred tank	$\frac{kl}{D} = 0.31\left(\frac{l^3 g\Delta\rho/\rho}{\upsilon^2}\right)\left(\frac{\upsilon}{D}\right)^{\frac{1}{3}}$	$\Delta\rho =$ density difference between bubble and fluid
Falling films	$\frac{kz}{D} = 0.69\left(\frac{zu}{D}\right)^{0.5}$	$z =$ position along film

a is packing area per bed volume; l is characteristic length (or bubble or drop diameter); ε is bed voidage; P/V is stirrer power per volume; u is superficial velocity (or drop velocity)
d_p is particle (packing material) diameter

- low volatility
- low viscosity
- nontoxic, nonflammable, and noncorrosive
- high thermal stability
- resistance to oxidation

Viscosity, to some degree, is also an important property associated with an appropriate solvent. If the solvent is too viscous it will be difficult to overcome the pressure drop of the column. In addition to achieving a lower pressure drop, a low-viscosity solvent will also lead to higher mass- and heat-transfer rates in the column because liquid-phase diffusion coefficients are inversely correlated with viscosity. To ensure safety it is also optimal for the solvent to be nontoxic, nonflammable, and noncorrosive. Examples of common solvents for CO_2 separation include, but are not limited to amines, carbonates, and ammonia.

In the stripping process, or solvent regeneration process, it is important that the binding of CO_2 can be "easily" reversed as to minimize the cost associated with reuse of the solvent. The most common stripping agent is steam. In the case of regeneration with steam, an additional separation step may be required since CO_2 would inevitably be mixed with steam. Operating temperature and pressure can be used to minimize stage requirements and/or solvent flow rate, which will influence the column volume required. Absorber pressures are usually dictated by the available pressure of the gas to be treated since it is usually too costly to add compression to operate at greater pressure. Stripper pressure and temperature may be optimized to minimize the compression energy required for CO_2 transport. The stripper temperature is limited by the thermal degradation of the solvent, while the absorber temperature is usually as close to ambient as the cooling sink will allow.

Key steps in the design of a column for absorption (or stripping) include determining parameters at the onset, *e.g.*, identification of feed components, concentrations, and flow rates; providing operating conditions when they may be known; selecting the solvent; and obtaining the appropriate equilibrium data. Also, before considering the separation process, the percentage of the solute in the gas mixture to be removed should be determined.

3.5 Design Parameters for Packed Columns

Design parameters such as absorber column height and diameter are required to accurately determine the capital cost of the separation process. Four primary factors influence the design of a separation process based upon mass transfer: 1) the allowable liquid flow rate, 2) that the contact area be proportional to the gas flow rate, 3) the energy requirements of the process, and 4) the number of *equilibrium stages*. To determine the number of stages required, the CO_2 concentration in the original gas mixture and the desired purity of the exiting gas mixture must be known. The energy requirement can be determined from material balance calculations and the relative quantities required in the gas and liquid phases, with the concentrations in

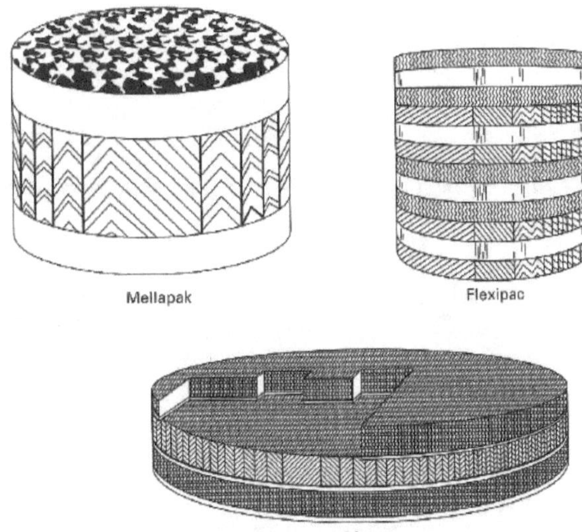

Fig. 3.10 Traditional structured packing materials used in a packed column. (Reprinted with permission of [3])

each phase determined from the equilibrium properties of the system in addition to the rate of CO_2 transfer from the gas to liquid phase. As previously discussed, the rate of mass transfer depends upon the properties of the phases in addition to the flow rates of the gas and liquid within the absorption column. Material balances and fluid dynamics both assist in determining the allowable liquid flow rate. The column cross-sectional area and subsequently its diameter can be determined with knowledge of the gas flow rate. The overall energy requirement is comprised of heat and mechanical energy requirements. Note that heat removal may be necessary in the absorption (exothermic) process depending upon the CO_2 loading and the material's ability to transfer heat away from the system. Additionally, heat is required for the solvent regeneration process. Finally, mechanical energy is required for fluid and gas transport or dispersion and for moving parts associated with the machinery of the system.

3.5.1 Trays vs. Packing

Most separation processes can be carried out using either trays or packing. A packed column can consist of either random or structured packing material as shown in Figs. 3.10 and 3.11. The factors favoring each for CO_2 capture applications will be discussed. In particular, the primary considerations include pressure drop, foaming, corrosion, liquid-gas distribution, and complexity.

The pressure drop in a packed column is lower than trayed columns since the open area within the collective packing material along a given stage approaches the area of the cross section of the column. With a trayed column, however, the open

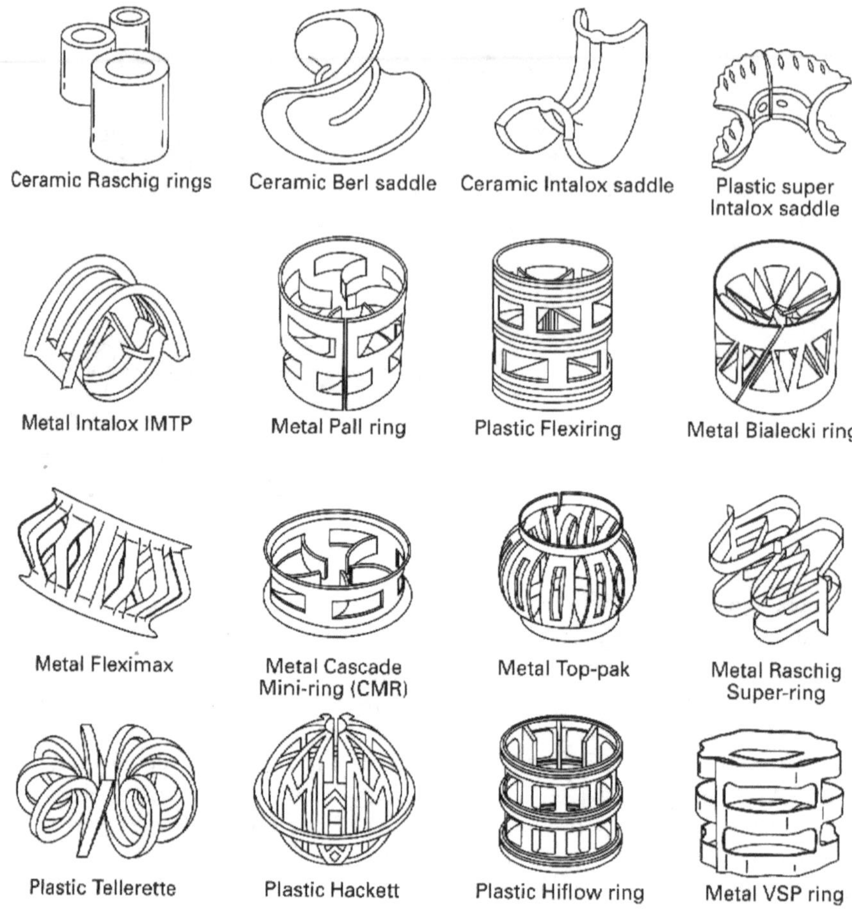

Fig. 3.11 Traditional random (or dumped) packing materials used in a packed column. (Reprinted with permission of [3])

area of the trays is only 8–15% of the column's cross-sectional area. Also, the trays themselves have an additional liquid pressure drop of approximately 55 mm of the liquid per tray, which is absent in packed columns. On average, the pressure drop in trayed columns is on the order of 10 mbar per theoretical stage versus 3–4 mbar in the case of random packing and 1.5–2 mbar per theoretical stage in the case of structured packing. Therefore, in the case in which gas is being moved by a fan or blower through the column, the pressure drop may be a controlling consideration. For absorption systems operating at 1 atm, such as postcombustion applications, the lower pressure drop in packed columns might favor the packing over the tray approach.

The large and open spacing associated with the random packing material in a column allows for improved foam dispersal over both structured packing and trays.

In structured packing, the solid walls of the material limit lateral movement of the foam, allowing foaming to propagate. Another consideration is corrosion. The range of materials used for random packing is greater than structured or trays. For instance, ceramic and plastic packing materials are affordable and corrosion-resistant. Thus far, the considerations of pressure drop, foaming, and corrosion have all favored random packing over structured packing or trays.

Advantages in which trays are favored include gas-liquid distribution and column complexity. To achieve a desired separation, it is crucial to maximize the extent of gas-liquid contact. If there is a maldistribution of gas and liquid throughout the column the separation may be ineffective. Maldistribution is a common problem in large-diameter columns, long beds, systems with low liquid flow rates, and small packing material. Another potential advantage of trays is the design complexity of the column. For instance, if cooling is required for an absorption process or heating for regeneration, the installation of these units is more difficult in a packed over a trayed column. Heat control for carbon-capture systems is an important consideration. In particular, for precombustion capture applications where the CO_2 concentration in the feed gas may be on the order of 40 mol% may result in significant exothermicity, which will have to be monitored and controlled for optimal solvent capacity. Additionally, in regeneration systems the control and distribution of heat is important for obtaining a desired CO_2 purity and effective separation for solvent recycling.

3.5.2 Pressure Drop

When a solvent flows down the column through packing material or over trays, it must overcome the pressure drop within the column due to fluid friction along the material. Examples of packing materials are shown in Figs. 3.10 and 3.11. Balance must exist between the gas-phase and liquid-phase mass flow rates (V and L, respectively); otherwise, if the gas velocity is too high it can prevent the solvent from falling down the packing material leading to liquid hold-up and subsequent flooding in the column. *Flooding velocity* is the gas velocity that leads to solvent flooding. The gas velocity should be chosen to be within 50–75% of the flooding velocity for safe operation. An empirical equation [67] for the pressure drop at flooding (ΔP_{flood}) is

$$\Delta P_{flood} = 0.115 F_p^{0.7} \tag{3.49}$$

such that F_p is a dimensionless term called the packing factor and typically ranges between 10 to 60 and ΔP_{flood} has units of inches of water per foot of packing. For higher packing factors, the pressure drop at flooding can be approximated at 2.0 inches of water per foot of packing (*i.e.*, 0.02 atm per m of packing).

3.5.3 Material Balances

Figure 3.12 shows the flows entering and exiting the column. The gas molar flow rate entering the bottom of the column is V_2 and the mole fraction of CO_2 in the gas

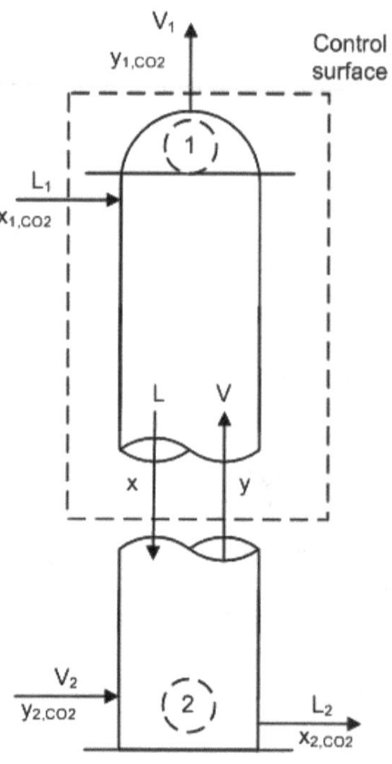

Fig. 3.12 Material balance components for packed column

feed is y_{2,CO_2}. The gas containing CO_2 flows upward through the column with the liquid solvent flowing downward. The role of the packing material within the column is to achieve sufficient contact between the gas and liquid phases. As discussed previously, CO_2 diffuses in the gas phase to the phase boundary, dissolves in the solvent (physical), then diffuses through the liquid phase, interacting with the binding component of a solvent chemically, which facilitates the continuous mass transfer from gas to liquid phases. The gas molar flow rate exiting the column, V_1, is ideally nearly void of CO_2 with a mole fraction of $y_{1,\ CO_2}$. The liquid molar flow rate entering the top of the column, L_1, will contain some fraction of CO_2, represented by the mole fraction, $x_{1,\ CO_2}$, since this stream is coming from the outlet of the stripping column, which will not be able to remove all of the CO_2. This highlights the importance of the stripper performance. The more CO_2 removed from the stripper, the more effective the mass transfer process from the gas to liquid phase will be. The solvent exiting the bottom of the absorber column will be loaded with CO_2 with mole fraction, $x_{2,\ CO_2}$ and liquid molar flow rate, L_2. Material balances may be carried out across the control surface of the absorber (and stripper) columns, as designated by the dotted line in Fig. 3.12. The goal behind carrying out the mass balances is to derive the operating line of the absorption process. From the operating line equation, the liquid-to-vapor ratio can be determined. The overall material balance equations based on the streams entering and exiting the column are:

$$\text{Total Material Balance: } L_1 + V_2 = L_2 + V_1 \tag{3.50}$$

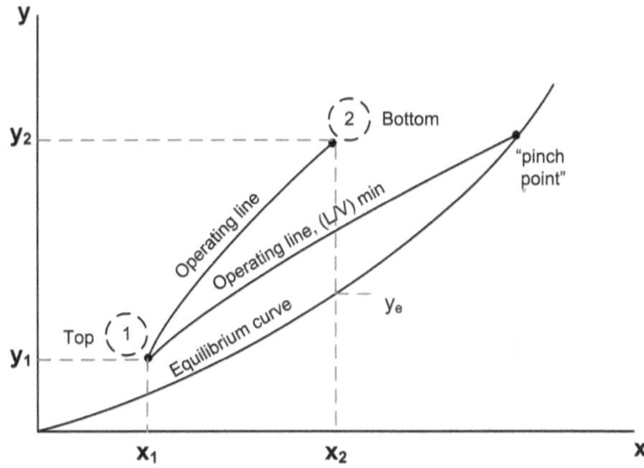

Fig. 3.13 Operating and equilibrium lines and limiting liquid-vapor ratio

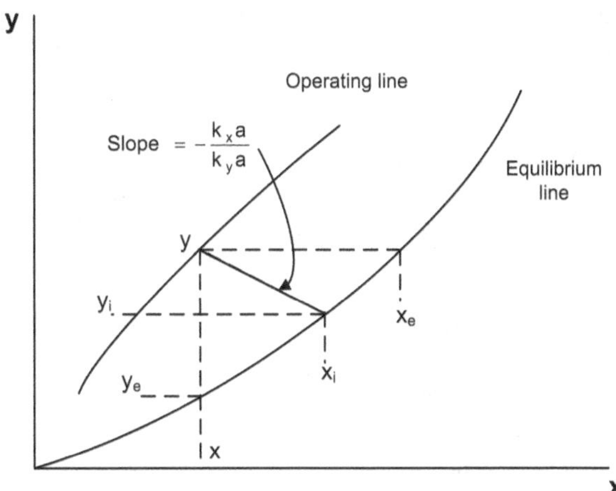

Fig. 3.14 Composition of interfacial and equilibrium points

$$CO_2 \text{ Balance: } L_1 x_{1,\,CO_2} + V_2 y_{2,\,CO_2} = L_2 x_{2,\,CO_2} + V_1 y_{1,\,CO_2} \tag{3.51}$$

These equations can be generalized to the following to represent any section of the column:

$$\text{Total Material Balance: } L_1 + V = L + V_1 \tag{3.52}$$

$$CO_2 \text{ Balance: } L_1 x_{1,\,CO_2} + V y_{CO_2} = L x_{CO_2} + V_1 y_{1,\,CO_2} \tag{3.53}$$

The relationship between x and y at any section of the column is the operating line and can be obtained by rearranging Eq. (3.53) for CO_2 yielding:

$$y = \frac{L}{V}x + \frac{V_1 y_{1,\,CO_2} - L_1 x_{1,\,CO_2}}{V} \tag{3.54}$$

The operating line can be plotted along with the equilibrium curve as shown in Figs. 3.13 and 3.14, which may be determined from experimental thermodynamic and

Table 3.12 Common mass-transfer coefficient definitions

Basic equation	Units of k[a]	Notes
$J = k\Delta c$	cm/sec	Common in model literature; convenient units for chemical reaction comparison
$J = k_p \Delta p$	mol/cm^{2}s Pa	Common for gas absorption processes; common in chemical engineering literature
$J = k_x \Delta x$ (or $k_y \Delta y$)	mol/cm^{2}s	Preferred for gas applications

[a] J has units of mol/cm^{2}s; c has units of mol/cm^3

solubility data associated with CO_2 in a given solvent. For the absorption process, the operating line must lie above the equilibrium curve, confirming the positive driving force for absorption, $(y - y_e)$, represented by the dotted vertical line in Fig. 3.13. Similarly, there is a driving force, $(x_e - x)$, which would be represented by a horizontal line. The opposite is true in the case of a stripping process. Within an absorption process, as CO_2 transfers from the gas to liquid phase throughout the height of the column, there is a decrease in the vapor flow rate, which manifests as a curved operating line as shown in Figs. 3.13 and 3.14. As the CO_2 content in a given gas mixture is more dilute this line loses its curvature. The average slope of the operating line from Eq. (3.54) is L/V, which is the ratio of liquid to vapor molar flow rates. This ratio influences the design parameters and hence, the economics of the process. An increase in L/V results in a greater driving force everywhere within the column with the exception of the very top, represented as point 1 in Fig. 3.13. As the slope, or L/V increases, a lower number of stages for effective separation is required, leading to a shorter column. However, there is a tradeoff since a higher L/V requires a larger quantity of solvent resulting in a more dilute liquid product, which will require additional energy in stripping. The optimal solvent flow rate for absorption is determined by balancing the operating costs of the absorber and stripper with the fixed capital equipment costs. A general rule of thumb is a liquid flow rate that is 1.1–1.5 times the minimum rate. The minimum possible liquid flow rate is that obtained from when the operating line just touches the equilibrium line, which is represented as the "pinch point" in Fig. 3.13. At this point an infinite number of stages is required since the concentration difference for mass transfer is zero at the bottom of the column.

The mass-transfer coefficient is often expressed in different forms throughout the literature, which makes understanding the complexities of absorption processes increasingly difficult. Table 3.12 lists the common definitions of mass-transfer coefficients. Up to this point the term "k" with units of cm/sec have been used; however, in most traditional chemical engineering textbooks, the term k_p with units of mol/cm^{2}s Pa or k_x with units of mol/cm^{2}s are more commonly used. Therefore, when discussing the rate of absorption in a staged process such as an absorber or stripping column, these latter definitions of mass-transfer coefficient are used in place of the previous "mass-transfer velocity" k term. Expressing the mass-transfer coefficient in the dimensions of velocity allows for an easier comparison to chemical reactions in addition to simultaneous heat and mass transfer. When this type of analysis is of interest, defining the mass-transfer coefficient in terms of k is more appropriate.

An additional complexity associated with the terminology of absorption processes is the use of individual versus overall mass-transfer coefficients. An individual mass-transfer coefficient is also termed as "local" and corresponds to a given stage within the column in contrast to the overall mass-transfer coefficient, which is also termed "average" and corresponds to the overall separation process in general. The final source of complexity in the terminology is the determination of the interfacial area, which would be required to determine the flux per area and corresponding mass-transfer coefficient. This unknown interfacial area is avoided by inherently including it within the mass-transfer coefficient as shown in the second two definitions in Table 3.12, and directly measuring the flux per column volume as demonstrated by Eqs. (3.55–3.58).

Rates of absorption may be expressed in terms of individual or overall gas- or liquid-phase mass transfer coefficients as follows:

$$r = k_y a(y - y_i) \tag{3.55}$$

$$r = k_x a(x_i - x) \tag{3.56}$$

$$r = k_y a(y - y_e) \tag{3.57}$$

$$r = k_x a(x_e - x) \tag{3.58}$$

such that $k_y a$ is the individual volumetric mass-transfer coefficient for the gas phase with units of mol/cm^{3}s, and is based upon a concentration difference driving force. The parameter a is the interfacial area per unit of packed volume and has units of cm^2/cm^3. The overall liquid-phase mass transfer coefficient, $K_x a$ has the same units as $k_x a$. For the gas-phase absorption rate expressions, a partial pressure driving force $(p - p_i)$ can replace the mole fraction driving force $(y - y_i)$ since concentration and partial pressure in the gas phase are directly proportional. The rate of absorption and corresponding gas-phase mass-transfer coefficients are often represented as:

$$r = k_p a(p - p_i) \tag{3.59}$$

$$r = k_p a(p - p_e) \tag{3.60}$$

such that $k_p a$ is the individual volumetric mass-transfer coefficient for the gas phase with units of mol/cm^{3}s atm, and is based on a partial-pressure difference driving force, and similarly for the overall gas-phase mass transfer coefficient, $K_p a$.

The composition at the interface between the liquid and gas phase is (y_i, x_i) and can be determined from the operating and equilibrium lines as shown in the y–x diagram of Fig. 3.14. Values associated with the interface are required for situations in which a given gas is rich in CO_2 or when the equilibrium line is strongly curved. Combining Eqs. (3.55) and (3.56) yields,

$$\frac{y - y_i}{x_i - x} = \frac{k_x a}{k_y a} \tag{3.61}$$

A line drawn from the operating line with slope $- k_x a / k_y a$ will intersect the equilibrium line at (y_i, x_i) as shown in Fig. 3.14. The overall driving forces are the vertical

Table 3.13 Selected overall gas-phase mass-transfer coefficients for CO_2 in various solvents[a] [13] (Packed-columns filled with Raschig rings)

Solvent	$K_y a$ (mol/s cm^{3}atm)	Reference
Water	2×10^{-7}	[68]
1-N sodium carbonate, 20% Na as bicarbonate	1×10^{-7}	[69]
3-N diethanolamine, 50% converted to carbonate	2×10^{-6}	[70]
2-N sodium hydroxide, 15% Na as carbonate	1.0×10^{-5}	[71]
2-N potassium hydroxide, 15% K as carbonate	1.7×10^{-5}	[72]
Hypothetical perfect solvent having no liquid-phase resistance and having infinite chemical reactivity	1.07×10^{-4}	[68]

[a]Basis: $L = 0.339$ g/s cm^2; $V = 0.041$ g/s cm^2; $T = 25°$C; $P = 1.0$ atm

(*i.e.*, $y - y_e$) or horizontal (*i.e.*, $x_e - x$) lines on the $y - x$ diagram of Fig. 3.14. The overall mass-transfer coefficients can be determined from $k_x a$ and $k_y a$, combined with the local slope of the equilibrium curve, m:

$$\frac{1}{K_y a} = \frac{1}{k_y a} + \frac{m}{k_x a} \tag{3.62}$$

$$\frac{1}{K_x a} = \frac{1}{k_x a} + \frac{1}{m k_y a} \tag{3.63}$$

Within Eqs. (3.62) and (3.63), the mass transfer can be described in terms of resistances in series, *e.g.*, the first term on the right-hand side of Eq. (3.62) represents the resistance to mass transfer in the gas phase, while the second term represents the resistance to mass transfer in the liquid phase.

Aqueous alkaline/based solvents such as potassium or sodium carbonate-bicarbonate or amines have also been used in the past to separate CO_2 from gas mixtures for ammonia synthesis processing. In ammonia synthesis, hydrogen is primarily produced from the partial oxidation of hydrocarbon, resulting in a mixture of CO_2, hydrogen, and nitrogen. The CO_2 is often removed in either a trayed or packed-column absorber. Table 3.13 provides several selected data illustrating the effect of increasing solvent alkalinity on the mass-transfer coefficient for a packed-column absorber. As the alkalinity of the solvent increases, the mass-transfer coefficient increases. In both absorption and stripping, the diffusion and chemical reactivity are closely coupled and must be treated simultaneously in determining the rate of absorption.

3.5.4 Column Height Estimation

The approach used to estimate the height of the column will be based upon the parameters outlined in the packed-bed absorption column depicted in Fig. 3.15. A

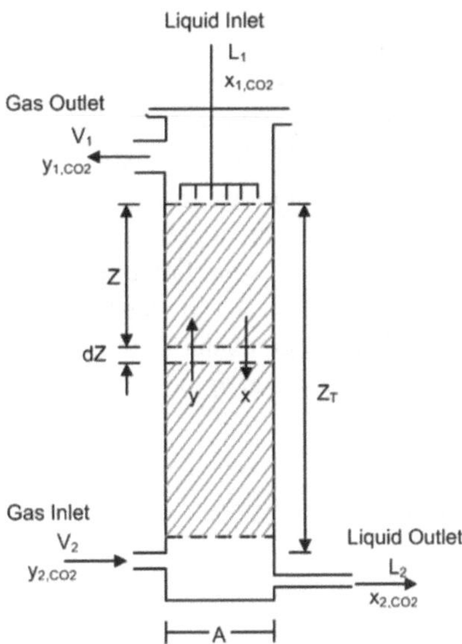

Fig. 3.15 Packed-bed absorption column for height estimation

similar approach, but for estimating the height of a trayed column is available in Appendix H. The cross-section of the column in Fig. 3.15 is A with a differential height of dz, and combined yield a volume, Adz. For dilute gases the molar flow rate change can be neglected, and the amount of CO_2 absorbed in section dz is Vdy, which is equivalent to the product of the rate of absorption and the differential volume as follows:

$$Vdy = K_y a(y - y_e)Adz \tag{3.64}$$

In Eq. (3.64) Vdy can be considered the loss of CO_2 from the gas phase into the solvent (liquid phase), while $K_y a(y - y_e)$ is the rate of absorption, in which $y - y_e$ is the mass-transfer driving force. Equation (3.64) can be rearranged and integrated as follows:

$$\frac{K_y aART}{V}\int_0^{z_T} dz = \int_2^1 \frac{dp}{p - p_e} \tag{3.65}$$

$$\frac{K_y aART Z_T}{V} = \int_2^1 \frac{dp}{p - p_e} \tag{3.66}$$

$$Z_T = \frac{V}{K_y aART}\int_2^1 \frac{dp}{p - p_e} \tag{3.67}$$

such that the height of the column is Z_T as expressed by Eq. (3.67).

The integral of Eq. (3.67) is the *number of mass-transfer units* (NTU), represented by the change in gas-phase partial pressure of CO_2 divided by the average driving force. The coefficient of Eq. (3.67) is the *height of a mass-transfer unit* (HTU). A straightforward design method is to determine NTU from a y–x diagram and multiply NTU by HTU obtained from the literature or from mass-transfer correlations, to ultimately determine the height of the column. The HTU can be thought of as a measure of separation effectiveness with a more efficient separation process resulting in a lower HTU. The NTU can be thought of as a measure of the difficulty associated with the separation process with a higher NTU required to increase the extent of CO_2 removal from a given gas mixture. For instance, to obtain a flue gas with 90% removal of CO_2 would require more transfer units than 50% removal.

Equation (3.67) is often rewritten in terms of the product of the number of transfer units (the integral value) and the height of each transfer unit (number in front of the integral with units of length) as:

$$Z_T = H_{OG} N_{OG} \tag{3.68}$$

or

$$Z_T = H_{OL} N_{OL} \tag{3.69}$$

such that H_{OG} is the height of the overall gas-phase mass-transfer unit:

$$= \frac{V}{K_y a A} \tag{3.70}$$

and N_{OG} is the number of overall gas-phase mass-transfer units:

$$= \int_2^1 \frac{dp}{p - p_e} \tag{3.71}$$

The relationships between the overall mass-transfer unit height to the individual liquid and vapor film mass-transfer units, H_L and H_G, respectively are:

$$H_{OL} = H_L + m\left(\frac{L}{V}\right)_m H_G \tag{3.72}$$

$$H_{OG} = H_G + m\left(\frac{L}{V}\right)_m H_L \tag{3.73}$$

such that m is the slope of the equilibrium line and $(L/V)_m$ is the slope of the operating line. Integration of Eq. (3.71) yields the number of transfer units. In cases where the solvent is dilute in CO_2, the number of mass-transfer units is:

$$N_{OG} = \int_2^1 \frac{dy}{y - y_e} = \frac{y_1 - y_2}{\Delta y_m} \tag{3.74}$$

such that Δy_m is the log-mean driving force expressed as:

$$\Delta y_m = \frac{\Delta y_1 - \Delta y_2}{ln\left(\frac{\Delta y_1}{\Delta y_2}\right)} \tag{3.75}$$

where

$\Delta y_1 = y_1 - y_e$ and
$\Delta y_2 = y_2 - y_e$.

In general, if the equilibrium curve and operating lines are approximately straight and the solvent feed is virtually CO_2-free, then the number of transfer units can be estimated by:

$$N_{OG} = \frac{1}{1 - m\left(\frac{V}{L}\right)_m} \ln\left[\left(1 - m\left(\frac{V}{L}\right)_m\right)\frac{y_1}{y_2} + m\left(\frac{V}{L}\right)_m\right] \quad (3.76)$$

This equation is plotted as a function of y_1/y_2 and $(L/V)_m$ in Fig. I.1 of Appendix I for a quick initial estimate. A rule of thumb from Colburn [73] is that optimal values for $m(V/L)_m$ will likely be between 0.7 and 0.8. For cases in which there may be slight curvature in the operating or equilibrium lines, this approach may be used as a rough estimate if each of the curves are separated into multiple straight-line segments.

If CO_2 is not sufficiently dilute in the gas mixture, full integration of Eq. (3.74) or the equivalent N_{OL} is required to determine the number of mass-transfer units. To determine the height of the column, the height of each mass-transfer unit must also be determined. The height of a mass-transfer unit can vary not just among CO_2 gas mixture applications, but also upon packing materials. The mass-transfer unit height of a given packing material will depend upon the gas and liquid flow rates and the uniformity of the liquid distribution throughout the column, which is dependent upon the column height and diameter. There are correlation methods available that provide reasonable estimates of mass-transfer unit heights and mass-transfer coefficients [74]. The Cornell Method [75] considered here is correlation-based and developed through a review of previously published data that explicitly take into account gas and liquid flow rates, in addition to the column diameter and height. The packing material used in the studies and reported here includes Berl saddles. Although it is commonly known that the mass-transfer efficiency of Pall rings and Intalox saddles are higher than an equivalent-sized Berl saddle, the correlations presented will serve well as a rough estimate. Cornell's correlations are:

$$H_{OG} = \frac{0.011\psi_n\,(Sc)_v^{0.5}\left(\frac{D_c}{0.305}\right)^{1.11}\left(\frac{Z_T}{3.05}\right)^{0.33}}{(L_w f_1 f_2 f_3)^{0.5}} \quad (3.77)$$

$$H_{OL} = 0.305\varphi_n\,(Sc)_L^{0.5}\,K_3\left(\frac{Z_T}{3.05}\right)^{0.15} \quad (3.78)$$

such that $(Sc)_v$ = gas Schmidt number = $\mu/\rho D$, $(Sc)_L$ = liquid Schmidt number = $\mu/\rho D$, D_c = column diameter (m), Z_T = column height (m), K_3 = percentage of flooding correction factor from Fig. I.2, $\psi_n = H_{OG}$ factor from Fig. I.3, $\varphi_n = H_{OL}$ factor from Fig. I.4, L_w = liquid mass flow rate per unit cross-sectional area in the column (kg/m^{2}s), f_1 = liquid viscosity correction factor = $(\mu/\mu_w)^{0.16}$, f_2 = liquid density correction factor $= (\rho/\rho_w)^{1.25}$, f_3 = surface tension correction factor $= (\sigma/\sigma_w)^{0.8}$, where the subscript "$w$" refers to the physical properties of water at 20°C.

The terms $D_c/0.305$ and $Z_T/3.05$ are included in Eqs. (3.77) and (3.78) to allow for the influence of column diameter and packed-bed height with the standard values of 0.305 m and 3.05 m used by Cornell, respectively. For purposes of design, for columns greater than 0.6 m in diameter, a fixed correction term of 2.3 should be used, and the correction for height should be included only when the distance between liquid redistributors is greater than 3 m. An estimate of N_{OG} can be obtained from Fig. I.1 as a first estimate, along with an H_{OG} of 1 m. To use Cornell's method, initial estimates of column height and diameter are required. To use Figs. I.2 and I.3 for determining the percentage flooding correction factor, an estimate of the column percentage flooding is required, which can be obtained from Fig. I.5, where a flooding line is included with the lines of constant pressure drop.

$$\text{Percentage flooding} = \left[\frac{K_4 \text{ at design pressure drop}}{K_4 \text{ at flooding}}\right]^{\frac{1}{2}} \tag{3.79}$$

such that K_4 is associated with the generalized pressure drop correlation of Fig. I.5 of Appendix I, and will be discussed in greater detail shortly.

3.5.5 Column Diameter Estimation

The cross-sectional area of a packed column determines its capacity. The column is designed to operate at the greatest pressure drop possible with respect to economic constraints to ensure reasonable liquid and gas distribution over the packing material. As a rule of thumb, for random packing the pressure drop should typically not exceed 80 mm of water per meter of packing height, at which point the gas velocity is approximately 80% of the flooding velocity. For absorber and stripping columns, recommended design values are between 15 to 50 mm water per m of packing [76]. If it is likely that the liquid may foam, then these values should be halved. The column cross-sectional area can be determined from the pressure drop correlation of Fig. I.5, which correlates the vapor and liquid flow rates, physical properties of the system, and packing characteristics, with respect to the vapor mass flow rate per unit cross-sectional area and constant pressure drop. The term K_4 from Fig. I.5 is a function of vapor flow, packing, and liquid properties as follows:

$$K_4 = \frac{13.1\left(V_w^*\right)^2 F_P\left(\frac{\mu}{\rho_L}\right)^{0.1}}{\rho_v(\rho_L - \rho_v)} \tag{3.80}$$

such that, $V_w{}^*$ = gas mass flow rate per unit column cross-sectional area (kg/m^{2}s), F_p = packing factor, characteristic of the size and type of packing, dimensionless, μ = liquid viscosity (N s/m^2), ρ_L, ρ_v = liquid and vapor density, respectively (kg/m^3).

In gas absorption, the liquid-to-vapor ratio is selected to yield the desired separation with the most economic solvent use. Provided the vapor (gas) flow rate is known and V_w* is determined by Eq. (3.80), the column area and related column diameter may be obtained.

Example 3.6 Consider an absorption column for CO_2 capture from coal-fired flue gas. The column is packed with Flexipac AQ Style 20 material (packing factor = 18) and is to be built to treat 35,000 mol/hr of flue gas with a CO_2 mole fraction of 0.119. A 35-wt% MEA solution will be used as the absorbent. The ratio of liquid-to-vapor is 4.15. The inlet temperatures of the entering flue gas and the solvent are 25°C and 40°C, respectively. (a) If the design pressure drop in the absorption column is 21 mm H_2O/m of packing, what should the velocity of the gas be? (b) What should the diameter of the column be?

Solution:
(a) The average molecular weight of the entering flue gas is (28.01)(0.881) + (44.01)(0.119) = 29.91 g/mol. Assuming the flue gas is ideal, the density of the entering flue gas at 25°C is:

$$\rho_v = \frac{pM}{RT} = \frac{(1)(29.91)}{(0.08206)(298.15)} = 1.22\frac{\text{g}}{\text{L}} = 1.22\frac{\text{kg}}{\text{m}^3}$$

The density and viscosity of the entering MEA solution is assumed to be similar to the density of water at 40°C and 1 atm ($\rho_L = 992\,\text{kg/m}^3$, $\mu = 0.7$ cP). The gas mass velocity can then be determined using the generalized pressure drop correlation (Fig. I.5) and Eq. (3.80). The abscissa value of Fig. I.5, when the liquid-to-vapor ratio is 4.15 and the pressure drop is 21 mm H_2O/m, is:

$$F_{LV} = \frac{L}{V}\sqrt{\frac{\rho_v}{\rho_L}} = 4.15\sqrt{\frac{1.22}{992}} = 0.15$$

At $F_{LV} = 0.15$, $K_4 = 0.8$. Eq. (3.80) is an empirical equation and the correct units must be used. The gas mass flow rate per unit column cross-sectional area is:

$$0.8 = \frac{[13.1(V_w^*)^2](18)\left(\frac{7\times10^{-4}}{992}\right)^{0.1}}{(1.22)(992 - 1.22)}$$

$$V_w^* = 4.09\frac{\text{kg}}{\text{m}^2\text{s}}$$

Therefore, the velocity of the entering flue gas is $\frac{V_w^*}{\rho_v} = \frac{4.11}{1.22} = 3.37\ \frac{\text{m}}{\text{s}}$

(b) The area of the absorption column is:

$$A = \frac{\dot{m}}{V_w^*}$$

where $\dot{m}$ is the mass flow rate of the entering flue gas in kg/s.

$$\dot{m} = \frac{(35000)(29.91)}{(3600)(1000)} = 0.29\ \frac{\text{kg}}{\text{s}}$$

Therefore, the area required for the absorption column is:

$$A = \frac{0.29}{4.09} = 0.07\ \text{m}^2$$

The corresponding column diameter is 0.30 m.

3.6 Work Required for Separation

Within the absorption and stripping processes gas transport is required to achieve desired flow rates. Typical gas movers consist of blowers, compressors, ejectors, exhausters, fans, vacuum pumps, and ventilators. Mechanically these gas movers operate in a similar fashion to liquid pumps; however, the greater volumes occupied by gases and the much lower viscosities of gases compared to liquids make the gas movers much larger and require operation at higher speeds, which leads to increased power usage and maintenance. In general, pumps require significantly less work than fans or blowers in terms of liquid and gas transport, respectively.

Gas Blowing Work Fans primarily operate near atmospheric pressure with pressure differentials less than 15 kPa. The power, P, required for moving a gas with a mass flow rate of $\dot{m}$, an average density, ρ, with a fan of efficiency, ε, over a pressure drop, Δp, is:

$$P = \frac{dw_f}{dt} = \frac{\dot{m}\Delta p}{\rho\ \varepsilon} \tag{3.81}$$

Blowers are more sophisticated and expensive than fans and can handle greater pressure drops up to 300 kPa. The difference between the work calculated for a fan versus a blower is that the density cannot be considered constant over large pressure drops. For adiabatic and reversible compression of an ideal gas, the blowing power is:

$$P = \frac{dw_b}{dt} = \frac{\dot{m}RTk}{M(k-1)\varepsilon}\left[\left(\frac{p_2}{p_1}\right)^{(k-1)/k} - 1\right] \tag{3.82}$$

such that T is the gas temperature, M is the molecular weight, p_1 is the initial gas pressure, p_2 is the final gas pressure, and k, the ratio of specific heats, c_p/c_v, which are available for selected gases in Appendix B. Typical fan and blower efficiencies range between 65 to 85%. To determine the work required to compress a gas from a pressure of p_1 to p_2 using either fan (Eq. 3.81) or blowing power (Eq. 3.82), the power expression can be divided by the molar flow rate of the gas, which will result in the work required per mole of gas compressed.

Solvent Pumping Work The solvent must be pumped to overcome pressure differences between the absorber and stripper, pressure drop in pipelines and nozzles, and elevation head when pumping to the top of contactors in the absorber and the stripper. The power required to pump a fluid with a volumetric flow rate, Q, over a pressure drop, Δp, with an intrinsic efficiency, ε_i is:

$$P = \frac{dw_p}{dt} = \frac{Q\Delta p}{\varepsilon_i} \tag{3.83}$$

The intrinsic efficiency is associated with friction and other energy losses within the pump and usually range between 40 to 85% [77]. For moderate pressure and relatively clean fluids having viscosities up to 0.5 Pa s, the intrinsic efficiency may be estimated from [78]:

$$\varepsilon_i = \left(1 - 0.12Q^{-0.27}\right)\left(1 - \mu^{0.8}\right) \tag{3.84}$$

in which the liquid volume flow rate Q has units of m^3/s and viscosity, μ has units of Pa s.

To determine the power consumption of the pump, the work in Eq. (3.83) must also be divided by the driver efficiency, ε_d, which may be approximated as 85% [77].

$$P = \frac{dw_p}{dt} = \frac{Q\Delta p}{\varepsilon_i \varepsilon_d} \tag{3.85}$$

In addition to blowing and pumping power, there are significant energy requirements associated with solvent regeneration. In fact, in the case of postcombustion capture applications, this portion represents the majority of the costs for CO_2 capture.

Regeneration Work Solvent regeneration typically requires the addition of heat to destroy the chemical bond formed upon absorption. In the case of physical absorption, the heat required is substantially less, but with the limitation of a significantly reduced mass-transfer rate due to a reduction in chemical reactivity with CO_2. The energy requirement associated with the release of CO_2 from a 30% MEA solution is used as an example with the methodology applicable to other solvents used for CO_2 capture. The use of MEA for postcombustion capture of CO_2 involves the circulation of the MEA solution between the absorber column and the stripping column, where CO_2 is released. The MEA solution that enters the absorber is the "lean" solution and the MEA solution entering the regeneration column is the "rich" solution. Not all of the CO_2 is released in the stripper, which operates between 100 and 140°C. For example, a lean 30% MEA solution may retain a loading of approximately 0.2 mol of CO_2 per mole of MEA, whereas the rich solution contains approximately 0.4 mol of CO_2 per mole of MEA [79].

The heat of absorption of CO_2 with a 30% MEA solution at 40°C (approximate temperature of the absorber column) is −84.3 kJ/mol of CO_2 [80]. However, this energy requirement is only part of the energy needed in the stripper.The heat capacity of a 30% MEA solution with 0.4 mol of CO_2 bound per mole of MEA has been measured as 3.418 J/g K [81]. The mass of the solution containing 1 mol of MEA also

Table 3.14 Selected heat capacities (J/g K) for amine-based CO_2 capture [76]

Solvent	0.0 mol CO_2/ mol solvent	0.4 mol CO_2/ mol solvent
20% MEA	3.911	3.648
30% MEA	3.734	3.418
20% DEA	3.915	3.740
30% MDEA	3.787	3.496
20% MEA/30% MDEA	3.410	3.135
20% DEA/30% MDEA	3.445	3.074

includes 0.4 mol of CO_2 as well as 7.9 mol of H_2O, resulting in a total of 221 grams per mole of MEA. Therefore, 756 J are required to heat 1 mol of MEA in solution 1°. Assuming the absorber column operates at 40°C and the stripper column at 120°C, the heat required to heat the 1 mol of MEA in solution is approximately 60.5 kJ. This requirement to heat up the solution is about 3.5 times the energy needed to release the 0.2 mol of CO_2 per mole of MEA. Of the 60.5 kJ, nearly 80% is needed to heat up the water. Table 3.14 provides a list of selected heat capacities for amine-based CO_2 capture.

While significant heat may be recovered from heating the MEA solution, for example by cross-exchanging the hot lean solution leaving the stripper column with the cooler rich solution entering the stripper, thermal losses of 10% or more can be anticipated. While the enthalpy to release the CO_2 is approximately 16.9 kJ per mole of MEA in solution, the thermal losses associated with the MEA solution cycle will be at least 6 kJ, or 35% of the heat required to release the CO_2. The same issue will arise for direct air capture if the release of the CO_2 involves heating "spectator molecules" such as water. Heat will have to be expended to heat up and potentially vaporize these molecules along with the bound CO_2 complex itself. However, it is important to note that a significant portion of the heat released upon absorption and heat used for solvent regeneration can be recovered through the use of heat exchangers. Subsequent regeneration cycles will only use a fraction of that of the initial cycle due to process heat integration.

Example 3.7 Determine the real work (as opposed to the thermodynamic minimum work) required to separate 90% of the CO_2 from the flue gas described in Example 3.6. Fan power, solvent pumping, and solvent heating for regeneration must be considered, along with appropriate efficiency estimates.

(a) Recall that the liquid-to-vapor ratio is an important parameter for the design of an absorption system. As the flue gas travels from the boiler outlet to the inlet of the absorption column, it loses energy along the way. To maintain the required gas flow into the column requires fan power. Assuming a fan with 80% efficiency is required to overcome a flue gas pressure drop of 14 kPa, what is the work required per mole of CO_2 captured?

(b) Solvent pumping is required to maintain the desired liquid flow into both the absorption and regeneration columns. Assume a required solvent flow rate of

40 mol/s into the absorption column. Calculate the flow rate of the rich solvent into the regeneration column assuming 90% capture of CO_2. Assuming that there is a total pressure drop of 300 kPa throughout the system (*i.e.*, solvent lines and across both packed columns), calculate the work per mole of CO_2 captured for pumping with a given pump efficiency of 75%.

(c) The work required per mole of CO_2 captured for the regeneration process depends heavily on the use of heat exchangers to reduce the heating duty, but assume the energy requirement is 159 kJ thermal/mol CO_2 captured [83], and a power plant efficiency of 33%. If the CO_2 must be compressed from .101 MPa to 15 MPa for pipeline transport, and the adiabatic single-stage compression power is given by:

$$P = \frac{\dot{m}RT}{M}\frac{k}{k-1}\left(r^{\frac{k-1}{k}} - 1\right)$$

such that $\dot{m}$ is the mass flow rate, T is the gas temperature, r is the compression ratio (*i.e.*, p_2/p_1), k is the ratio of heat capacities (*i.e.*, c_p/c_v), and M is the molecular weight of the gas, calculate the regeneration and compression energy required per mol of CO_2 captured, as well as the total work of the absorption-regeneration process for 90% capture per mole of CO_2 captured.

(d) What is the 2nd-Law efficiency of this capture process?

(e) If the equivalent net power generated is 230 kW before the carbon capture process, what would the energy penalty be?

Solution:
(a) From Example 3.6, the mass flow rate, $\dot{m}$, is 0.29 kg/s and the density, ρ, is 1.22 kg/m^3. From Eq. (3.81),

$$\dot{w}_f = \frac{(0.29)(14000)}{(1.22)(0.8)} = 4.16\ \frac{\text{kJ}}{\text{s}}$$

Recall that the total molar flow rate is 35000 mol/hr, so the desired capture rate of CO_2 is:

$$\dot{n}_{CO_2} = (0.9)(0.119)(35000\,\text{mol/hr})/(3600\,\text{s/hr}) = 1.04\,\text{mol/s}$$

Therefore, the fan work required per mole of CO_2 captured is:

$$w_f = \frac{\dot{w}_f}{\dot{n}_{CO_2}} = \frac{4.16}{1.04} = 4.0\ \frac{\text{kJ}}{\text{mol}}$$

(b) The flow rate of the rich solvent into the regeneration column is the mixing of the lean solvent and the flow rate of the capture CO_2. From part (a), the flow rate of the captured CO_2 is 1.04 mol/s. Thus, the flow rate of the rich solvent is 40 + 1.04 = 41.04 mol/s.

The solvent flow rate is not constant through the system (lean vs. rich). Assume that the flow rate through the system can be taken as the average of the rich and lean flow rates, 40.52 mol/s. By maintaining the assumption that the solvent has the fluid properties of water, the volumetric flow rate is:

$$Q = \frac{\dot{n}M}{\rho} = \frac{(40.52)(18.01)}{1} = 729\ \frac{\text{cm}^3}{\text{s}} = 7.29 \times 10^{-4}\ \frac{\text{m}^3}{\text{s}}$$

The work required for pumping the solution is:

$$w_p = \frac{(7.29 \times 10^{-4})(300000)}{(0.75)} = 291.6\ \frac{\text{J}}{\text{s}}$$

If the driver efficiency of the pump is 85%, the total power consumption of the pump (Eq. 3.85) is:

$$P = \frac{w_p}{\varepsilon_d} = \frac{291.6}{0.85} = 0.343\,\text{kW}$$

Thus, the actual pump work required per mole of CO_2 captured is 0.343/ 1.04 = 0.33 kJ/mol CO_2 captured.

(c) The work required for the regeneration process, that is, the equivalent power that could have been produced by the thermal energy used in the regenerator, is (159 kJ thermal/mol)(0.33 work output/thermal input) = 52 kJ/mol. The compression energy can be calculated by dividing the compression power by the molar flow rate and compressor efficiency. Assuming a temperature of 40°C = 313 K, a compression ratio of r = (15 MPa)/(0.101 MPa) = 149, and a ratio of heat capacities from Appendix B as 1.29 (average of values for 289 K and 339 K), yields:

$$w_c = \frac{P_{ad}}{\frac{\dot{m}}{M}} = RT\frac{k}{k-1}\left(r^{\frac{k-1}{k}} - 1\right)$$

$$= (8.31)(313)\left(\frac{1.29}{0.29}\right)\left(149^{\frac{0.29}{1.29}} - 1\right) = 24.1\frac{\text{kJ}}{\text{mol}}$$

Therefore the total work of the absorption-regeneration process is

$$w_{Tot} = w_f + w_p + (w_{regen} + w_c) = 4.0 + 0.33 + 52 + 24.1 = 80.4\frac{\text{kJ}}{\text{mol}}$$

Greater than 90% of the total work required for 90% CO_2 capture is the work related to the regeneration and compression processes.

(d) The 2nd-law efficiency is determined from:

$$\eta_{2nd} = \frac{W_{min}}{W_{real}}$$

where W_{min} is the minimum thermodynamic work calculated for the separation with 90% CO_2 captured from the flue gas and the W_{real} is the actual work for the same process. Given a desired CO_2 purity of 95% and that all streams are at ~25°C, the minimum thermodynamic work for separation is:

$$W_{min} = R\Big[T_R\left(y_R^{CO_2} ln\left(y_R^{CO_2}\right) + y_R^{N_2} ln\left(y_R^{N_2}\right)\right) + T_P\Big(y_P^{CO_2} ln\left(y_P^{CO_2}\right) + y_P^{N_2} ln\left(y_P^{N_2}\right)\Big) - T_F\left(y_F^{CO_2} ln\left(y_F^{CO_2}\right) + y_F^{N_2} ln\left(y_F^{N_2}\right)\right)\Big]$$

W_{min} = 6.78 kJ/s or 6.51 kJ/mol CO_2 captured based upon a capture rate of approximately 1.04 mol/s.

Therefore, the second law efficiency is $\eta_{2nd} = 6.51/(4+0.33+52) = 0.115$ or 11.5%. Note that we did not include the energy for compression, since this is not included in the minimum work calculation.

(e) The energy used by the capture and compression process is: (80.4 kJ/mol) (1.04 mol/s) = 83.6 kW, so the energy penalty is 83.6 kW/230 kW = 36%.

3.7 Problems

Problem 3.1 The gas-phase diffusivity, as determined by Eq. (3.8), is dependent on temperature ($T^{\frac{3}{2}}$), pressure ($1/p$) and the square root of the sum of the inverse molecular weights of the 2 components, $\left(\frac{1}{M_1} + \frac{1}{M_2}\right)^{\frac{1}{2}}$. Using the data from Table 3.2, fit the data for D to $\left(\frac{1}{M_1} + \frac{1}{M_2}\right)^{\frac{1}{2}}$, where M_1 is 44.01 g/mol for CO_2 and M_2 is molecular weight of the second component. Comment on the effect that using data at varying temperatures has on the relationship.

Problem 3.2 The liquid-phase diffusivity, as determined by Eq. (3.11) is dependent on temperature, solvent viscosity and molecular weight, the association parameter of the solvent, and the molar volume occupied by the gas in solution. Determine the diffusivity of CO_2 in the following solvents using the solvent viscosity data of Table 3.3 and compare the empirical-based prediction to the experimental liquid-phase diffusivity data available in Table 3.3. Assume the molar volume of CO_2 is 34.0 cm^3/mol solvent and the provided solvent association parameters.

Water, $\phi = 2.2$; Water, $\phi = 2.6$; Ethanol, $\phi = 1.5$; Heptane, $\phi = 1.0$

Problem 3.3 Assume a constant flue gas mixture comprised of 12 mol% CO_2 and 500 ppm SO_2 is to be scrubbed in an absorber using a 1 M amine-based solvent, in which the primary mechanism is assumed to be reaction of the acid gas with a secondary amine. Neglect possible side reactions toward carbonate (or sulfate) species and assume the following reaction takes place in solution:

$$AO_2 + R_2NH + H_2O \leftrightarrow R_2NAO_2{}^- + H_3O^+$$

such that A = C or S, with the reaction proceeding as second-order.

If the activation barrier of the reaction of SO_2 with the secondary amine in solution is lower than the corresponding CO_2 reaction by 10 kJ/mol of AO_2 reacted, determine the ratio of the rate constants *and* the ratio of reaction rates between the SO_2 and CO_2 reactions at 298 K. Assume the Henry's Law constants of CO_2 and SO_2 in the solvent are approximately 30 atm L/mol and 10 atm L/mol, respectively.

Problem 3.4 Using the equilibrium constant data available in Appendix G and Eq. (3.15), reproduce the plot in Fig. 3.6. Assume a solubility of CO_2 in water (*i.e.*, the reciprocal of the Henry's Law constant, H) of 0.039 mol/L atm at 30°C. In addition, estimate the pH of "pristine rainwater" given a CO_2 concentration in air of 390 ppm. How would the pH of rainwater change with a CO_2 concentration in air of 500 ppm?

$$CO_2 \overset{H_{CO_2}}{\Longleftrightarrow} CO_{2(g)} \tag{R.1}$$

$$CO_2 + H_2O \overset{K_1}{\Longleftrightarrow} H_2CO_3 \tag{R.2}$$

$$CO_2 + OH^- \overset{K_2}{\Longleftrightarrow} HCO_3^- \tag{R.3}$$

$$H^+ + CO_3^{2-} \overset{K_3}{\Longleftrightarrow} HCO_3^- \tag{R.4}$$

$$H^+ + HCO_3^- \overset{K_4}{\Longleftrightarrow} H_2CO_3 \tag{R.5}$$

$$H^+ + OH^- \overset{K_5}{\Longleftrightarrow} H_2O \tag{R.6}$$

Problem 3.5 Consider the transport of CO_2 from a coal-fired flue gas across the gas-liquid interface of a solvent fluid-coated Raschig ring in a packed absorption column. The diffusive flux associated with the transport of CO_2 **from the bulk gas** phase to the gas-liquid interface is equivalent to

$$J_{G,CO_2} = \frac{D_{G,CO_2}}{\delta}(c_{\infty,CO_2} - c_{i,CO_2})$$

while the diffusive flux of CO_2 **from the gas-liquid** interface to the bulk fluid is equivalent to:

$$J_{L,CO_2} = \frac{D_{L,CO_2}}{\delta}(c_{i,CO_2} - c_{\infty,CO_2})$$

For the following two 20 wt% MDEA solutions [82]:

Temperature	ρ (g/cm^3)	μ (cP)	H_{CO_2} (atm L/mol)
25°C	1.0152	1.941	32.30
50°C	1.0047	1.051	55.47

Determine the ratio of the liquid-to-gas phase:

a. film thicknesses, assuming that the film thickness is proportional to the ratio of the fluid viscosity to fluid density, and
b. fluxes, assuming the bulk concentration of CO_2 in the liquid phase is negligible.

Problem 3.6 Repeat Problem 3.5 for a fuel gas mixture from an integrated gasification combined cycle (IGCC) power plant. State the major differences and explain.

Problem 3.7 Using the Film Model, estimate the enhancement factor, E, and the resultant flux, J_{L,CO_2}, for the following cases:

a. postcombustion capture system using a 5-M MEA solution (approximately 30 wt% MEA),
b. direct air capture system using a 1-M NaOH solution, and
c. postcombustion capture system using a 10-M unobtainium solution, which has the reaction rate of carbonic anhydrase, $z = 1$, and $H = 20$ atm.

For all, assume $k_L = 10^{-1}$ cm/s, $D = D_B = 1 \times 10^{-5}$ cm^2/s. Compare your results to Example 3.4 in the text.

Problem 3.8 Consider an absorption column for CO_2 capture from a natural gas-fired flue gas. The column is packed with Flexipac AQ Style 20 material (packing factor = 18) and is to be built to treat 35,000 mol/hr of flue gas with a CO_2 mol fraction of 0.06. A 35 wt% MEA solution is used as the absorbent. The ratio of liquid-to-gas is 4.15. The inlet temperatures of the entering flue gas and the solvent are 25 and 40°C, respectively. Determine the following:

a. If the design pressure drop in the absorption column is 21 mmH_2O/m of packing, what should the velocity of the gas be?
b. What should the diameter of the column be?
c. How does the design of this column compare with that of Example 3.6?

Problem 3.9 Using data from Table 3.14, determine the energy required to heat a 20 wt% MEA solution loaded with 0.4 mol CO_2/mol solvent from 40°C (the absorber column temperature) to 120°C (the stripper column temperature). What percentage of this energy is used to heat up the water in the solution? Repeat for a 30 wt% MDEA solution.

References

1. Gas-Liquid Reactions, Danckwerts, P.V., Copyright (1970), with permission from McGraw-Hill Companies, Inc
2. J Phys Chem A, McCann N, Phan D, Wang X, Conway W, Burns R, Attalla M, Puxty G, Maeder M (2009) Kinetics and mechanism of carbamate formation from CO_2 (aq), carbonate species, and monoethanolamine in aqueous solution. Copyright 2009, American Chemical Society
3. John Wiley & Sons, Inc., Seader JD, Henley EJ (2006) Separation Process Principles
4. van Krevelen DW, Hoftijzer PJ (1948) In Chimie et Industrie, Numero Speciale du XXIe. Congress International de Chimie Industrielle, Brussels
5. Setschenow M (1892) Action de líacide carbonique sur les solutions dessels a acides forts. Etude absortiometrique Ann Chim Phys 25:226–270
6. (a) Barrett PVL (1966) Gas absorption on a sieve plate. Cambridge University, Cambridge; (b) Danckwerts PV (1970) Gas–liquid reactions. McGraw-Hill, New York, p 276
7. Onda K, Sada E, Kobayashi T, Kito S, Ito K (1970) Salting-out parameters of gas solubility in aqueous salt solutions J Chem Eng Jpn 3(1):18–24
8. Houghton G, McLean AM, Ritchie PD (1957) Compressibility, fugacity, and water-solubility of carbon dioxide in the region 0–36 atm. and 0–100 °C Chem Eng Sci 6(3):132–137
9. Muldoon MJ, Aki SNVK, Anderson JL, Dixon JNK, Brennecke JF (2007) Improving carbon dioxide solubility in ionic liquids J Phys Chem B 111(30):9001–9009
10. Jacquemin J, Husson P, Mayer V, Cibulka I (2007) High-pressure volumetric properties of imidazolium-based ionic liquids: effect of the anion J Chem Eng Data 52(6):2204–2211
11. (a) Cadena C, Anthony JL, Shah JK, Morrow TI, Brennecke JF, Maginn EJ (2004) Why is CO_2 so soluble in imidazolium-based ionic liquids? J Am Chem Soc 126(16), 5300–5308; (b) Huang X, Margulis CJ, Li Y, Berne BJ (2005) Why is the partial molar volume of CO_2 so small when dissolved in a room temperature ionic liquid? Structure and dynamics of CO_2 dissolved in $[Bmim^+][PF_6^-]$ J Am Chem Soc 127(50)17842–17851
12. (a) Blanchard LA, Gu Z, Brennecke JF (2001) High-pressure phase behavior of ionic liquid/CO_2 systems J Phys Chem B 105(12):2437–2444; (b) Huang J, Rüther T (2009) Why are ionic liquids attractive for CO_2 absorption? An overview Aust J Chem 62(4):298–308
13. Sherwood TK, Pigford RL, Wilke CR (1975) Mass transfer. McGraw-Hill, New York, p 677
14. (a) Mandal BP, Kundu M, Padhiyar NU, Bandyopadhyay SS (2004) Physical solubility and diffusivity of N_2O and CO_2 into aqueous solutions of (2-amino-2-methyl-1-propanol +diethanolamine) and (N-methyldiethanolamine + diethanolamine) J Chem Eng Data 49(2):264–270; (b) Mandal BP, Kundu M, Bandyopadhyay SS (2005) Physical solubility and diffusivity of N_2O and CO_2 into aqueous solutions of (2-amino-2-methyl-1-propanol + monoethanolamine) and (N-methyldiethanolamine + monoethanolamine) J Chem Eng Data 50(2):352–358; (c) Mathonat C, Majer V, Mather AE, Grolier JPE (1997) Enthalpies of absorption and solubility of CO_2 in aqueous solutions of methyldiethanolamine Fluid Phase Equilib 140(1–2):171–182
15. Chapman S, Cowling TG (1991) The mathematical theory of non-uniform gases: an account of the kinetic theory of viscosity, thermal conduction, and diffusion in gases. Cambridge University Press, Cambridge
16. Wilke CR, Chang PC (1955) Correlations of diffusion coefficients in dilute solutions Am Inst Chem Eng 1:264
17. Le Bas G (1915) The molecular volumes of liquid chemical compounds. Longmans, Green, New York
18. Hayduk W, Laudie H (1974) Prediction of diffusion coefficients for non-electrolytes in dilute aequous solutions Am Inst Chem Eng 20:611
19. Lusis MA, Ratcliff GA (1971) Diffusion of inert and hydrogen-bonding solutes in aliphatic alcohols Am Inst Chem Eng 17:1492
20. Akgerman A, Gainer JL (1972) Diffusion of gases in liquids Ind Eng Chem Fund 11:373
21. Gainer JL, Metzner AB (1965) Transport phenomena 6, vol In AIChE- Inst Chem Eng Joint Meet, London

22. Hiss TG, Cussler EL (1973) Diffusion in high viscosity liquids Am Inst Chem Eng 19:698
23. Treybal RE (1981) Mass transfer operations, vol 2. McGraw-Hill, Singapore
24. Mandal BP, Kundu M, Bandyopadhyay SS (2003) Density and viscosity of aqueous solutions of (N-methyldiethanolamine + monoethanolamine), (N-methyldiethanolamine + diethanolamine), (2-amino-2-methyl-1-propanol + monoethanolamine), and (2-amino-2-methyl-1-propanol + diethanolamine) J Chem Eng Data 48(3)703–707
25. Gubbins KE, Bhatia KK, Walker RDJ (1966) Am Inst Chem Eng 2:548
26. Ratcliff GA, Holdcroft JG (1963) Trans Inst Chem Eng 41:315
27. Shiflett MB, Yokozeki A (2005) Solubilities and diffusivities of carbon dioxide in ionic liquids: [bmim][PF_6] and [bmim][BF_4] Ind Eng Chem Res 44(12):4453–4464
28. Morgan D, Ferguson L, Scovazzo P (2005) Diffusivities of gases in room-temperature ionic liquids: data and correlations obtained using a lag-time technique Ind Eng Chem Res 44(13):4815–4823
29. Tokuda H, Tsuzuki S, Susan MABH, Hayamizu K, Watanabe M (2006) How ionic are room-temperature ionic liquids? An indicator of the physicochemical properties J Phys Chem B 110(39):19593–19600
30. Guerrero Sanchez C, Lara Ceniceros T, Jimenez Regalado E, Raşa M, Schubert US (2007) Magnetorheological fluids based on ionic liquids Adv Mater 19(13):1740–1747
31. Bonhôte P, Dias AP, Papageorgiou N, Kalyanasundaram K, Grätzel M (1996) Hydrophobic, highly conductive ambient-temperature molten salts Inorg Chem 35(5):1168–1178
32. Camper D, Bara J, Koval C, Noble R (2006) Bulk-fluid solubility and membrane feasibility of Rmim-based room-temperature ionic liquids Ind Eng Chem Res 45(18):6279–6283
33. Brubaker DW, Kammermeyer K (1954) Collected research papers for 1954 Ind Eng Chem 46:733
34. Davies GA, Ponter AB, Crains K (1967) The diffusion of carbon dioxide in organic liquids Can J Chem Eng 45:372
35. Walker NA, Smith FA, Cathers IR (1980) Bicarbonate assimilation by fresh-water charophytes and higher plants: I. Membrane transport of bicarbonate ions is not proven J Membr Biol 57(1):51–58
36. Glasstone S, Lewis D (1960) Elements of physical chemistry. Macmillan, London, p 760
37. Hou Y, Baltus RE (2007) Experimental measurement of the solubility and diffusivity of CO_2 in room-temperature ionic liquids using a transient thin-liquid-film method Ind Eng Chem Res 46(24):8166–8175
38. Palmer DA, Van Eldik R (1983) The chemistry of metal carbonato and carbon dioxide complexes Chem Rev 83(6):651–731
39. (a) Peng Z, Merz Jr KM (1993) Theoretical investigation of the $CO_2 + OH^- \rightarrow HCO_3^-$ reaction in the gas and aqueous phases J Am Chem Soc 115(21):9640–9647; (b) Nemukhin AV, Topol IA, Grigorenko BL, Burt SK (2002) On the origin of potential barrier for the reaction $OH^- + CO_2 \rightarrow HCO_3^-$ in water: studies by using continuum and cluster solvation methods J Phys Chem B 106(7):1734–1740
40. Li WK, McKee ML (1997) Theoretical study of OH and H_2O addition to SO_2 J Phys Chem A 101(50):9778–9782
41. Glasstone S, Laidler KJ, Eyring H (1941) The theory of rate process. McGraw-Hill, New York
42. Wynne Jones WFK, Eyring H (1935) The absolute rate of reactions in condensed phases J Chem Phys 3:492
43. Evans MG, Polanyi M (1936) Further considerations on the thermodynamics of chemical equilibria and reaction rates. T Faraday Soc 32:1333–1360
44. Bell RP (1937) Relations between the energy and entropy of solution and their significance T Faraday Soc 33:496–501
45. Pinsent BRW, Pearson L, Roughton FJW (1956) The kinetics of combination of carbon dioxide with hydroxide ions T Faraday Soc 52:1512–1520
46. Caplow M (1968) Kinetics of carbamate formation and breakdown J Am Chem Soc 90(24):6795–6803

47. (a) Crooks JE, Donnellan JP (1989) Kinetics and mechanism of the reaction between carbon dioxide and amines in aqueous solution J Am Chem Soc Farad T 2 1989(4):331–333; (b) da Silva EF, Svendsen HF (2004) Ab initio study of the reaction of carbamate formation from CO_2 and alkanolamines Ind Eng Chem Res 43(13):3413–3418
48. (a) Jensen MB, Jorgensen E, Fourholt C (1954) Reactions between carbon dioxide and amino alcohols. I. Monoethanolamine and diethanolamine Acta Chem Scand 8(7):1137; (b) Danckwerts PV, Sharma MM (1966) The absorption of carbon dioxide into solutions of alkalis and amines: (with some notes on hydrogen sulphide and carbonyl sulphide) Chem Eng 1966:CE244–CE279; (c) Penny DE, Ritter TJ (1983) Kinetic study of the reaction between carbon dioxide and primary amines J Am Chem Soc Farad T 1 79(9):2103–2109
49. McCann N, Maeder M, Hasse H (2011) Prediction of the overall enthalpy of CO_2 absorption in aqueous amine systems from experimentally determined reaction enthalpies Energy Proc 4:1542–1549
50. (a) Alper E (1920) Reaction mechanism and kinetics of aqueous solutions of 2-amino-2-methyl-1-propanol and carbon dioxide Ind Eng Chem Res 29(8):1725–1728; (b) Yih SM, Shen KP (1988) Kinetics of carbon dioxide reaction with sterically hindered 2-amino-2-methyl-1-propanol aqueous solutions Ind Eng Chem Res 27(12):2237–2241; (c) Rinker EB, Sami SA, Sandall OC (1995) Kinetics and modelling of carbon dioxide absorption into aqueous solutions of N-methyldiethanolamine Chem Eng Sci 50(5):755–768
51. (a) McCann N, Maeder M, Attalla M (2008) Simulation of enthalpy and capacity of CO_2 absorption by aqueous amine systems Ind Eng Chem Res 47(6):2002–2009; (b) Jou FY, Otto FD, Mather AE (1994) Vapor–liquid equilibrium of carbon dioxide in aqueous mixtures of monoethanolamine and methyldiethanolamine Ind Eng Chem Res 33(8):2002–2005
52. McCann N, Phan D, Wang X, Conway W, Burns R, Attalla M, Puxty G, Maeder M (2009) Kinetics and mechanism of carbamate formation from CO_2 (aq), carbonate species, and monoethanolamine in aqueous solution J Phys Chem A 113(17):5022–5029
53. Bishnoi S, Rochelle GT (2000) Absorption of carbon dioxide into aqueous piperazine: reaction kinetics, mass transfer and solubility Chem Eng Sci 55(22):5531–5543
54. Khalifah RG (1971) The carbon dioxide hydration activity of carbonic anhydrase J Biol Chem 246(8):2561
55. (a) Harned HS, Owen BB (1958) The physical chemistry of electrolytic solutions, vol 1 Reinhold Publishing Corporation, New York; (b) Puxty G, Rowland R (2011) Modeling CO_2 mass transfer in amine mixtures: PZ-AMP and PZ-MDEA Environ SciTechnol 45:2398–2405
56. Albert A, Serjeant EP (1962) Ionization constants of acids and bases: a laboratory manual. Methuen, London
57. Jensen BS (1959) The synthesis of 1-phenyl-3-methyl-4-acyl-pyrazolones-5 Acta Chem Scand 13(8):1668–1670
58. Bates RG, Bower VE (1962) Revised standard values for pH measurements from 0 to 95 °C J Res Natl Bur Stand 66A(2):179–184
59. Versteeg GF, Van Dijck LAJ, Van Swaaij WPM (1996) On the kinetics between CO_2 and alkanolamines both in aqueous and non-aqueous solutions. An overview Chem Eng Commun 144:113–158
60. Silverman DN (1994) In: Pradier JP, Pradier CM (eds) Carbon dioxide chemistry: environmental issues. Woodhead Publishing, Cambridge, p 406
61. Lewis WK, Whitman WG (1924) Principles of gas absorption Ind Eng Chem 16(12):1215–1220
62. Whitman WG (1923) The two-film theory of gas absorption Chem Metall Eng 29:146–148
63. Nernst W (1904) Theory of reaction velocity in heterogenous systems Z Phys Chem 47:52–55
64. Higbie R (1935) Rate of absorption of a gas into a still liquid Trans Am Inst Chem Eng 31:365–389
65. Danckwerts PV (1951) Significance of liquid-film coefficients in gas absorption Ind Eng Chem 43(6):1460–1467
66. (a) Richards GM, Ratcliff GA, Danckwerts PV (1964) Kinetics of CO_2 absorption–III: First-order reaction in a packed column Chem Eng Sci 19(5):325–328; (b) Danckwerts PV, McNeil

KM (1967) The effects of catalysis on rates of absorption of CO_2 into aqueous amine-potash solutions Chem Eng Sci 22(7):925–930
67. Kister HZ, Scherffius J, Afshar K, Abkar E (2007) Realistically predict capacity and pressure drop for packed columns Chem Eng Prog 103(7):28–38
68. Sherwood TK, Holloway FAL (1940) Performance of packed towers-liquid film data for several packings Trans Am Inst Chem Eng 36:39–70
69. Furnas CC, Bellinger F (1938) Operating characteristics of packed columns I Trans Am Inst Chem Eng 34:251
70. Cryder DS, Maloney JO (1941) The rate of absorption of carbon dioxide in diethanolamine solutions Trans Am Inst Chem Eng 37:827–852
71. Tepe JB, Dodge BF (1943) Absorption of carbon dioxide by sodium hydroxide solutions in a packed column Trans Am Inst Chem Eng 39:255–276
72. Spector NA, Dodge BF (1946) Removal of carbon dioxide from atmospheric air Trans Am Inst Chem Eng 42:827–848
73. Colburn AP (1939) The simplified calculation of diffusal processes. General considerations of two-film resistance Trans Am Inst Chem Eng 35:211–236
74. (a) Onda K, Takeuchi H, Okumoto Y (1968) Mass transfer coefficients between gas and liquid phases in packed columns. J Chem Eng Jpn 1(1):56–62; (b) Vital TJ, Grossel SS, Olsen PI (1984) Estimating separation efficiency Hydrocarbon Process 63(11):147–153
75. Cornell D, Knapp WG, Close HJ, Fair JR (1960) Mass transfer efficiency-packed columns Chem Eng Prog 56:68
76. Towler G, Sinnott R (2008) Chemical engineering design principles, practice and economics of plant and process design. Elsevier, Burlington, p 1245
77. Ulrich GD, Vasudevan PT (2004) Chemical engineering process design and economics a practical guide, 2nd edn. Process Publishing, Durham, p 706
78. Tan SH (1981) centrifugal-pump power needs Oil Gas J
79. Ramezan M, Skone TJ, Nsakala N, Liljedahl GN (2007) Carbon dioxide capture from existing coal-fired power plants; National Energy Technology Laboratory (NETL). U.S. Department of Energy, Pittsburgh
80. Kim I, Svendsen HF (2007) Heat of absorption of carbon dioxide (CO_2) in monoethanolamine (MEA) and 2-(Aminoethyl) ethanolamine (AEEA) solutions Ind Eng Chem Res 46(17):5803–5809
81. Weiland RH, Dingman JC, Cronin DB (1997) Heat capacity of aqueous monoethanolamine, diethanolamine, N-methyldiethanolamine, and N-methyldiethanolamine-based blends with carbon dioxide J Chem Eng Data 42(5):1004–1006
82. Al-Ghawas HA, Hagewiesche DP, Ruiz-Ibanez G, Sandall OC (1989) Physicochemical properties important for carbon dioxide absorption in aqueous methyldiethanolamine J Chem Eng Data 34(4):385–391
83. Singh D, Croiset E, Douglas PL, Douglas MA (2003) Techno-economic study of CO_2 capture from an existing coal-fired power plant: MEA scrubbing vs. O_2/CO_2 Recycle Combustion 44(19):3073–3091

Chapter 4
Adsorption

In an adsorption process a gas mixture contacts small porous particles, which can selectively adsorb or complex with CO_2 for its effective removal from the gas mixture. Sorbent technologies may also be developed to capture CO_2 indirectly by focusing on the selective adsorption of other gases in a given gas mixture, *e.g.*, N_2, O_2, CH_4, H_2, etc. Adsorption is particularly known for its effectiveness in the separation of dilute mixtures. Molecules of CO_2 may be held loosely by weak intermolecular forces, termed *physisorption* or strongly via covalent bonding, termed *chemisorption*. Generally, physisorption occurs when the *heat of adsorption* is less than approximately 10–15 kcal/mol, while chemisorption occurs with heats of adsorption greater than 15 kcal/mol [10]. These are rules of thumb, however, and exceptions do exist. For instance, the heat of physisorption of CO_2 in some zeolites [12] has been reported to be as high as 50 kcal/mol, with heats of chemisorption known to extend from as low as 15 kcal/mol to over 100 kcal/mol. The heat of adsorption is a direct measure of the binding strength between a fluid molecule and the surface. Typically, physisorption is characterized by a low heat of adsorption (typically less than 2–3 times the latent heat of evaporation), by monolayer or multilayer coverage, and is a rapid non-activated and reversible process, and although *polarization* is possible, no electron transfer occurs. Chemisorption is characterized by high heats of adsorption (typically greater than 2–3 times the latent heat of evaporation), highly specific sites of adsorption, monolayer coverage only, and possible dissociation. In addition, chemisorption is a slow process due to the electron transfer leading to bonding between the adsorbate and surface and the required activation barrier that has to be overcome for the formation of the bound complex.

A molecule in the fluid phase prior to adsorption is referred to as an *adsorptive*, once on the surface it is defined as an *adsorbate*, with the surface defined as the *adsorbent*, or more general to include both adsorption and desorption processes, *sorbent*. In the case of CO_2 capture, adsorption and desorption are of equal importance since any material designed for capture will have to be regenerated due to the sheer adsorbent volume required for any significant CO_2 mitigation. For adsorption to be a thermodynamically favorable process, it must be exothermic. The relationship between the change Gibbs free energy (ΔG), enthalpy (ΔH), and entropy (ΔS) leads to an understanding of why this is the case. The entropy difference between

J. Wilcox, *Carbon Capture*,
DOI 10.1007/978-1-4614-2215-0_4,

the condensed surface-bound phase and fluid (gas or vapor) phase is negative since the number of translational degrees of freedom is reduced from 3 in the fluid phase to no more than 2 in the adsorbed phase, with the number of rotational degrees of freedom also decreasing upon adsorption. For any significant adsorption to occur the process must be spontaneous, *i.e.*, the Gibbs free energy must be negative. Since, $\Delta G = \Delta H - T\Delta S$, it follows that the adsorption process must be exothermic (*i.e.*, $\Delta H < 0$). Although chemisorption and physisorption processes are both possible for CO_2 capture, due to the volume of CO_2 to be handled and captured it is probable that physisorption will be the primary mechanism in terms of material design for this application, to minimize the energy required for regeneration. In general, the heat of adsorption provides an indication of the extent of chemical affinity an adsorbate has to a surface, with greater amounts of heat corresponding to stronger adsorbate -sorbent bonding. Due to the exothermicity of the adsorption process, it is important to keep in mind the extent of heat release at the scales required for significant CO_2 capture to determine whether heat removal throughout the process is required.

Figure 4.1 shows a schematic outlining the structure of the chapter. As discussed in Sect. 1.6 of Chap. 1, the costs can be divided into non-technical and technical. The factors that influence the cost of the technology associated with an adsorption separation process are discussed in this chapter and hence, the compression of CO_2 required for transport is not included in the schematic. Figure 4.1 is a simplistic view of the relationship between cost, equipment, material, and process parameters

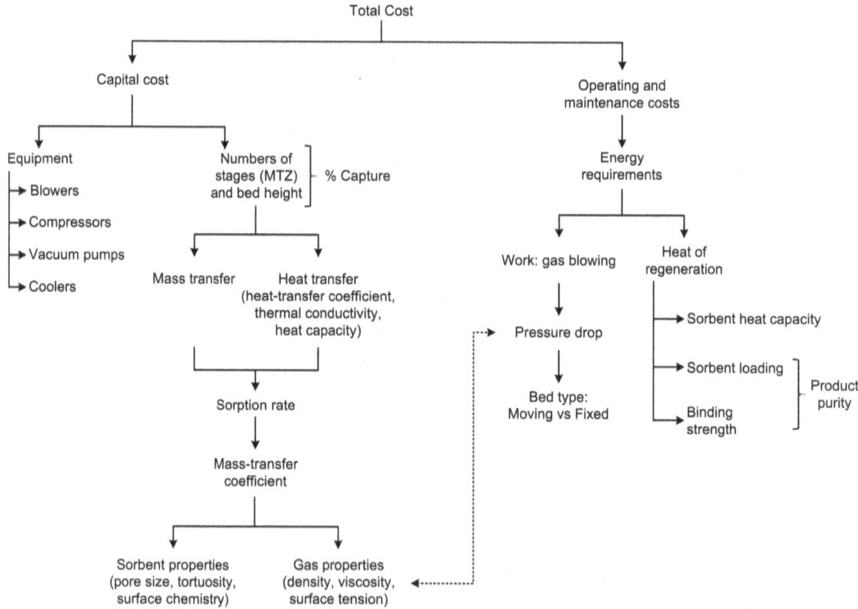

Fig. 4.1 Schematic outlining the components that comprise the cost of the CO_2 adsorption separation process

to emphasize that these aspects are not disconnected, but rather, highly dependent and intertwined in a complex fashion. This schematic is not meant to be inclusive of all connections and components of adsorption, but rather a general overview of the primary constituents. In addition to the equipment that may be required in adsorption processes, such as compressors, blowers, and vacuum pumps, the length of the adsorber bed also dictates the capital cost, which is determined from the mass-transfer behavior and corresponding rate of adsorption. The time scale associated with a sorbent reaching sufficient capacity prior to regeneration is based in part upon its pore size distribution, since the pore size influences the time it takes the solute species to travel into the sorbent particles and into the pore structure representative of the void space. For instance, adsorptive species may travel more easily in sorbent particles comprised primarily of larger-sized pores. However, a network comprised of pores that are too large, may limit the available surface area and pore volume required for achieving reasonable capacities. The pore size distribution also plays a role in determining the extent of sorbent loading, which is necessary for assessing the heat required to regenerate the sorbent. The heat requirements for regeneration along with the required gas blower power for overcoming the pressure drop through a sorbent bed, collectively dictate the operating and maintenance costs associated with the adsorption process for CO_2 capture. Significant attention in the chapter is focused on sorbent characterization, since this is a crucial component to understanding sorbent capacities and rates of adsorption, which are phenomena underlying the operating and maintenance and capital costs of the adsorption separation process.

4.1 Adsorption Fundamentals

4.1.1 Physical Adsorption

Forces and Adsorption Energies Physisorption processes are predominantly associated with van der Waals forces, also known as dispersion-repulsion forces and electrostatic forces, which are sourced from polarization, dipole, quadrupole, and higher-pole interactions. The van der Waals forces are present in all systems, but the electrostatic interactions are only present in systems that contain charge, such as zeolites, metal-organic frameworks (MOFs), and sorbents with surface functional groups and surface defects. The dispersion-repulsive forces between two isolated systems can be investigated by considering each type of interaction with its own potential energy term. For instance, the attractive potential from the dispersion forces, Φ_D between two isolated systems 1 and 2 of distance r_{12} is:

$$\Phi_D = -\frac{A_1}{r_{12}^6} - \frac{A_2}{r_{12}^8} - \frac{A_3}{r_{12}^{10}} \tag{4.1}$$

such that A_1, A_2, and A_3 are constants. The first, second, and third terms in Eq. (4.1) represent the instantaneous induced dipole, induced dipole-induced quadrupole, and the induced quadrupole-induced quadrupole interactions, respectively. When two systems approach one another, there is a short-range repulsive energy, Φ_R associated with the finite size of the systems that can be represented semiempirically by:

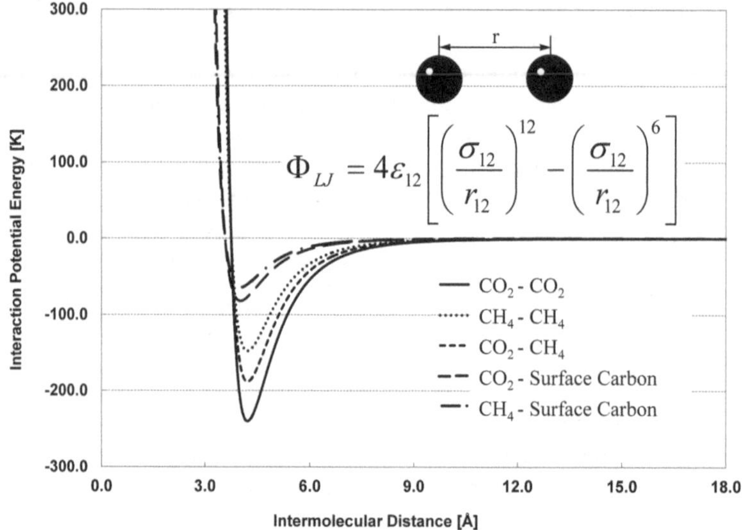

Fig. 4.2 Lennard-Jones potential for CO_2–CO_2, CO_2-methane, and CO_2-graphite basal plane interactions

$$\Phi_R = -\frac{B}{r_{12}^{12}} \tag{4.2}$$

Often times, the total dispersion-repulsion expression is truncated by removing the higher-order dispersion energy contributions leading to the well-known Lennard-Jones potential:

$$\Phi_{LJ} = 4\varepsilon\left[\left(\frac{\sigma}{r}\right)^{12} - \left(\frac{\sigma}{r}\right)^{6}\right] \tag{4.3}$$

This expression is represented in Fig. 4.2 for various CO_2 interactions. The force constants ε and σ are a function of temperature, pressure, and type of species under consideration. Tabulated data for CO_2 and the gas species expected to interact with CO_2 in typical gas mixtures are available in Appendix F.

The Lennard-Jones parameters can be "averaged" to represent the interaction potential between two different systems, more specifically, by the geometric mean of ε and the arithmetic mean of σ as follows:

$$\varepsilon_{12} = \sqrt{\varepsilon_1\varepsilon_2} \quad \text{and} \quad \sigma_{12} = \frac{1}{2}(\sigma_1 + \sigma_2) \tag{4.4}$$

In addition to the dispersion-repulsion, electrostatic forces exist in the case of charged systems such as MOFs and zeolites. In charged systems in which there exists a significant electric field in the neighborhood of the surface, additional energy contributions can be sourced from polarization (Φ_P), field-dipole (Φ_μ), and field gradient-quadrupole (Φ_Q) interactions, which are given by the following expressions:

$$\Phi_P = -\frac{1}{2}\alpha E^2; \quad \Phi_\mu = -\mu E; \quad \Phi_Q = \frac{1}{2}Q\frac{\partial E}{\partial r} \tag{4.5}$$

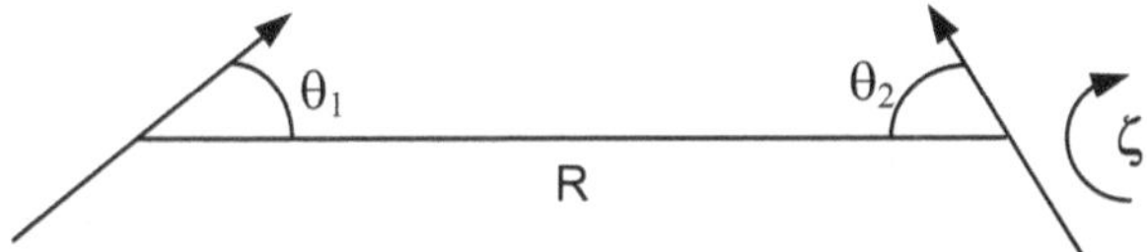

Fig. 4.3 Configuration of a pair of interacting axially-symmetric charge distributions, with the arrows indicating the dipolar directions

such that E is the electric field, α is the polarizability, μ is the *dipole moment*, and Q is the *quadrupole moment* defined by:

$$Q = \frac{1}{2}\int q(\rho,\theta)(3\cos^2\theta - 1)\rho^2 \mathrm{d}V \tag{4.6}$$

such that $q(\rho, \theta)$ is the local charge density at the point, (ρ, θ) with the origin defined as the center of the system with the integration carried out over the entire system. For a charged system, such as a zeolite or MOF, the overall potential is given by the sum of the dispersion-repulsion and electrostatic interactions as:

$$\Phi = \Phi_D + \Phi_R + \Phi_P + \Phi_\mu + \Phi_Q + \Phi_S \tag{4.7}$$

such that Φ_S is the contribution from sorbate-sorbate interactions on the surface at high (surface) coverage. As an example, Yang et al. have shown that CO_2-MOF adsorption energies may contain up to a 30% contribution from electrostatic interactions at low to moderate pressures, with decreasing contribution (*e.g.*, less than 3%) at higher pressure [13].

The Lennard-Jones parameters may be derived from theory using quantum chemical calculations or from experiment using detailed deviation measurements from the ideal gas law based upon the second virial coefficient, $B(T)$, which may be expressed by:

$$\frac{P\widetilde{V}}{RT} = 1 + \frac{B(T)}{\widetilde{V}} + \frac{C(T)}{\widetilde{V}^2} + \cdots \tag{4.8}$$

Equation (4.8) is known as the virial equation of state and $B(T)$ represents the first deviation from the ideal-gas law, such that $\widetilde{V}$ is the molar volume, with the deviation from ideality related to the molecular-pair interactions. Figure 4.3 illustrates the two axially-symmetric charge distributions of molecules 1 and 2, with the angle ζ representative of the space formed between the planes of the axes of molecules 1 and 2 and their line of centers. The interaction potential energy, u_{12} between molecules 1 and 2 can be expressed by the summation of Eqs. (4.1) and (4.2) or by the Lennard-Jones potential of Eq. (4.3). If u_{12} is the interaction potential energy between two axially-asymmetric molecules 1 and 2, then the classical statistical-mechanical expression for $B(T)$ is:

$$B(T) = \frac{N_A}{4}\int_0^\infty R^2 \mathrm{d}R \int_0^\pi \sin\theta_1 \mathrm{d}\theta_1 \int_0^\pi \sin\theta_2 \mathrm{d}\theta_2 \int_0^{2\pi} \mathrm{d}\zeta[1 - \mathrm{e}^{-u_{12}/kT}] \tag{4.9}$$

such that R, θ, and ζ are defined in Fig. 4.3, and N_A is Avogadro's number.

Table 4.1 Kinetic and electrostatic properties of common gases

Molecule	Kinetic diameter (nm) [15]	Dipole moment (Debye) [16]	Quadrupole moment (10^{-40} Coulomb·m^2) [17]	Polarizability (10^{-24} cm^3) [16–17, 17d]
CO_2	0.330	0	−13.71, −10.0	2.64, 2.91, 3.02
N_2	0.364	0	−4.91	0.78, 1.74
O_2	0.346	0	−1.33	1.57, 1.77
H_2O	0.280	1.85	6.67	1.45, 1.48
SO_2	0.360	1.63	−14.6	3.72, 3.89, 4.28
NO	0.317	0.16	−6.00	1.7
NO_2	0.340	0.316	Unknown	3.02
NH_3	0.260	1.47, 5.10	−7.39	2.22, 2.67, 2.81
HCl	0.346	1.11, 3.57	13.28	2.63, 2.94
CO	0.376	0.11, 0.37	−8.33, −6.92	1.95, 2.19
N_2O	0.317	0.16, 0.54	−12.02, −10.0	3.03, 3.32
Ar	0.340	0	0	1.64, 1.83
H_2	0.289	0	2.09, 2.2	0.81, 0.90
CH_4	0.380	0	0	2.6

The Lennard-Jones parameters may also be obtained experimentally from transport properties such as viscosity or thermal conductivity measurements.

Effective separation takes place through a preferred surface interaction with CO_2 caused by differences in polarity or surface reactivity. Highly porous solid particles with interconnected pores allow for large surface areas, which are required for the separation of large volumes of gas. Specific surface areas of typical sorbent materials range from 500 and up to 6200 [14] in the case of some MOFs in units of m^2/g of sorbent. Aside from direct surface interactions, separation can also take place based upon differences in molecular weight or kinetic diameters of the various gas components in a given mixture, which lead to different rates of diffusion through the material. It is important to recognize that adsorption may be used to selectively separate gas species other than CO_2 in a gas mixture; however, typical gas-mixture components include N_2, O_2, and methane and consequently CO_2 tends to be more reactive than these species due to its stronger quadrupole moment (see Table 4.1) making it an easier target for selective capture. The electrostatic and kinetic properties of some common gases are listed in Table 4.1. These parameters can be used as selective criteria for separation via adsorption or diffusion through a bed of sorbent particles. Notice that the kinetic diameter of CO_2 is similar to most of the gas species that would be present in gas mixtures of interest. This implies that it would be difficult to separate CO_2 from N_2 for instance, on size alone. The other three parameters in Table 4.1 in part represent the inherent reactivity of CO_2. It is important to notice that the quadrupole moment of CO_2 is greater than most of the other gases. This is the parameter that allows for the high selectivity in terms of the separation of CO_2 from N_2 using a zeolite sorbent. However, in flue gas water vapor is present at approximately 8–10% molar volume and has a very strong dipole moment, which CO_2 does not possess, as shown in Table 4.1. This is the reason that zeolite-based sorbents and MOFs have difficulty in achieving high CO_2 selectivity over water. The mechanism of adsorption in these systems is based primarily upon charge selectivity.

Table 4.2 IUPAC pore size classifications [7]

Type	Pore size (Å)
Micropore	< 20
Mesopore	20–500
Macropore	> 500

Adsorption Isotherms An *adsorption isotherm* represents the equilibrium relationship between the CO_2 molecules in the fluid phase and adsorbed phase at a given temperature. The gas-phase concentration of CO_2 in the fluid is traditionally expressed in the form of partial pressure or mole percent. The concentration of CO_2 adsorbed on a surface is traditionally expressed in terms of the mass adsorbed per unit mass of sorbent material. The shape of the adsorption isotherm is dependent upon the interactions of CO_2 with other fluid species in the confined pore space, interactions with CO_2 and the pore walls, and in the case of *micropores*, the wall-wall interactions of the pore may also influence sorption behavior. The International Union of Pure and Applied Chemistry (IUPAC) as shown in Table 4.2, have defined pore size classifications, such that micropores have an internal diameter of less than 2 nm, *mesopores* between 2 and 50 nm, and *macropores* greater than 50 nm.

The mechanism of adsorption differs in micropores versus pores of diameter greater than 20 Å (2 nm). For instance, in micropores adsorption is governed by CO_2-surface interactions with wall-wall interactions also playing a significant role. Within mesopores the fluid-fluid interactions become more important leading to capillary condensation in the pores. Condensation in the pores is represented by the formation of a liquid-like phase at a pressure lower than the *saturation pressure* of the bulk fluid. In the mesopore, the conditions of the bulk phase transition (to condensed phase) are shifted, taking place at lower pressures and/or higher temperatures due to the fluid-wall interactions. In macropores, the fluid-wall interactions play less of a role and the density of the fluid in macropores can be equated to the bulk density of the fluid at a given temperature and pressure.

In 1985, IUPAC classified six types of *sorption* isotherms as pictured in Fig. 4.4 [18]. The term "sorption" is used when both adsorption and desorption are referred to Type I is a reversible isotherm that is obtained when adsorption is limited to single layer or very few molecular layers. Materials that are microporous will often exhibit adsorption isotherms of this type. Type II isotherms are typically obtained in the case of macroporous sorbents in which there is an unrestricted extent of multilayer adsorption. The inflection point at location B represents the stage at which monolayer coverage is complete and multilayer coverage begins to take place. Isotherms of Type III are known as unfavorable and are not common, but occur in cases in which fluid-wall interactions are weak, such as the case of water vapor on a non-defective graphitic carbon surface. Type IV isotherms are common in the case of mesopores and possess the characteristic hysteresis loop, which is accounted for by condensation in the pore. The initial plateau feature of this isotherm, similar to Type II, represents monolayer coverage, with the second plateau reaching an upper limit of multi-layer coverage. Hysteresis occurs in this case due to the energetically irreversible condensation effect that takes place in the pore. Type V isotherms also exhibit both

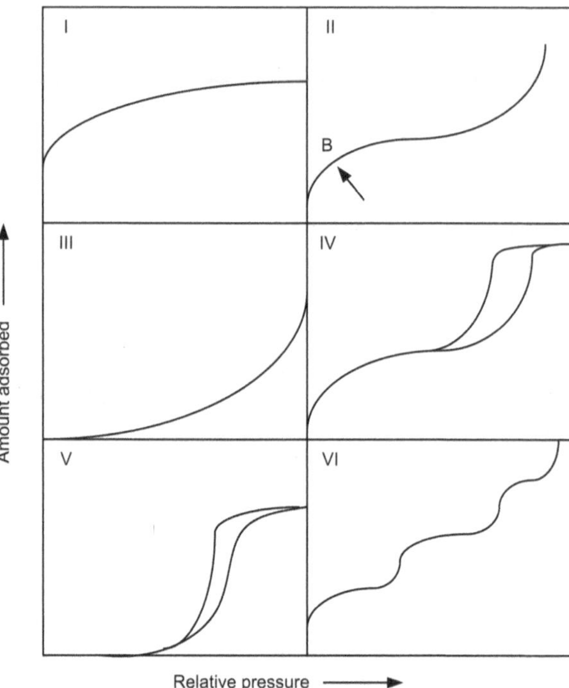

Fig. 4.4 Sorption isotherm types by IUPAC classification. The Henry's law regime is the low-loading region where the isotherm becomes linear. (Reproduced with permission [7])

condensation and hysteresis, but with the first-stage adsorption phenomena similar to Type III, in which the fluid-surface interactions are weak and the second-stage adsorption phenomena similar to Type IV in which the fluid-fluid interactions play a dominant role far from the pore surface. Type VI isotherms represent a stepwise adsorption process, in which successive two-dimensional phase transitions of the "gas-liquid-solid" and "commensurate-incommensurate" type take place [19] and occur in the case of argon [20] and krypton [21] adsorption on graphitic carbon surfaces at 77 K.

Langmuir and BET Theories "For his discoveries and investigations in surface chemistry," Irving Langmuir was awarded the Nobel Prize in Chemistry in 1932. Based upon kinetic theory, Langmuir [22] was able to describe the Type I adsorption isotherm based on the assumptions that only a monolayer of adsorbate species adsorb to the surface, lateral interactions do not take place, and that the surface is homogeneous with identical adsorption sites. The Langmuir equation for Type I isotherms is:

$$\frac{W}{W_m} = \frac{Kp}{1 + Kp} \tag{4.10}$$

such that p is the adsorbate pressure, W is the sorbate loading, W_m the maximum sorbate loading based upon the monolayer assumption, and K, the adsorption equilibrium constant per site defined as:

$$K = \frac{k p_c}{N_m \nu e^{-E/RT}} \tag{4.11}$$

with

$$k = \frac{N_A}{\sqrt{2\pi MRT}} \tag{4.12}$$

such that N_A Avogadro's constant, M is the molecular weight, R is the ideal gas constant, T is the temperature, and in Eq. (4.11), p_c is the condensation probability, representing the likelihood that a fluid molecule's collision with the pore wall will result in adsorption, N_m is the number of adsorbate molecules comprising a complete monolayer per unit area, ν is the vibrational frequency of the adsorbate normal to the surface in its adsorbed state, and the Boltzmann distribution term, $e^{-E/RT}$ is the probability that an adsorbed molecule contains sufficient energy to overcome the net attractive potential of the surface. From experimental adsorption measurements, $1/p$ and $1/W$ of Eq. (4.10) can be obtained and plotted with the resulting straight line determining both K and W_m parameters using the y-intercept, $1/W_m$ and the slope, $1/KW_m$ of the equation, respectively. The calculation of W_m leads to the determination of the sorbent surface area, S from:

$$S = \frac{W_m N_A A}{M} \tag{4.13}$$

such that A is the cross-sectional area of the adsorbate and M the adsorbate molecular weight. Surface area measurements obtained from Langmuir-type adsorption isotherms are not able to provide indication of the adsorption mechanism, *i.e.*, whether physisorption or chemisorption is occurring. For instance, in the case of chemisorption, the adsorbate chemically binds to only active sites on the surface, leaving some portion of the surface uncovered. In the case of physisorption, which is primarily associated with pore filling, there is not a clearly defined monolayer region within the width of the pore. In addition, the concept of a monolayer breaks down for very small (*i.e.*, micro) pores where the pore diameter approaches the diameter of the adsorbed molecules. In such systems, the isotherm commonly has the form of Eq. (4.10), but the saturation capacity corresponds to total filling of the pore volume rather than monolayer coverage. Also, the Langmuir model is based on the assumption of localized adsorption without interaction between adsorbed molecules. In the case of higher coverage, the adsorbate-adsorbate interaction cannot be ignored, leading to deviations from the simple Langmuir model.

Example 4.1 The following experimental data for the equilibrium adsorption of pure CO_2 on MOF Cu-BTC at 25°C were obtained by Liang et al. [83]:

Pressure (bar)	Adsorption capacity (mol/kg)
0.1	0.47
0.3	0.78
0.5	1.35
1.0	2.31
1.5	2.94
2.0	3.56
3.1	4.26
4.1	4.77
5.1	5.08

Fit the data to a Langmuir isotherm.

Solution: By using the linearized form of the Langmuir equation (Eq. 4.10), linear regression can be used to obtain the constants, K and W_m. The values obtained are $K = 0.64\ \text{bar}^{-1}$ and $W_m = 5.80$ mol/kg. The Langmuir equation is then:

$$\frac{W}{5.80} = \frac{0.64p}{1 + 0.64p}$$

The predicted values are:

Pressure (bar)	Adsorption capacity (mol/kg)	Langmuir prediction (mol/kg)
0.1	0.47	0.45
0.3	0.78	0.88
0.5	1.35	1.41
1.0	2.31	2.31
1.5	2.94	2.88
2.0	3.56	3.31
3.1	4.26	3.86
4.1	4.77	4.22
5.1	5.08	4.47

In 1938 Stephen Brunauer, Paul Hugh Emmett and Edward Teller [23] extended Langmuir 's monolayer adsorption theory to include multilayer adsorption, which is traditionally termed BET theory. The assumption underlying BET theory is that the uppermost molecules of the adsorbed layers are in dynamic equilibrium with the vapor phase. The expression for BET multilayer adsorption is:

$$\frac{1}{W[(p/p_0) - 1]} = \frac{1}{W_m C} + \frac{C - 1}{W_m C}\left(\frac{p}{p_0}\right) \tag{4.14}$$

with C expressed as,

$$C = \frac{p_{c1}\nu_1}{p_{c2}\nu_2} e^{(E_1 - E_L)/RT} \tag{4.15}$$

such that p_{c1} and p_{c2} are the probability of a fluid particle condensing upon collision with layer 1 and layer 2, respectively, ν_1 and ν_2 are the vibrational frequencies of the adsorbate normal to layer 1 and layer 2, respectively, E_1 is the activation barrier required to overcome the potential associated with adsorption on layer 1, while E_L is energy associated with condensation to the liquid phase. In Eq. (4.14), p_0 is the saturation pressure of the bulk phase of the pure adsorbate. Plotting $1/W[p/p_0 - 1]$ versus p/p_0 traditionally yields a straight line in the range of $0.05 \leq p/p_0 \leq 0.35$ [24]. The slope of the BET plot is $(C - 1)/W_mC$ with y-intercept, $1/W_mC$. Solving the two equations provide estimates of C and W_m, which can be used to calculate the available surface area using Eq. (4.13). Assuming the liquid is comprised of spheres having 12 nearest neighbors with 6 neighbors in a given plane, and with a hexagonal-close-packed configuration, the adsorbate cross-sectional area may be approximated by [25]:

$$A = 1.091\left(\widetilde{V}/N_A\right)^{2/3} \times 10^{16} \text{Å}^2 \tag{4.16}$$

such that $\widetilde{V}$ is the molar volume of the bulk fluid phase.

Details of the derivation associated with both Langmuir and BET isotherm equations are omitted here, but can be found in Lowell et al. [25]. Assumptions of BET theory include equality among all adsorption sites in addition to the exclusion of lateral adsorbate interactions, with the heat of adsorption at the second and higher layers equivalent to the heat of liquid-phase condensation. It is well known that BET is generally valid between *relative pressures* (*i.e.*, p/p_0) of 0.05 and 0.35. BET theory is applicable to nonporous and mesoporous materials, but is not generally applicable to micropores since these pores have unusually high adsorption energies due to the overlapping potentials of the pore walls; therefore, it is difficult to determine the extent of monolayer and multilayer adsorption versus pore filling. In the application of BET theory to micropores, the calculated surface area is not representative of the true micropore surface area. For this reason, more sophisticated models are recommended for the investigation of adsorption in micropores and are discussed in detail in Sect. 4.2.2 on Pore Characterization.

Hysteresis in Mesopores In mesopores (*i.e.*, pores between 2 and 50 nm) both fluid-wall and fluid-fluid interactions play a role in the adsorption mechanism. Capillary (or pore) condensation takes place when a gas in a pore condenses to a liquid-like phase at pore pressures less than the saturation pressure (p_0) of the bulk liquid. In addition to multilayer adsorption and pore condensation, hysteresis is also a common mesopore phenomenon. During the multilayer adsorption process, the film thickness will be limited by its stability, which is governed by the attractive fluid-wall interactions, and the *surface tension* and curvature of the liquid-vapor interface. The difference in chemical potential between the adsorbed phase (μ) and that of the boundary (μ_0)

can be expressed by:

$$\mu - \mu_0 = \Delta\mu = -\alpha l^{-m} - \frac{\sigma}{a\Delta\rho} \tag{4.17}$$

such that α is the fluid-wall interaction parameter, l is the pore surface film thickness, with the l^{-m} term resulting from the long-range van der Waals interactions between the fluid and the wall. Typical values for m are between 2.5 and 3, with the lower range applicable to strongly attractive sorbents such as graphite. Also in Eq. (4.17), σ is the surface tension of the adsorbed liquid-like film, a is the core radius of the pore, such that $a = r - l$, where r is the pore radius, and $\Delta\rho$ is the density difference between the liquid-like film and the gas phase. At some critical distance from the pore wall, or critical layer thickness, l_c, core condensation controlled by the intermolecular forces in the fluid will take place. This condensation represents a phase transition from a gas to a liquid-like phase occurring at a chemical potential less than that of the gas-liquid coexistence of the bulk fluid. Adapted from Lowell et al. [25], these phenomena are illustrated in Fig. 4.5. Mesopore filling commences with monolayer formation (A), and is followed by multilayer adsorption (B), which leads to the start of capillary condensation at the core of the pore after reaching a critical film thickness (C to D). The plateau of the adsorption isotherm (D) is representative of a completely filled pore in which the gas phase is separated from a hemispherical meniscus. As the pressure is reduced to less than the pore condensation pressure, the meniscus is reduced (E), with the closure of the hysteresis loop corresponding to a multilayer film in equilibrium with the gas phase (F). Notice that in the adsorption isotherm plot of Fig. 4.5, in the relative pressure range between A $\leftrightarrow$ F, the sorption phenomena are reversible.

The Kelvin approach may also be used to describe the coexistence of liquid and gas phases occurring in pores of fairly uniform shape and width. The Kelvin equation relates the chemical potential difference of adsorbed and gas-liquid boundary fluid molecules to macroscopic properties such as surface tension and bulk densities as follows:

$$\Delta\mu = \mu - \mu_0 = -RT \ln\left(\frac{p}{p_0}\right) = -\frac{2\sigma\cos\theta}{r\Delta\rho} \tag{4.18}$$

such that σ is the surface tension of the liquid, θ is the *contact angle* of the liquid against the pore wall, r is the average curvature radius corresponding to the pore radius, and $\Delta\rho$ is the difference in density of the liquid bulk at coexistence and the bulk gas. The contact angle, θ, can be interpreted as a measure of the relative strength of fluid-wall versus fluid-fluid interactions. In general, understanding hysteresis can aid in material and process optimization for low-energy sorbent regeneration.

4.1.2 Chemical Adsorption

The previous discussion on adsorption forces was focused primarily on physisorption processes. In the case of chemisorption in which covalent bonds are formed upon

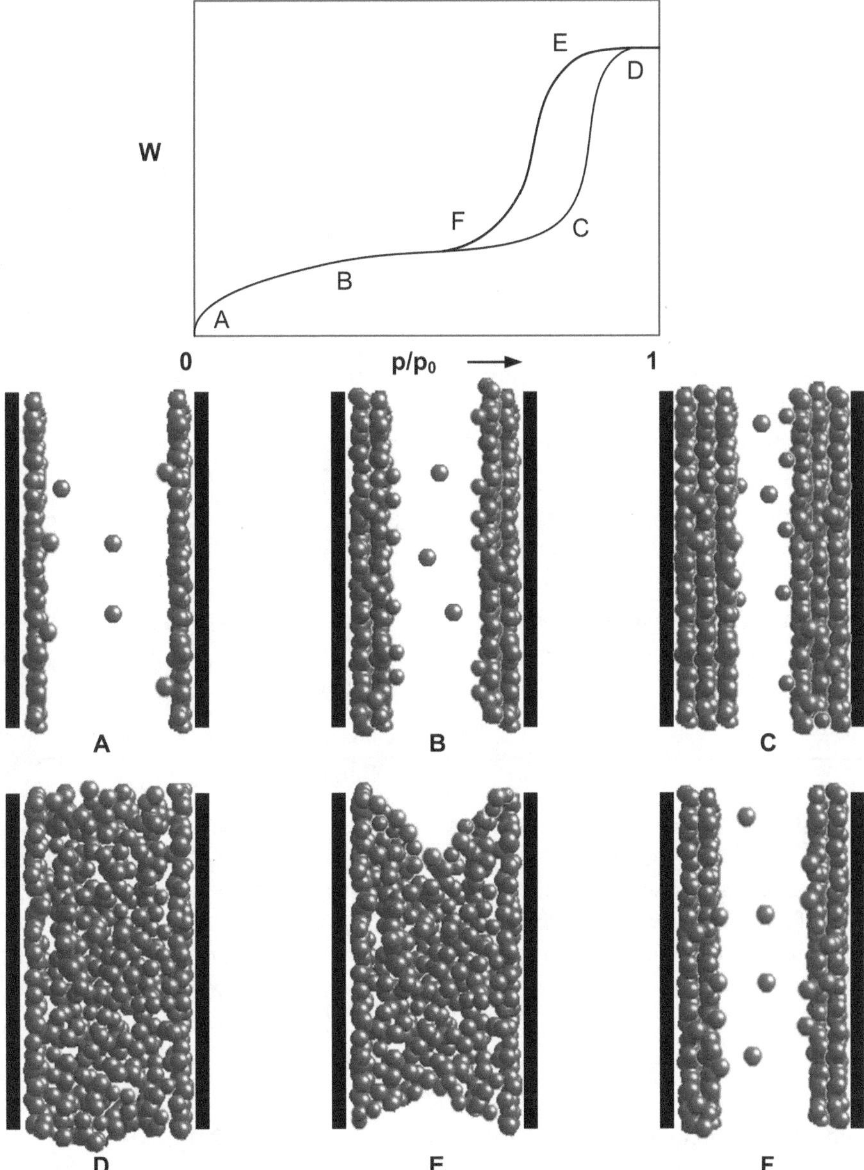

Fig. 4.5 Schematic representation of multilayer adsorption and pore condensation in a slit-shaped pore. (With kind permission from [5])

adsorption there is increased likelihood of greater heat release that may influence sorbent uptake. The dimensionless parameter, which measures the relative rates of reaction and diffusion within the pores is termed the Thiele modulus (ϕ) [26], or by some, the Jüttner modulus [27], and is defined as:

$$\phi = R_p \sqrt{\frac{k}{D_e}} \tag{4.19}$$

such that R_p is the sorbent particle radius, k is an intrinsic reaction rate constant that is related to the surface rate constant by, $k = sk_s$, in which s is the surface area per unit particle volume, and D_e is the effective diffusivity. The Thiele modulus can be derived, similar to that carried out by Ruthven [84], from considering the steady-state differential mass balance for a shell element of a microporous particle as follows:

$$\frac{\mathrm{d}^2 c}{\mathrm{d}r^2} + \frac{2}{r}\frac{\mathrm{d}c}{\mathrm{d}r} = \frac{k}{D_e}c \tag{4.20}$$

Assuming: that the effective diffusivity is independent of concentration and substitution of Eq. (4.19) into Eq. (4.20), in dimensionless form leads to:

$$\frac{\mathrm{d}^2 C}{\mathrm{d}R^2} + \frac{2}{R}\frac{\mathrm{d}C}{\mathrm{d}R} = \frac{R_p^2 k}{D_e}C = \phi^2 C \tag{4.21}$$

such that $C = c/c_0$, $R = r/R_p$, and ϕ is the Thiele modulus. Assuming the following boundary conditions: $r = R_p \rightarrow R = 1.0$; $c = c_0 \rightarrow C = 1.0$; $r = 0 \rightarrow R = 0$; and $\mathrm{d}c/\mathrm{d}r = \mathrm{d}C/\mathrm{d}R = 0$, yields a relatively simple solution of,

$$C(R) = \frac{c(r)}{c_0} = \frac{\sinh(\phi R)}{R \sinh \phi} \tag{4.22}$$

such that the concentration of the diffusing species is dependent upon its diffusion depth within the particle. To determine the effect that diffusion into a pore has on the rate of reaction we can define the effectiveness factor as the ratio of the average rate of reaction in the particle to the average rate of reaction without the inclusion of diffusion limitations. Mathematically, the effectiveness factor, η, may be defined as,

$$\eta = \int_{R=0}^{R_p} \frac{4\pi R^2 C(R)\mathrm{d}R}{\left(\frac{4}{3}\pi R_p^3\right) c_0} \tag{4.23}$$

and upon integration of Eq. (4.23) after substitution of Eq. (4.22) yields,

$$\eta = \frac{3}{\phi}\left[\frac{1}{\tanh \phi} - \frac{1}{\phi}\right], \tag{4.24}$$

which is reduced to $\eta \approx 3/\phi$ for large ϕ, when the Thiele modulus is small, the effectiveness factor approaches unity, resulting in a minimal change in reactant concentration over the depth of the particle, indicating that the reaction rate is kinetically-controlled rather than diffusion-controlled. When the Thiele modulus is large, the reactant concentration drops significantly at the surface and is nearly zero at the particle's center, resulting in a diffusion-limited reaction rate. Since the Thiele modulus is directly proportional to the particle radius from Eq. (4.19), larger particles tend to exhibit diffusion-limited behavior ($\phi \to \infty$), while smaller particles tend to exhibit reaction-controlled behavior ($\phi \to 0$). In the case of physisorption processes, since the mechanism of adsorption is not activation barrier-controlled, intraparticle diffusion resistance dominates adsorption.

4.2 Common Types of Sorbents

Table 4.3 lists the physical characteristics and adsorption properties of common sorbents applied to CO_2 capture. Low-temperature sorbents investigated for CO_2 capture include activated carbon, ion-exchange resins, silica gel, activated alumina, and surface functionalized nanoporous materials based on silica and carbon. High-temperature sorbents investigated include metal oxides (*e.g.*, CaO), which can withstand temperatures of up to 1000°C, hydrotalcites, which are naturally occurring anionic clays that can operate up to 400°C, and lithium zirconate (Li_2ZrO_3) that can withstand temperatures up to 750°C with capacity increase through the addition of potassium carbonate (K_2CO_3). Several of these sorbent types are discussed next in further detail.

Activated Carbon The synthesis of activated carbon takes place by preoxidation in air for several days at approximately 270°C, followed by thermal treatment of the carbonaceous material through O_2, CO_2 or steam-activation at temperatures between 700 and 1100°C [34]. The purpose of the thermal treatment is to remove tar-like components, which leads to pore opening and increased pore connectivity. Additionally, the thermal gas treatment allows for the removal of the aliphatic components of the material through the formation of CO and CO_2 gases, leaving behind a framework primarily of aromatic character. Figure 4.6 shows the evolution of micro and mesopores followed by multiple-day preoxidation treatment. Activated carbon is traditionally synthesized from carbonaceous material including various-ranked coal, coconut shell, and wood.

Figure 4.7 illustrates how the pore size distribution can be influenced by the type of activated carbon, based upon the structure and morphology of the original source material. The morphology of activated carbon is based upon microcrystalline graphite components stacked in a random orientation that allows for the formation of micropores. Pore size distributions of activated carbon tend to span a broad range and are often trimodal. Typical size ranges are listed in Table 4.4.

The pore surfaces within activated carbons are inherently nonpolar with slight polarity arising from chemisorbed oxygen functional groups from thermal treatment. Due to their inherent nonpolarity, activated carbons tend to be hydrophobic and organophilic. Understanding the extent of water adsorption on sorbents is crucial to the design and optimization of a sorbent with selectivity toward CO_2 since water may often compete with CO_2 and is present in most gas streams from which CO_2 is to be separated. In the case of poly(vinylidene chloride)-based activated carbon,

Table 4.3 Properties for traditional commercial sorbents

Sorbent	Pore diameter (nm)	Sorbent density (kg/m^3)	Sorbent porosity	BET surface area (m^2/g)	H_2O cap. wt%, 25°C, 4.6 mmHg	CO_2 cap. wt%, 25°C, 250 mmHg	Regeneration temp, °C
Activated carbon [17a]							
Small-pore	1–2.5	500–900	0.4–0.6	400–1200	1	5	–
Large-pore	> 3	600–800	–	200–600	–	7	–
Zeolites [17a]							
3A	0.3	670–740	0.2	700	20	–	> 350
4A	0.4	660–720	0.3	700	23	13	120–350
5A	0.5	670–720	0.4	650	21	15	120–350
13X	0.8	610–710	0.5	600	25	16	120–350
Mordenite	0.3–0.4	720–800	0.25	700	9	6	–
Chabazite	0.4–0.5	640–720	0.35	650	16	12	–
Silica gel [17a]							
Small-pore	2.2–2.6	1000	0.47	800	11	3	130–280
Large-pore	10–15	620	0.71	320	–	–	130–280
Activated alumina [17a]	1–7.5	800	0.50	320	7	2	150–315
MOFs [14a, 28]	0.4–2.4	200–1000	0.79–0.90	150–6200	9	4–14[f], 15	25–80
Ion exchange resins	< 1–12	1100[a], 1270[b]	0.2–0.5	15–120	–	–	60
Hollow fibers [29a]	2.5–11	1250	0.3–0.8	450–1100	–	–	100–150
CMS [29a, 30]	0.3–0.9	640–1000	0.5	400	> 20	1.2–2.5	100–200
Amine-based [31]	8–40	1000–1500		5–500	–	5–14	80–120
Hydrotalcites [32]	2–20	150–550	0.15–0.5	16–290	–	–	120–400
Chemisorbents [32c, 33]	0.2–20	2000		250–1250	–	–	700–920[c], 150–500[d], 350[e]

[a]Anion resin, [b]Cation resin, [c]CaO-based, [d]MgO-based, [e]Al_2O_3-based, [f]At conditions of 25°C, CO_2 partial pressure of 1 atm

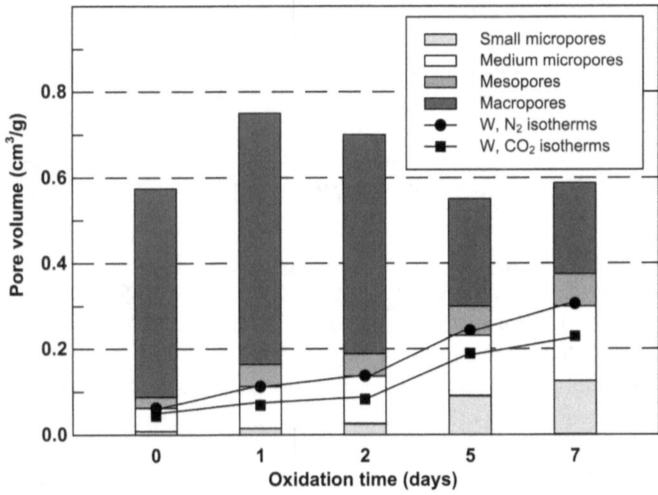

Fig. 4.6 Evolution of micro and mesopores with coal preoxidation and the volume (loading, *W*) occupied by N_2 versus CO_2. (Reprinted with permission from Elsevier [6])

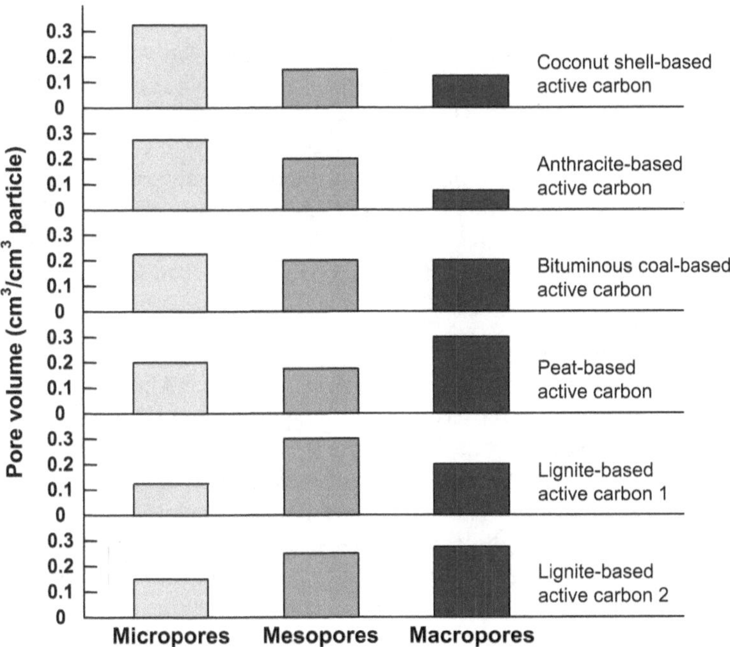

Fig. 4.7 Pore size distribution in selected activated carbons derived from various sources. (Reprinted with permission of John Wiley & Sons, Inc. [3])

Table 4.4 Pore characteristics of typical[a] activated carbons [16]

	Micropores	Mesopores	Macropores
Diameter (Å)	<20	20–500	>500
Pore volume (cm^3/g)	0.15–0.5	0.02–0.1	0.2–0.5
Surface area (m^2/g)	100–1000	10–100	0.5–2

[a]Particle density 0.6–0.9 g/cm^3; porosity 0.4–0.6

water adsorption at low pressure is significantly reduced after the removal of surface oxygen functional groups through degassing at 1000°C. After this treatment, water adsorption occurs only at pressures high enough to induce *capillary condensation* in the mesopores [35]. Treatment of activated carbon by exposure to H_2 is also a method for increasing the hydrophobicity of activated carbon. It is important to keep in mind, however, that these treatments may also lead to decreased adsorption of CO_2, depending on the mechanism of CO_2-surface attraction and subsequent adsorption, *i.e.*, whether the mechanism of interaction is through the acidic carbon atom or basic oxygen atoms.

Carbon Molecular Sieves Through careful treatment of carbon-based materials, they can be made to contain narrow pore size distributions, thereby termed carbon molecular sieves (CMS). The earliest example feedstock for the synthesis of CMS is polyvinylidene dichloride, with a wider variety of materials being used more recently [36]. Due to the wide pore size distribution of activated carbons, they show low selectivity for molecules with similar diameters. Generally, commercially available CMS materials are synthesized by controlled oxidation and subsequent thermal treatment of high-rank coal or anthracite [37]. Additional steps to CMS synthesis include controlled hydrocarbon cracking within the micropore network and well-controlled partial oxidation or gasification [38]. These CMS materials can be synthesized with micropore diameters between 4 to 9 Å. The largest market for CMS currently is air separation. Additional details of these materials and their history are available in a review by Jüntgen [39].

Zeolites These materials are a class of porous crystalline aluminosilicates comprised of periodic arrays of Si- or Al-oxide tetrahedra, and have been traditionally used as molecular sieves for gas separation applications [40]. The substitution of silicon atoms within the zeolite framework by aluminum atoms creates *anions* of aluminum at these sites that are counterbalanced by exchangeable anions, leading to a polar structure that preferentially adsorbs polar (dipolar, quadrupolar, etc.) gas molecules. As previously discussed, water has a large permanent dipole moment. Although CO_2 has no permanent dipole, it has a strong quadrupole moment. Since, the dipole interaction is stronger than the quadrupole interaction, water is adsorbed preferentially over CO_2. Many zeolites can also be produced in pure silica form, the most common example being "silicalite," which is the pure silica analog of ZSM-5. Such materials have very little affinity to polar molecules and are therefore classed as "hydrophobic." However, this lack of polarity means that their affinity to CO_2

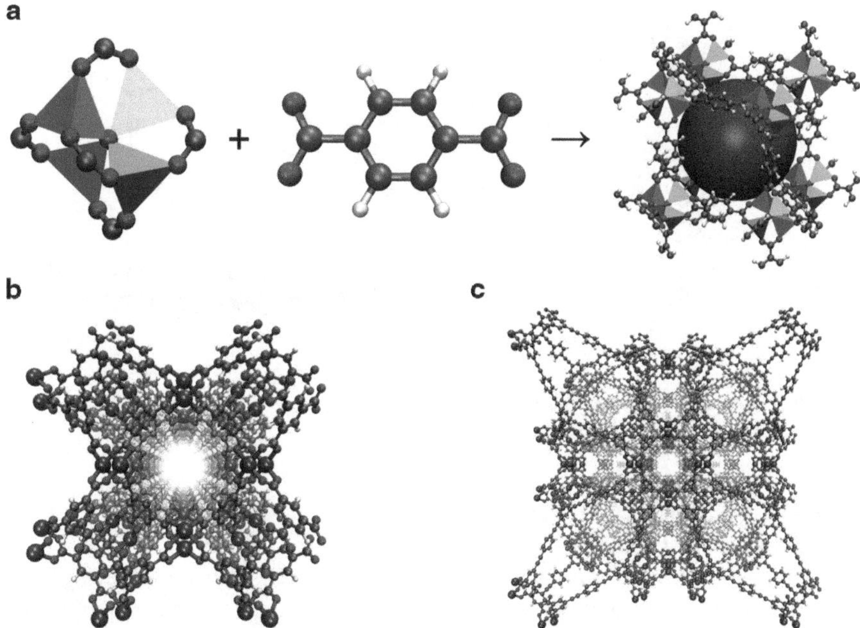

Fig. 4.8 **a** Construction of the MOF-5 framework. *Left*, the $Zn_4(O)O_{12}C_6$ cluster with the ZnO_4 tetrahedral indicated in cyan (O, *red*; C, *grey*); *Middle*, organic linker (benzenedicarboxylic acid); *Right*, one of the cavities in the $Zn_4(O)(BDC)_3$, MOF-5, framework. Eight clusters constitute a unit cell and enclose a large cavity, indicated by a purple sphere of diameter 18.5 Å. **b** $[Cu_3(TMA)_2(H_2O)_3]_n$ polymer framework viewed down the [100] direction, showing nanochannels with fourfold symmetry (O, *red*; C, *grey*, Cu, *brown*). **c** Structural features of NU-100 MOFs (O, *red*; C, *grey*, Cu, *brown*). (Courtesy of [9])

is also greatly reduced. Water competition is a major challenge associated with the application of zeolites for CO_2 capture.

Metal Organic Frameworks Metal-containing vertices and organic-based linkers comprise the framework of MOFs as illustrated in Fig. 4.8 [41]. These sorbents are designer materials with controlled and narrow pore size distributions in the molecular-size region. The choice of metal in addition to the chemical composition and length of linker makes these sorbents tunable and thereby potentially selective toward CO_2 capture. The mechanism by which MOFs separate CO_2 from gas mixtures is similar to that of zeolites. For instance, polar functionality can be added to the linker groups by including polar functional groups such as $-NO_2$ (*i.e.*, ZIF-78) and $-CN$ (*i.e.*, ZIF-82) [42] that can interact with the strong permanent quadrupole moment of CO_2. In particular, MOFs containing these polar functional groups showed higher CO_2/CH_4, CO_2/N_2, and CO_2/O_2 selectivities compared to others since the $-NO_2$ and $-CN$ groups have greater permanent dipole moments than other functional groups. Many reviews are available on these emerging designer materials [43]. Common MOF densities and surface areas can range from 0.2 to 1.0 g/cm^3 and 500 to approximately 6200 m^2/g, respectively [14].

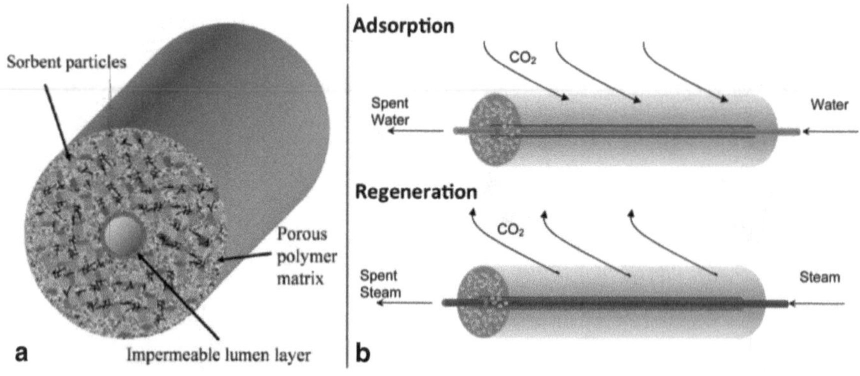

Fig. 4.9 **a** Schematic of hollow-fiber sorbent and **b** adsorption and regeneration modes of hollow-fiber sorbent. (Reprinted with permission from [8])

Hollow-Fiber Sorbents These materials are a unique class developed by Jones and Koros [44] that combine molecular- and process-level requirements for optimal CO_2 capture. The motivation behind these materials is based upon achieving rapid temperature-swing adsorption with low pressure drop and thermal requirements. These are hybrid materials consisting of polymer-anchored sorbents within a porous polymeric hollow fiber matrix. The inner tube within the hollow fiber structure is comprised of impermeable polyvinylidene chloride that allows for the internal transport of cooling water, to achieve rapid heat dissipation and hence maximum CO_2 loading. The inner tube also allows for the transport of steam to achieve rapid heating during regeneration. A schematic of the hollow fiber sorbent framework along with an illustration of the adsorption and regeneration processes are shown in Figs. 4.9a and 4.9b, respectively. Additionally, the porosity of the polymer matrix is similar to that of a membrane with a porosity gradient decreasing from the outer porous membrane wall to the inner impermeable polymer-based layer, which aids in enhancing uptake of gas into pores of the polymer matrix.

Example 4.2 Using data from Table 4.3, determine: (a) the volume fraction of pores in zeolite 5A filled with adsorbed CO_2 when its partial pressure is 250 mmHg at 25°C and (b) whether the amount of CO_2 adsorbed is equivalent to more than a monolayer. At these conditions, the partial pressure is considerably below the CO_2 vapor pressure of (48011 mmHg). The area of an adsorbed CO_2 molecule is given by Eq. (4.16).

Solution: (a) Take 1 kg of zeolite 5A as a basis and the sorbent particle density, $\rho_p = 700\ \text{kg/m}^3$ (zeolite 5A). The pore volume can be calculated using the relation $V_p = \varepsilon/\rho_p$, such that ε is the *void fraction*, and the data in Table 4.3 as.

$$V_p = 0.4/(700\ \text{kg/m}^3) = 5.7 \times 10^{-4}\ \text{m}^3/\text{kg}$$

For 1 kg, the pore volume is 5.7×10^{-4} m^3. Based on the capacity value in Table 4.3, the mass of adsorbed CO_2 is 0.15 kg.

Assuming the molar volume of CO_2 in the zeolite packing space is 6.28×10^{-5} m^3/mol, the volume of adsorbed CO_2 is:

$$\frac{(6.28 \times 10^{-5}\ \mathrm{m^3/mol})(0.15\ \mathrm{kg})}{0.044\ \mathrm{kg/mol}} = 2.42 \times 10^{-4}\ \mathrm{m^3}$$

and the fraction of pores filled with CO_2 is $2.14 \times 10^{-4}/5.7 \times 10^{-4} = 0.375$.

(b) From Table 4.3, the surface area of 1 kg zeolite 5A is 650×10^3 m^2. From Eq. (4.16):

$$A = 1.091\left(\frac{6.28 \times 10^{-5}}{6.023 \times 10^{23}}\right)^{2/3} = 2.42 \times 10^{-19}\ \mathrm{m^2/molecule}$$

The number of CO_2 molecules adsorbed is:

$$\frac{(0.15\ \mathrm{kg})(6.023 \times 10^{23}\ \mathrm{molecule/mol})}{0.04401(\mathrm{kg/mol})} = 2.05 \times 10^{24}\ \mathrm{molecules}$$

The number of CO_2 molecules in a monolayer of 650×10^3 m^2 of surface area is:

$$650 \times 10^3/2.42 \times 10^{-19} = 2.69 \times 10^{24}\ \mathrm{molecules}$$

In the 5A zeolite, $2.05 \times 10^{24}/2.69 \times 10^{24} = 0.76$ monolayers are adsorbed within a given pore.

4.2.1 Water Adsorption

Pore filling with water is dependent upon *relative humidity*, and may substantially influence the mass-transfer and subsequent adsorption of CO_2 in micro and meso-porous materials. The relative humidity, $\mathcal{H}_R$, is defined as the ratio of the partial pressure of water vapor to the vapor pressure of water at a given temperature and is usually described as a percentage. The total extent of water adsorption is dependent on the volume of smaller pores rather than simply the surface area. In polar molecular sieves, water is held strongly with its adsorption nearly irreversible. Traditionally, the amount of water adsorbed at a given partial pressure decreases significantly with increasing temperature. In the case of molecular sieves this is not always true since the water molecules are bound so tightly on the surface, trapped by the wall-wall potential forces of the micropores. In the case of slightly larger pore systems, that is with pore sizes large enough to hold multiple molecular water layers, capillary effects will dominate allowing the water molecules located away from the walls to essentially evaporate out as temperature increases. For air with 1% water vapor at 20°C, $\mathcal{H}_R$ is equal to 7.6 mm Hg/17.52 mm Hg × 100 = 43.4%. At this relative humidity, silica gel adsorbs 0.26 kg/kg of sorbent. For the same concentration (or partial pressure of water) at 40°C, $\mathcal{H}_R$ is equal to 7.6 mm Hg/55.28 mmHg × 100 = 13.7%, and silica gel adsorbs significantly less, *i.e.*, 0.082 kg/kg sorbent.

The effects of water on CO_2 adsorption require careful attention. In some sorbents, the presence of water might enhance CO_2 adsorption, while in others inhibition exists due to direct competition for active adsorption sites. Whether water will assist in adsorption or limit it, depends on the mechanism of adsorption, *i.e.*, whether CO_2 is binding via an ionic interaction or covalently through a carbamate or carbonate bond. As previously discussed, in the case of charged sorbents, such as MOFs, zeolites, and tertiary-amine -functionalized sorbents, CO_2 will interact with active sites by physical means dependent upon van der Waals and electrostatic interactions due to its permanent quadrupole moment. If water is present it will interact preferentially with the cations of these systems due to water's enhanced electrostatic contribution, *i.e.*, its permanent dipole moment. In the case of tertiary-functionalized systems, a proton from water will interact to form positively charged nitrogen atoms, which will interact with CO_2 via a zwitterion mechanism. Additionally, this will result in an increase in hydroxyl concentration leading to the potential formation of bicarbonate. In this case, the sorption behavior is comprised of two unique contributions. The mechanism by which CO_2 adsorbs in the case of amine-functionalized sorbents is quite similar to the mechanism of binding in amine-based solvents, as described in detail in Sect. 3.3 of Chap. 3. In the case of zeolites, in which cations are present and preferentially adsorbing water, the interaction can lead to a locally ordered hydrogen-bonding network, which may allow for oxygen atoms of some of the outer molecules of the hydration shell to be available for carbonate formation with CO_2. However, in general the adsorption capacities of CO_2 on charged systems such as zeolites and MOFs are limited by the presence of water due to direct competition of the cation adsorption sites.

4.2.2 Pore Characterization

Sorbent characterization is an important aspect when considering the design and tuning of new materials for CO_2 capture applications. Different assumptions and approaches are used to determine the surface area of micro and mesoporous materials. Characterization experiments can provide information such as surface chemistry, surface area, porosity, pore connectivity, and pore size distribution. In the previous Langmuir and BET models for meso and macropore characterization, surface layering is assumed, and with the kinetic theory of gases serving as a foundation, a dynamic equilibrium is expected to exist between the fluid particles condensed on the surface and those present in the gas phase. Mesopores, in particular, fill via pore condensation representative of a gas-liquid phase transition, while micropores reflect a continuous filling process due to the overlapping potential fields of the pore walls.

The micropore range can be further divided into sub catagories, *e.g.*, pores smaller than 7 Å are known as *ultramicropores* and pores between 7 and 20 Å are known as *supermicropores*. The properties (*e.g.*, density and packing) of the condensed surface phase in micropores are assumed to be similar to those of the fluid particles of the bulk liquid phase. At temperatures less than the fluid's *critical temperature*, T_c, (*e.g.*, $T \leq 0.8T_c$), it is assumed that the pore volume is strictly a liquid with an approximate volume of $V = W/\rho$, such that W and ρ are the fluid weight and density, respectively. The energy, E, associated with micropore filling is equivalent to the isothermal work required to compress a gas from its equilibrium pressure, p, to its vapor pressure, p_0 in the micropore, *e.g.*, $E = RT \ln p_0/p$. Details of the derivation are omitted here, but these two relations can be combined to arrive at the following Dubinin-Radushkevich [45] ("DR") equation:

$$\ln W = \ln (V\rho) - k\left[\ln\left(\frac{p_0}{p}\right)\right]^2 \tag{4.25}$$

such that,

$$k = 2.303K\left(\frac{RT}{\beta}\right)^2 \tag{4.26}$$

where V is the total micropore volume, K is a constant that can be determined by the pore size distribution, and β is the energy ratio with the reference corresponding to the vapor-phase value of a single adsorbate and is often taken as unity. If the pore surface is heterogeneous in terms of chemical functionality or roughness, this model may fail and the use of a more generalized form, *i.e.*, the Dubinin-Astakhov [46] ("DA") model is recommended. The equation for the DA model is identical to Eq. (4.25) except that the exponent "2" is replaced by an additional variable, n. This model was further improved to include adsorbate-adsorbate surface interactions and their impact on pore filling and adsorption isotherm shape, and is known as the Horvath-Kawazoe [47] model. Additional details associated with adsorption mechanisms of micropores are available in Lowell et al. [25].

If the fluid weight of the micropore filling is expressed as a function of pore radii, r, then the DA equation can be written as [11]:

$$\theta = \exp\left[-\left(\frac{k}{E}\right)^n r^{-3n}\right] \tag{4.27}$$

where θ represents the extent of micropore filling, n is the fitting parameter in the DA equation determined by linear regression of ln (W) versus $\left[\ln\left(\frac{p_0}{p}\right)\right]^n$, and k is the interaction constant of fluid molecule interactions in units of J Å^3. The DR and DA equations may be used to determine pore size distributions in microporous material, but do have the limitation in that this approach will not capture bi- or multiple-modal pore size distributions. An excellent review is available that discusses the use of physical adsorption as a method to characterize nanoporous materials [48]. Additional approaches exist, such as nonlocal density functional theory, that allow

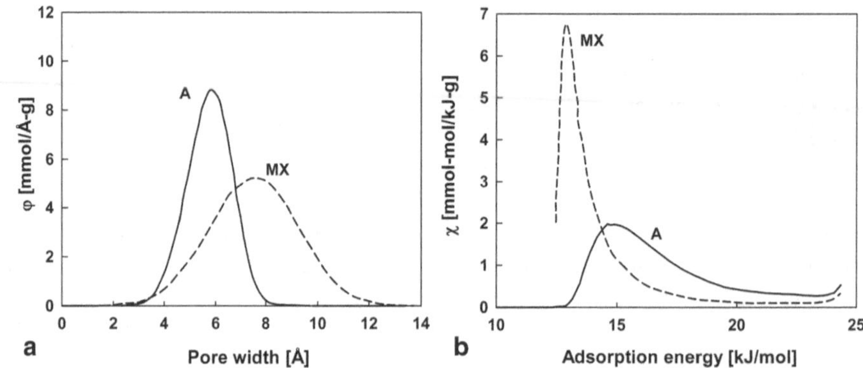

Fig. 4.10 **a** Micropore size distribution CMS-A and a superactivated MX and **b** corresponding adsorption energy distribution. (Copyright with permission from Elsevier [4])

for pore size determinination in general nanoporous material that contain both micro and meso-scale regimes.

Pore filling for characterization of microporous materials such as zeolites, CMS, and MOFs is difficult since it requires relative pressures of 10^{-7} to 10^{-5} to fill pores with diameters between 0.5 and 1 nm. Rates of diffusion and adsorption equilibration into pores at these relative pressures can also be very slow. At approximately 87 K, argon fills micropores of these diameters at a higher relative pressure compared to nitrogen, *i.e.*, $10^{-5} < p/p_0 < 10^{-3}$, which leads to faster diffusion and equilibration times, and an accurate pore size analysis can therefore be achieved over the complete micro- and mesopore size ranges [25]. Challenges with these low-pressure experiments include the need for a turbomolecular pump and potentially restricted diffusion into ultramicropores. The use of CO_2 as a probe molecule has been shown to overcome these challenges [49]. The saturation pressure of CO_2 at 273 K is approximately 26,400 Torr and to achieve low relative pressures (down to $p/p_0 = 10^{-7}$) a turbomolecular pump is not necessary. At the relatively high temperature and pressure conditions possible with CO_2, diffusion challenges are minimal and equilibration can occur much faster. On average, a micropore analysis with nitrogen at 77 K can take up to 24 h, while the same analysis with CO_2 at 273 K takes an average of 5 h. For micropore volumes with limited access to CO_2 at room temperature, helium adsorption at approximately 5 K may be used [50]; however, a cryostat, turbomolecular pump, and low-pressure transducers are required.

It is important to discuss briefly the relationship between adsorption energy and the structural heterogeneity of sorbent materials, which will have implications associated with sorbent regeneration. Figure 4.10 shows the micropore size distribution of two carbons, *i.e.*, CMS-A with a BET surface area of approximately 1100 m^2/g and Maxsorb (MX) superactivated carbon (Kansai Coke & Chemicals) with a BET surface area of approximately 2200 m^2/g. As expected and can be seen from Fig. 4.10a, the pore size distribution in the activated carbon is broader than that of the CMS. From Fig. 4.10b, a distribution in adsorption energy is evident. Recalling the energy contributions associated with adsorption, the smaller pores will have wall-wall interactions that will influence the extent of adsorption. Specifically, the sum of the interactions energies equates to the adsorption energy, and since the interaction potential includes the wall-wall interaction in smaller pores, the adsorption energy will be influenced by the pore size. In this case, the DR equation with the assumption of a Gaussian pore size distribution was used to fit the adsorption isotherms to determine the micropore size distribution. The adsorption energies associated with CMS are higher and with a broader tail than for activated carbon, which is consistent with the enhanced interactions in smaller pores.

Example 4.3 Given the following parameters associated with low-pressure CO_2 adsorption on a carbon nanotube (CNT)-based sorbent, determine: (a) the specific pore volume V(cm^3(STP)/g) of the CNT sorbent and (b) with the assumption of a Gaussian distribution, determine the micropore distribution of this CNT sorbent.

Parameters	Value
Temperature, T (K)	273
CNTs weight, m (g)	0.1596
CO_2 density @ STP, ρ (kg/m^3)	0.001839
CO_2 liquid density @ 273K, ρ (kg/m^3)	1.023
Interaction constant of CO_2, k (J · Å^3)	3145000
Affinity coefficient of CO_2, β (dimensionless)	0.3510

Relative Pressure (p/p_0)	Quantity Adsorbed (cm^3/g STP)	Relative Pressure (p/p_0)	Quantity Adsorbed (cm^3/g STP)	Relative Pressure (p/p_0)	Quantity Adsorbed (cm^3/g STP)
8.977E-05	0.4008	0.001430	4.120	0.008156	16.31
0.0001859	0.7564	0.001557	4.425	0.009099	17.77
0.0002731	0.9986	0.001726	4.835	0.01004	19.05
0.000356	1.300	0.001904	5.197	0.01102	20.44
0.0003807	1.361	0.002086	5.607	0.01214	21.93
0.0004031	1.446	0.002310	6.122	0.01335	23.53
0.0004483	1.538	0.002546	6.616	0.01459	25.11
0.0004823	1.661	0.002795	7.160	0.01618	27.07
0.0005426	1.821	0.003089	7.765	0.01784	29.00
0.0005887	1.986	0.003399	8.351	0.01961	31.10
0.0006605	2.162	0.003742	9.033	0.02167	33.41
0.0007145	2.332	0.004133	9.735	0.02383	35.80
0.0008011	2.516	0.004547	10.46	0.02624	38.31
0.0008657	2.736	0.005008	11.31	0.02889	41.08
0.0009722	2.961	0.005528	12.18	0.03208	44.21
0.001051	3.186	0.006077	13.13	0.03506	47.22
0.001173	3.476	0.006717	14.13		
0.001282	3.796	0.007377	15.19		

Solution: (a) Optimize the value n in the Dubinin-Astakhov equation from a fit of the low-pressure adsorption isotherm using a linear regression.

$$\ln(W) = \ln(V\rho) - K\left[\ln\left(\frac{p_0}{p}\right)\right]^n$$

such that n is 1.2356, and the square of the sample correlation coefficient, R^2, is 0.9999. Plotting $\ln(W)$ as a function of $\left[\ln\left(\frac{p_0}{p}\right)\right]^n$ yields:

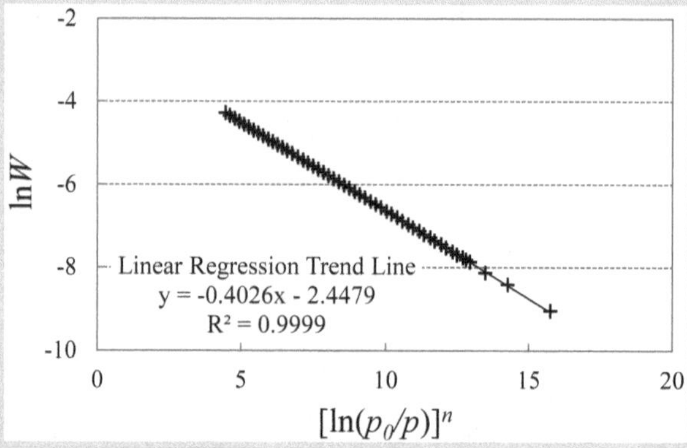

The slope of the line, $-K = -0.4026$, and the y-intercept of the line, $\ln(V\rho) = -2.4479$. Therefore, the pore volume V is:

$$V = \frac{e^{-2.447}}{\rho} = \frac{e^{-2.447}}{1.023} = 0.085 \text{ cm}^3$$

such that ρ is the adsorbed liquid density of CO_2 at 273 K. Then, the specific pore volume (*i.e.*, per mass basis) of this CNT sample is:

$$\frac{V}{m} = \frac{0.085 \text{ cm}^3}{0.1596 \text{ g}} = 0.5296 \text{ cm}^3/\text{g}$$

(b) Based upon a Gaussian distribution assumption, and assuming that no perturbation takes place in the adsorbed phase, from Eq. (4.27):

$$\frac{dV}{V dr} = \frac{d\theta}{dr} = \frac{d\theta}{d(\frac{d}{2})} = 3n\left(\frac{k}{E}\right)^n r^{-(3n+1)} \exp\left[-\left(\frac{k}{E}\right)^n r^{-3n}\right]$$

Such that θ represents the extent of micropore filling. The quantities V, E, and n have been determined from the linear regression in part (a), such that n is 1.235 and

$$E = RTK^{-\frac{1}{n}} = \left(8.315 \frac{\text{J}}{\text{mol K}}\right)(273 \text{ K})\left(0.4026^{-\frac{1}{1.235}}\right) = 6986 \text{ J/mol}$$

Such that K is the negative slope of ln (W) versus $\left[\ln\left(\frac{p_0}{p}\right)\right]^n$, as shown in the solution of part (a) and the resulting pore size distribution is:

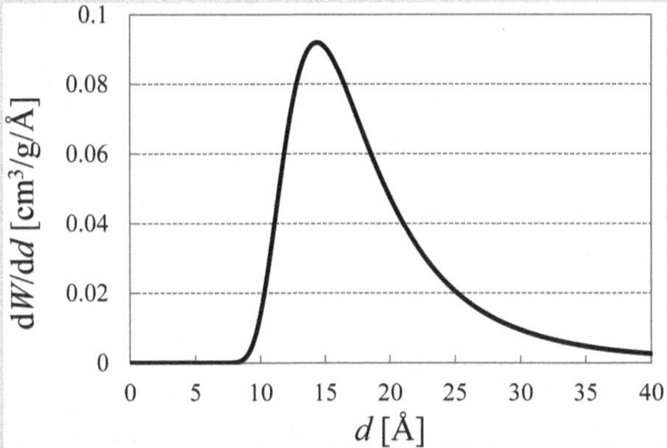

The details of the micropore size distribution approach can be found in the previous work by Medek et al. [11]. A limitation with this approach is that only a single average pore size may be determined, indicating the need for a separate model approach for the analysis of sorption data for systems with bi- or tri-modal pore size distributions.

4.2.3 *Pelletization*

Pelletization is a crucial process to the synthesis of sorbents for commercial application since sorbent particles with a reasonable porosity are desired to obtain high volumetric capacities. Additionally, aspects of the pelletization process, such as the choice of binder and pore structure, can influence the mechanical strength of the sorbent. Mechanical strength is in particular important for achieving a high number of sorbent cycles, and for particle fluidization processes such as fluidized bed configurations for adsorption and chemical looping combustion [51].

Although sorbents are comprised primarily of micro and mesoporous channels, to avoid excessive pressure drop through the adsorbent bed (and consequent excessive power requirements), it is crucial that the sorbent particles themselves are not too small. Figure 4.11 shows examples of industrial-based sorbents, which are pelletized composites of smaller microparticles held together with a binder, creating a system of macropores of similar diameter to the particles themselves. Within each of the microparticles, the micro and/or mesopore structure exists containing the majority of the surface area and corresponding active sites for adsorption. In Fig. 4.11a the microparticles are comprised of an ordered and homogeneous pore network that

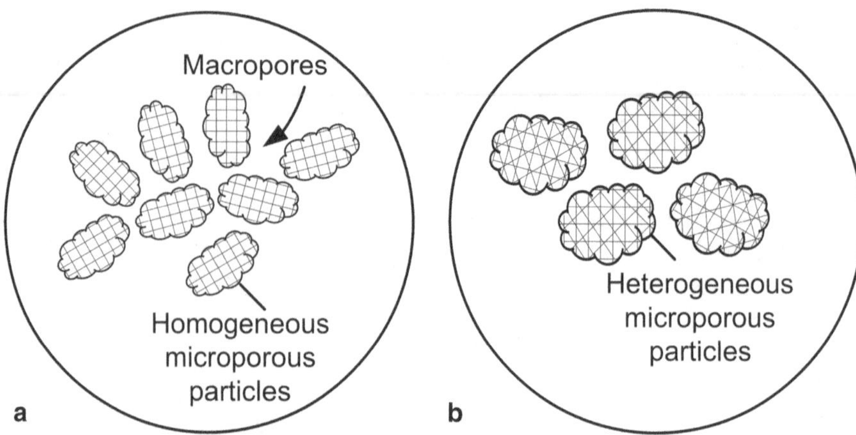

Fig. 4.11 Examples of pelletized industrial sorbents with multiple-modal pore size distributions comprised of particles with **a** homogeneous micro or mesopores with narrow pore size distribution such as zeolites, CMS, or MOFs and **b** heterogeneous micro and mesopores with wide pore size distribution such as activated carbon

includes pores of a narrow size distribution, which is found in ordered crystalline materials such as zeolites, MOFs, and CMS. In Fig. 4.11b the microparticles are comprised of heterogeneous pore networks, in which the pore size distribution is broad, as is generally the case for activated carbons and other amorphous sorbents.

4.3 Adsorption Kinetics

4.3.1 Diffusion in Pores

Capturing CO_2 from a gas mixture via an adsorption process requires very high surface areas for optimal capacity and activity, which further requires the sorbent materials to have very fine pores in the micro and/or mesoporous range. Transport through these pores occurs through various diffusion mechanisms depending on the pore size, and understanding these mechanisms is essential for the effective design of sorbents for CO_2 capture since transport rates define limits on the cycle times that may be possible. This has also been shown in previous experiments of Yue et al. [52] and Song et al. [53] regarding amine-functionalized silica-based and "molecular basket" sorbents, respectively. Yue et al. used thermogravimetric analysis and found sorbent capacity to increase from 308 to 373 K. Typically, as temperature increases, due to the exothermicity associated with adsorption, the sorbent capacity decreases. In this system, due to diffusion limitations in the smaller pores, an increase in temperature results in an increase in the diffusion rate, leading to enhanced capacity at higher temperature. Song et al., also using thermogravimetric analysis, found sorbent capacity to increase from 323 to 348 K, where it exhibited a maximum, at

373 K, and subsequent decrease thereafter. When pore size becomes restrictive, as it is for some microporous sorbents, and certainly for amine-functionalized silica-based sorbents, the pore openings become restricted, with diffusional limitations becoming more important than thermodynamic limitations.

The division of micro, meso, and macropores is primarily based upon the differing forces that control the mechanism of adsorption in each of the pore size ranges. Surface forces are dominant in the case of micropores in which the fluid particle never escapes the surface force field even at the pore center. Capillary forces become dominant in mesopores, while macropores barely contribute to the adsorption capacity, but play a central role in bulk fluid transport. *Steric hindrance* plays an important role in micropores due to the relative size of the molecule in comparison to the micropore itself and can result in diffusion becoming an activated process, in which the fluid diffuses across the surface by hopping between adjacent "potential-well" sites located on the surface with an activation barrier associated with each hop. Within the mesopores, Knudsen diffusion may play a dominant role, in which fluid-surface interactions occur more frequently than fluid-fluid interactions, although, in this size regime surface diffusion and capillary effects may also influence the transport behavior. The role of the pore surface is relatively minor in larger pores, so macropore diffusion occurs primarily through bulk or molecular diffusion mechanisms since interactions between fluid particles occur far more frequently than fluid-surface interactions.

Considering straight cylindrical pores can provide insight into the various diffusion mechanisms taking place; however, it is important to keep in mind that a real system is comprised mainly of a random three-dimensional network of interconnected pores of varying shapes, diameters, and orientations with the exception of microcrystalline ordered systems such as MOFs and zeolites. As expected, transport in micropores is dominated by surface diffusion and heavily influenced by steric hindrance of the fluid particles within the pore. The primary diffusion mechanisms involved in meso and macropores are *molecular diffusion*, *Knudsen diffusion*, and *Poiseuille flow*. When the diameter of the pore is greater than the *mean free path* of the fluid particles, the diffusing molecules will interact with each other more than with the pore walls, thereby minimizing the wall effects on the transport. In this case, the pore diffusivity for a straight cylindrical pore is essentially the same as the molecular diffusivity, D_m. In mesopores, the mean free path becomes comparable to the pore diameter, leading to an increase in collisions between the fluid particles and the wall. When a fluid particle strikes the wall, there is an energy exchange between the surface atoms or molecules and those comprising the fluid particle, with the result that the fluid particle rebounds from the surface in a purely random direction. In the case of Knudsen diffusion, the molecular diffusion component is negligible since the momentum transfer between the fluid particle and the wall greatly exceeds the momentum transfer between diffusing fluid particles. Knudsen diffusion, D_K of a fluid particle with molecular weight, M in units of g/mol, within a pore of radius, r, at a temperature, T is approximated by:

$$D_K = 9700r\sqrt{\frac{T}{M}} \tag{4.28}$$

Table 4.5 Relative importance of molecular and Knudsen diffusion, and Poiseuille flow for air at 20°C in a straight cylindrical pore [57]

p (atm)	D_m (cm²/s)	r (cm)	D_K (cm²/s)	D (cm²/s)	D_{Pois} (cm²/s)	D_{total} (cm²/s)	$\frac{D_{Pois}}{D_{total}}$
1.0	0.2	10^{-6}	0.03	0.027	0.0007	0.027	0.026
		10^{-5}	0.3	0.121	0.07	0.19	0.37
		10^{-4}	3.0	0.19	7.0	7.2	0.97
10	0.02	10^{-6}	0.03	0.012	0.007	0.019	0.37
		10^{-5}	0.3	0.019	0.7	0.719	0.97
		10^{-4}	3.0	0.020	70	70	1.0

such that r has units of cm, T is in K, and D_K has units of cm²/s. It is evident that Knudsen diffusion is independent of pressure (since the mechanism is not dependent upon molecular collisions) and is only weakly dependent upon temperature.

Consider the diffusion of a binary gas mixture (A and B) in the transition region of the mesopore range, in which momentum transfer between fluid particles and the wall both play a role. The following expression [54] provides a relationship between the overall diffusivity D and the individual contributions from molecular and Knudsen diffusion:

$$\frac{1}{D} = \frac{1}{D_K} + \frac{1}{D_{AB}}[1 - y_A(1 + N_B/N_A)] \tag{4.29}$$

such that D_{AB} is the diffusion of fluid A in fluid B, y_A is the mole fraction of fluid A, and N_B and N_A are the diffusive flux of fluid A and B, respectively. In the dilute case, *i.e.*, when y_A is small, this expression reduces to:

$$\frac{1}{D} = \frac{1}{D_K} + \frac{1}{D_{AB}} \tag{4.30}$$

If there exists a pressure gradient exists across the cylindrical pore there will be laminar (bulk) flow, *i.e.*, Poiseuille flow, across the pore represented by:

$$D_{Pois} = \frac{pr^2}{8\mu} \tag{4.31}$$

such that p is the pressure and μ is the fluid viscosity. Modeling the diffusion effects collectively is difficult and as such, Maxwell [55] developed a simplified model termed the "dusty gas model," in which the fluid particles interact with the pore wall, represented by "dust." Details of this model are available in the literature [56], but the model indicates that each of the transport contributions can be added to represent the total diffusion. Order of magnitude estimations have been calculated for air at ambient conditions to provide a sense of the relative contributions of each transport mechanism as a function of pore size. As seen from Table 4.5, flow in the Poiseuille regime becomes important in larger pores; however, it is important to note that this contribution becomes more significant at higher pressures, *i.e.*, at 10 atm this phenomena accounts for 37% of the flow in a 20-nm pore.

Capillary condensation can also affect diffusion in mesopores. Through the effect of surface tension the equilibrium vapor pressure of a fluid in a pore is lower than in the bulk phase resulting in condensation within the pore (capillary). The vapor pressure of the fluid in a confined pore is well below the vapor pressure of the free liquid in its bulk phase. One may think that when capillary condensation occurs within the pores of a three-dimensional pore network that diffusion would be substantially reduced. In fact, the opposite has been known to occur in systems exhibiting a wide pore size distribution. A qualitative explanation is that through capillary condensation in pores of a certain diameter, diffusion takes place through other potentially shorter channels, effectively lowering the *tortuosity* of the system. This counterintuitive result was first observed by Carman and Haul [58], with a detailed analysis provided by Weisz [59].

Tortuosity Although a great deal regarding adsorption phenomena can be understood by focusing on single-pore systems, it is crucial to investigate pore networks and connectivity in three dimensions to ultimately capture the interplay between adsorption and transport phenomena. Tortuosity is defined as the deviation from the linear path of a particle moving from point A to point B in a three-dimensional pore system. In modeling the diffusion of fluids in porous solids, the bulk fluid diffusivity, D_b differs from the effective, or measured diffusivity, D_e, depending upon the extent of tortuosity in the structure of the solid as,

$$D_e = D_b \cdot \varepsilon / \tau \tag{4.32}$$

such that ε is the void fraction (porosity) and τ is the tortuosity factor of the solid. The tortuosity factor combines all deviations of straight-line paths into a single dimensionless parameter, L_{eff}/L, such that L_{eff} is the traveled distance through the porous solid and L is the simple straight-line path through the solid. Typical tortuosity factors range from 1 for straight non-intersecting pores to values up to 5 for complex or low-porosity solids.

4.3.2 Mass Transfer

The selective transport of CO_2 from a gas mixture to the active binding site on the surface of a porous particle takes place through a series of mass-transfer processes as shown in Fig. 4.12, beginning with external film transport. In total, there are three mass-transfer regimes, and hence, three unique mass-transfer resistance contributions. These include mass-transfer resistance (1) in the liquid film surrounding the pellet, (2) in the macropores, which are formed from the interparticle spacing of the bound microparticles (see Sect. 4.2.3 on pelletization) that comprise the pellet, and (3) in the micropores of the microparticles. Mass-transfer within the liquid film surrounding the pellet is termed external mass-transfer, with the mass-transfer within the micro and macropore networks are termed internal. The pore size distribution of the microparticles may include mesopores as well, but micropores are the focus for simplicity.

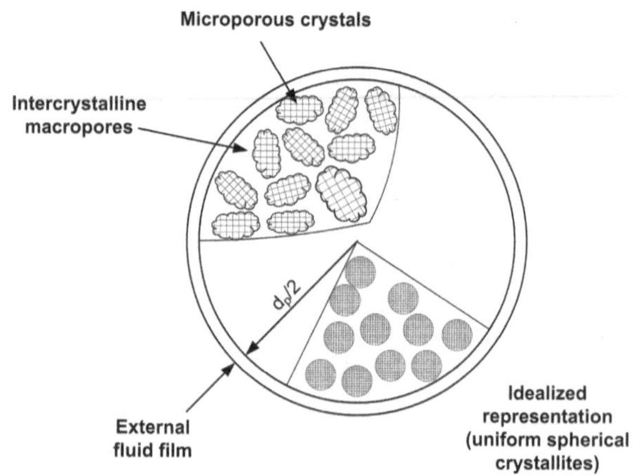

Fig. 4.12 Schematic of sorbent pellet displaying the three primary resistances to mass transfer. (Reprinted with permission of John Wiley & Sons, Inc. [3])

External Mass Transfer Whether capture of CO_2 is taking place from a flue gas or directly from the air, water vapor will be present to some extent. Therefore, the surface of a sorbent particle will always be surrounded by a fluid film, through which transport occurs via molecular diffusion. The extent of diffusion resistance depends upon the thickness of the boundary layer of the fluid film, which in turn depends upon the hydrodynamic conditions of the system. The external resistance may be correlated in terms of a mass-transfer coefficient, k_f with the molar flux of CO_2, defined as:

$$J_{f,\,CO_2} = D_f \frac{(c_{i,CO_2} - C_{S,CO_2})}{\delta} = K_f(C_{i,CO_2} - C_{S,CO_2}) \tag{4.33}$$

such that J_{f,CO_2} is the molar flux of CO_2 at the interface of the gas and fluid film, D_f is the diffusivity of CO_2 in the film, δ is the film thickness, c_{i,CO_2} is the concentration of CO_2 at the gas-film interface and c_{S,CO_2} is the concentration of CO_2 at the pellet surface, and k_f is the mass-transfer coefficient of the film. The concentration of CO_2 throughout the film is not necessarily constant.

Since it is difficult to assess the film thickness, correlations between the hydrodynamic conditions and particle size of the sorbent material with the mass-transfer coefficient are useful. For the ideal case of a unique isolated spherical sorbent particle in a stagnant fluid, it has been shown that the Sherwood number (Sh) is equivalent to:

$$Sh = \frac{k_f d_p}{D} \approx 2.0 \tag{4.34}$$

such that d_p is the particle diameter and D is the overall diffusivity. Under conditions of flow, which would most often be the case in a CO_2 separation process, Sh may be much greater than 2.0. The hydrodynamic conditions can be characterized using the Schmidt (Sc) and Reynolds (Re) numbers, as previously defined in Sect. 3.4.2 of Chap. 3. For a variety of well-defined fluid-solid contacting scenarios, the following correlation has been developed for flow through a packed-bed:

$$Sh = 2.0 + 1.1 Re^{0.6} Sc^{1/3} \tag{4.35}$$

This relationship is the standard mass-transfer correlation for packed beds with *Re* between 3 and 10^4 [60].

Internal Mass Transfer In addition to external mass-transfer within the film surrounding the sorbent pellet, internal mass-transfer can take place within the micro and/or macropore networks. If the mass-transfer resistance in the micropores is dominant, the CO_2 concentration throughout the pellet will be essentially uniform and the rate of adsorption is independent of the size of the particles comprising the pellets. However, if the mass-transfer resistance is dominant in the macropores that comprise the void space between the particles, the concentration of CO_2 will be uniform through the microparticle (*i.e.*, within the micropore network of the microparticle), but nonuniform throughout the pellet, with the rate of adsorption is dependent upon the particle size [16b].

Consider the ideal case of isothermal single-component adsorption with micropore diffusion control. The sorbent uptake for this system can be modeled based upon CO_2 adsorption on a single microparticle (*e.g.*, crystalline zeolite microparticle), since the concentration of CO_2 is considered to be uniform throughout the pellet when the system is modeled as a collection of identical microparticles (as illustrated in the bottom right of the pellet in Fig. 4.12). Assuming a spherical microparticle, the concentration of CO_2 in the adsorbed phase, W, within the microparticle as a function of time t and particle radius r can be described in spherical coordinates as:

$$\frac{\partial W}{\partial t} = D_{p,CO_2}\left(\frac{\partial^2 W}{\partial r^2} + \frac{2}{r}\frac{\partial W}{\partial r}\right) \tag{4.36}$$

such that D_{p,CO_2} is the diffusivity of CO_2 in the micropore network of the microparticle and is considered constant in this simple example. Equation (4.36) may be solved for various initial and boundary conditions to plot the CO_2 uptake as a function of time and microparticle radius.

A similar approach can be used to investigate macropore-controlled mass-transfer. In this case, the concentration of CO_2 is uniform throughout the microparticles, but non-uniform throughout the pellet. The rate of uptake can be determined by assuming equilibrium between the CO_2 in the fluid phase of the macropores and the CO_2 in the adsorbed phase of the macroparticles. Assuming the diffusivity of CO_2 in the macropore, D_p, is independent of the CO_2 concentration, a mass balance can be performed on a spherical shell element of the porous pellet as:

$$(1-\varepsilon)\frac{\partial W}{\partial t} + \varepsilon\frac{\partial c}{\partial t} = \frac{\varepsilon D_p}{2}\left(\frac{\partial^2 c}{\partial {d_p}^2} + \frac{2}{d_p}\frac{\partial c}{\partial d_p}\right) \tag{4.37}$$

such that ε is the porosity of the pellet, c is the concentration of CO_2 in the fluid phase of the macropores, W is the concentration of CO_2 in the adsorbed phase, and $1/2d_p$ is the pellet radius. At equilibrium conditions, W in Eq. (4.37) may be substituted for the appropriate adsorption isotherm relationship.

In 1955, Eugen Glueckauf published his work on the formulae for diffusion into spheres and their application to chromatography [61]. Based upon a linear-driving force approximation, he determined that the rate of concentration change for systems in which the adsorption isotherm is linear or slightly curved and conditions are maintained close to equilibrium, for the case of CO_2 capture may be adequately approximated by:

$$\frac{\partial c}{\partial t} = k(c - c_{S,CO_2}) = \frac{15D_e}{r^2}(c - c_{S,CO_2}) \tag{4.38}$$

such that c is the gas concentration within the porous spheres, c_{S,CO_2} is the concentration of CO_2 in the adsorbed phase, k is the mass-transfer coefficient, D_e is the effective fluid diffusivity, and r is the pore radius [61]. The resemblance to Eq. (4.33) is clear. The general mass-transfer coefficient, k, can be expressed as the product of the mass-transfer coefficient of a pore, k_p, and the ratio of external area to particle volume, which is $3/r$ for a spherical particle. The mass-transfer coefficient, k_p, associated with diffusion into the pores decreases with time as the fluid particles penetrate deeper into the sorbent to reach the adsorption sites. Based upon the mass-transfer coefficient derived from Glueckauf in Eq. (4.38), the following expression may be used as an approximation of the internal mass-transfer coefficient for uptake in spherical particles:

$$k_p \approx \frac{10D_e}{d_p} \tag{4.39}$$

such that D_e is the effective fluid diffusivity, d_p is the is the sorbent particle diameter.

Total Mass Transfer The previous discussion focused on external and internal mass-transfer, which are dependent purely on the particles themselves, and not the bed in which they are housed. Mass-transfer resistance can also take place with transport throughout the bed. A solute material balance [62] for some section length of bed, dL, can be carried out to understand the mass-transfer process through a bed filled with sorbent material as shown in Fig. 4.13, with the difference between the feed and the outlet flows equal to the rate of accumulation in the fluid and sorbent as follows:

$$\varepsilon\frac{\partial c}{\partial t} + (1-\varepsilon)\rho_p\frac{\partial W}{\partial t} = -u_0\frac{\partial c}{\partial L} \tag{4.40}$$

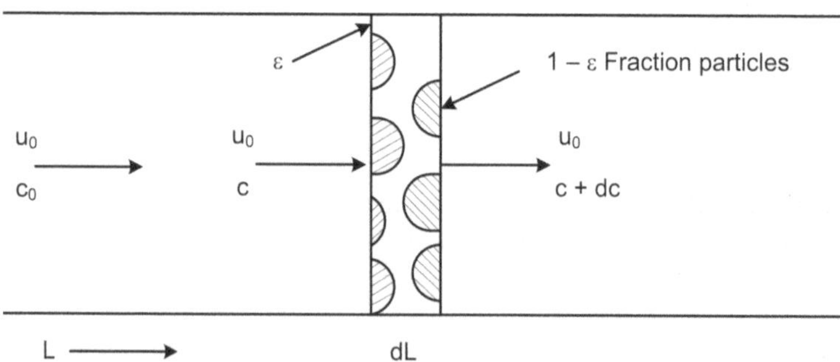

Fig. 4.13 Mass balance on a section, dL of a fixed bed. (With permission from McGraw-Hill Companies, Inc. [2])

such that ε is the external void fraction of the bed, and the solute dissolved in the pore fluid is included with the particle fraction $1-\varepsilon$. Additional terms include the particle density ρ_p, the concentration of CO_2 in the adsorbed phase W, and the superficial velocity, u_0 of the fluid through the packed bed. It is important to note that the term, "$(1-\varepsilon)\rho_p$" is the bed density, ρ_b. For simplicity, the velocity is assumed constant, which may not be valid for a non-dilute system, such as CO_2 in flue gas (*i.e.*, ~12 mol%). In the case of CO_2 capture from a gas mixture, the first accumulation term in Eq. (4.40) is negligible compared to the second term, which represents accumulation of CO_2 within the solid. In this simple case, axial dispersion is neglected, but will be discussed shortly.

As previously described, the mechanism of CO_2 transfer from a gas mixture to a solid sorbent includes diffusion through the fluid film surrounding the pellet (or sorbent) particle and diffusion through the pore network to the surface sites for CO_2 adsorption. Physical adsorption can be thought of as an instantaneous process with equilibrium assumed to exist between the adsorbed phase and fluid phase within the pore. Similar to Eq. (4.38), the mass-transfer process may be approximated by:

$$(1-\varepsilon)\rho_p\frac{\partial W}{\partial t} = ka(c - c_{S,CO_2}) \tag{4.41}$$

such that k is the overall mass-transfer coefficient, a is the external surface area of the sorbent particles, which is $3(1-\varepsilon)/r$ for spherical particles of radius, r, and c_{S,CO_2} is the concentration of CO_2 in the adsorbed phase in equilibrium with the average concentration of CO_2, W, within the solid sorbent.

The overall mass-transfer resistances can be represented by the reciprocal of the total mass-transfer coefficient, K_c, represented as,

$$\frac{1}{K_c} \approx \frac{1}{k_f} + \frac{d_p}{10D_e} \tag{4.42}$$

such that k_f is the resistance due to the fluid film surrounding the pellet, and $d_p/10D_e$ is the internal resistance due to either micro, macro, or some combined pore transport resistance. The relative importance between micro and macropore resistances is dependent on the system and the conditions, but also on the ratio of the diffusion time constants, which is dependent on the square of the sorbent particle radius [16b]. This provides a quick and easy approach to estimating the controlling resistance between these two pore size regimes.

Solutions to the mass balance of Eq. (4.37) exist for a number of different isotherm shapes and controlling steps with a dimensionless time parameter, $\bar{t}$ and a parameter, N, representing the number of mass-transfer units. More specifically, these parameters can be defined as:

$$\bar{t} = \frac{u_0 c_0 (t - L\varepsilon/u_0)}{\rho_p L(1-\varepsilon)(W_{sat} - W_0)} \tag{4.43}$$

$$N \equiv \frac{K_c a L}{u_0} \tag{4.44}$$

such that $\bar{t}$ is the ratio of the time to the ideal time, t^*, with mass transfer absent. The adsorber unit may be operated with complete removal of CO_2 in a time $\bar{t}$ equal to unity and a step-function increase in concentration from zero to c/c_0 equal to unity. The term, $L\varepsilon/u_0$ in Eq. (4.43) is the fluid displacement time from external voids in the packed bed, while the term $\rho_p(1-\varepsilon)$ represents the bed density, ρ_b. For a finite mass-transfer rate, breakthrough will occur at some time $\bar{t}$ less than unity, with the steepness of the breakthrough curve dependent on both the number of transfer units (height of the adsorber) and shape of the equilibrium curve. Breakthrough is discussed in additional detail in Sect. 4.4.2.

4.3.3 Heat-Transfer Effects

In an adsorption process, the solid sorbent is typically held in a fixed bed with the gas stream continuously passing over the bed until saturation is reached, at which point the flow is switched to a second bed, with sorbent regeneration carried out in the saturated bed by desorbing the adsorbed gas. Adsorption and regeneration are traditionally carried out in a cyclic fashion. Similar to absorption, the adsorption process is exothermic and, depending upon the volume of gas adsorbed and its extent of heat release upon adsorption, cooling may be important to consider when designing an adsorption system. The two effects in a nonisothermal system is the influence of temperature on sorbent loading and diffusivity, which can be minimized by reducing the concentration step size over which uptake is measured [16b]. When heat-transfer controls the uptake of the sorbent, the uptake curves show an initially rapid uptake, then a slow approach to equilibrium.

Lee and Luss [63] and Carberry [64] have shown in application to the nonisothermal behavior of catalysts, that the major resistance to heat transfer lies in the external fluid film surrounding the particle, rather than throughout the particle itself. The Biot number (Bi) is a useful dimensionless parameter for characterizing the ratio of internal to external mass- and heat-transfer gradients [16b]. The Biot numbers for mass- and heat-transfer, respectively are defined by:

$$Bi_{\mathrm{m}} = \frac{k_f d_p}{6\varepsilon D_p}, \quad Bi_{\mathrm{h}} = \frac{h d_p}{6\lambda_s} \tag{4.45}$$

such that k_f is the external fluid film mass-transfer coefficient, D_p is the pore diffusivity, d_p is the sorbent particle diameter, h is the overall heat-transfer coefficient between the particle and the fluid, and λ_s is the thermal conductivity of the solid.

The Biot numbers can be expressed in terms of the Sherwood number for mass-transfer and Nusselt number (Nu) for heat transfer as:

$$Bi_{\mathrm{m}} = \frac{Sh}{6}\frac{D_m}{\varepsilon D_p}, \quad Bi_{\mathrm{h}} = \frac{Nu}{6}\frac{\lambda_g}{3\lambda_s} \tag{4.46}$$

such that D_m is the molecular diffusivity and λ_g is the thermal conductivity of the fluid.

Since Sh within the film is on the order of 2.0 and $D_p \leq D_m/\tau$, the minimum value of $\mathrm{Bi_m}$ is expressed as $\tau/3\varepsilon$, which is approximately 3.0 with standard tortuosity and porosity values [16b, 57, 65]. Therefore, the internal concentration gradient is substantially greater than the external gradient. Additional mass-transfer resistance from Knudsen or intercrystalline diffusion will decrease D_p further, implying that the intraparticle resistance is of greater importance than the film resistance. In terms of heat transfer, the relationship between internal and external resistance is the exact opposite. For gas-phase systems, λ_s/λ_g is on the order of 10^2 to10^3, implying that $Bi_\mathrm{h} \ll 1.0$, leading to an external temperature gradient much greater than that within the particle. Hence, the model of an isothermal sorbent particle in which all heat-transfer resistance takes place in the external film and all mass-transfer resistance takes place within the particle itself is a realistic representation.

4.3.4 Axial Dispersion

During fluid transport in a packed bed, axial mixing is possible, but not desired since it reduces the efficiency of the separation process [16b]. Flow through a packed bed with the inclusion of axial dispersion results in the modification of Eq. (4.40) with an additional dispersion term and corresponding dispersion coefficient, D_L:

$$-D_L\frac{\partial^2 c}{\partial L^2} + u_0\frac{\partial c}{\partial L} + \varepsilon\frac{\partial c}{\partial t} + (1-\varepsilon)\rho_p\frac{\partial W}{\partial t} = 0 \tag{4.47}$$

Minimizing the extent of axial dispersion is a primary concern when designing a sorbent separation process. The two major driving forces for axial dispersion are (1) molecular diffusion and (2) turbulent mixing, which takes place when the flows are divided and recombined around sorbent particles. To a first approximation, the two effects are additive [16b], with the dispersion coefficient equal to:

$$D_L = \gamma_1 D_m + \gamma_2 d_p u \tag{4.48}$$

such that γ_1 and γ_2 are constants with typical values of 0.7 and 0.5, respectively, and D_m is the molecular diffusivity, d_p is the sorbent particle diameter, and u is the bulk fluid velocity. Expressed in terms of the axial Peclet number, this becomes:

$$\frac{1}{Pe} = \frac{D_L}{ud_p} \tag{4.49}$$

such that Pe is the axial Peclet number for flow around particles of diameter d_p. Langer et al. investigated the variation of Pe with the particle diameter for flow through packed beds and found that for particles sizes less than 3 mm in diameter, Pe is reduced, thereby enhancing the extent of axial dispersion [66].

4.4 Column Dynamics

4.4.1 Mass-Transfer Zone

In a fixed-bed adsorption process, the CO_2 fluid and solid phase concentrations change with time and bed position in the case of a CO_2-selective adsorption process. An example of a concentration profile is shown in Fig. 4.14a, in which c/c_0 is the ratio of the fluid to feed concentrations. After several minutes, the sorbent material at the beginning of the bed is nearly saturated, with the majority of the mass-transfer taking place farther from the inlet. The region where the majority of the mass-transfer takes place is termed the *mass-transfer zone* (*MTZ*) with typical limits on c/c_0 of 0.95 to 0.05. The curve in Fig. 4.14a is called a *breakthrough curve*, which is used to investigate the dynamic behavior of adsorption processes and also aids in determining when flow should be stopped or diverted to a fresh sorbent bed. At times t_1, t_2, and t_3, the exit concentration is nearly zero CO_2, while at the *breakthrough time* t_b, the relative concentration is dependent upon the given system, but in the example of Fig. 4.14 from McCabe et al. [62]. the relative concentration is approximately 0.1. In this example, Continuing the adsorption process beyond the break point results in a rise in concentration to approximately 0.5 followed by a slower approach to unity. The shape of the adsorption isotherm influences the shape of the breakthrough curve as well. For instance, in the case of a favorable isotherm (concave down), the breakthrough curves tend to exhibit a uniform pattern, becoming "self-sharpening" as they advance down the length of the column. Therefore, for a given mass-transfer mechanism and equilibrium data, a single theoretical breakthrough curve can represent all operating conditions. Conversely, in the case of an unfavorable isotherm

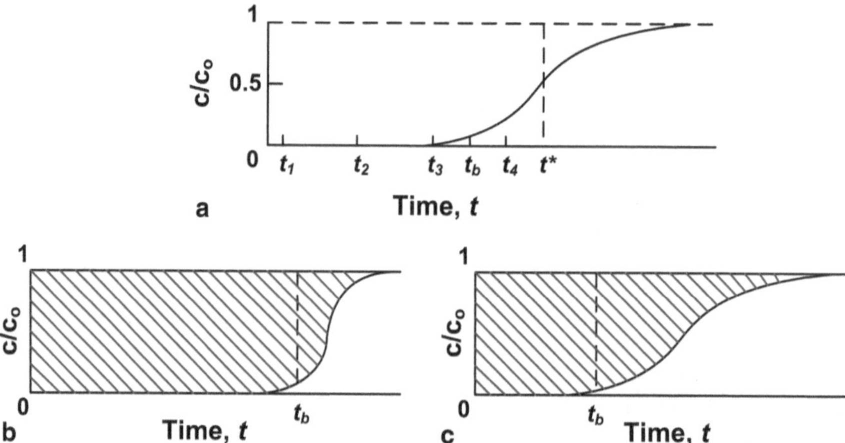

Fig. 4.14 Breakthrough curves for **a** fixed-bed adsorption **b** narrow and **c** wide mass-transfer zone. (With permission from McGraw-Hill Companies, Inc. [2])

(concave up), the breakthrough curves do not sharpen, but rather become stretched as they advance down the length of the column [67].

The area between the S-shaped curve and the horizontal line at $c/c_0 = 1$ in Fig. 4.14 is proportional to the total adsorbed, assuming a constant flow throughout the bed and that equilibrium is reached with respect to the feed. Ideal adsorption is represented by a vertical breakthrough curve, in which the total amount adsorbed is proportional to the area of the rectangle with width t^* and height 1.0. By a material balance, the movement of the adsorption front through the bed and the effect of process variables on t^* can be determined. With CO_2 as the adsorbate, the CO_2 feed rate F_{CO_2} can be represented by,

$$F_{CO_2} = u_0 c_0 \tag{4.50}$$

such that u_0 is the superficial velocity and c_0 the initial feed concentration of CO_2. For an ideal adsorption process, all of the CO_2 in the feed stream would be adsorbed at time t^*, with the concentration in the porous sorbent increasing from an initial value W_0 to an equilibrium saturation value, W_{sat}, achieving an overall balance of,

$$u_0 c_0 t^* = L\rho_b(W_{sat} - W_0) \tag{4.51}$$

or in terms of the ideal adsorption time,

$$t^* = \frac{L\rho_b(W_{sat} - W_0)}{u_0 c_0} \tag{4.52}$$

such that L is the length of the packed bed and ρ_b is the bed density. For fresh sorbent material, $W_0 = 0$, but due to the high cost, sorbents are typically not fully regenerated. Figures 4.14b and 4.14c represent examples of narrow and wide mass-transfer zones, respectively, with the actual amount adsorbed at breakthrough determined through by integrating above the curve up to time t_b, as represented by the shaded region behind the vertical dotted line of Figs. 4.14b and 4.14c. The amount adsorbed at saturation is determined by integrating the shaded region above the breakthrough-curve from time zero to infinity. The goal of an optimal sorbent is to achieve a narrow mass-transfer zone, making the most efficient use of the sorbent and thereby reducing regeneration costs. In fact, an ideal sorbent would have a vertical breakthrough curve, which would be representative of zero mass-transfer resistance and minimal axial dispersion.

4.4.2 Breakthrough

The way in which the CO_2 concentration changes in the gas phase as a function of distance throughout the packed bed will influence the process design. Figure 4.15 illustrates the movement of the CO_2 concentration wave throughout the bed, in addition to the CO_2 outlet concentration as a function of time for an essentially

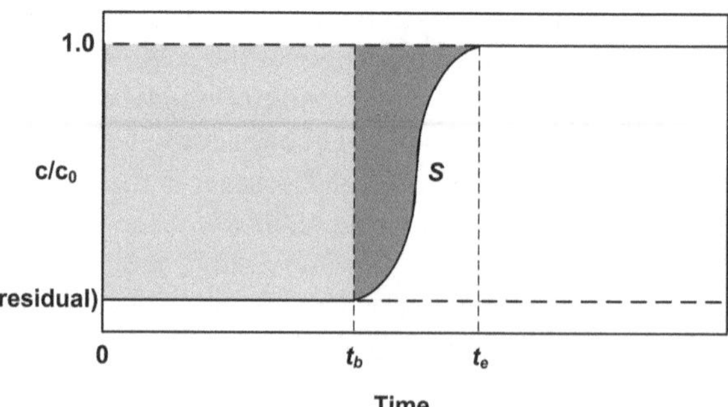

Fig. 4.15 Time trace of adsorptive concentration in an adsorber, with nonzero initial concentration after the first cycle. (With permission from McGraw-Hill Companies, Inc. [2])

isothermal bed. The following three zones exist within the bed: (1) the zone nearest the inlet where the sorbent is fully saturated and in equilibrium with the CO_2 feed concentration, (2) deeper within the bed a zone exists, in which the adsorbed phase concentration approximates an S-shape, and (3) a zone in which no adsorption takes place since the majority of the CO_2 in the feed is removed in the first two zones. The width of the mass-transfer zone (S curve) directly affects the operation of the sorbent bed. As the ratio of the mass-transfer zone width to the bed length increases, the bed use per cycle decreases, with more sorbent required to process a given feed rate. For a linear adsorption isotherm, widening of the mass-transfer zone is typical, unlike the case of Langmuir or Freundlich[1] isotherms, in which widening does not occur, but rather the mass-transfer zone remains fairly constant. Collins [68] has developed a technique for determining the bed length of a full-scale adsorption process based upon bench-scale breakthrough curves, and upon the condition of a constant mass-transfer zone. Within this approach the fraction of the bed saturated, θ at the breakthrough time, t_b is determined. This fraction can be accurately determined from numerical integration of the breakthrough curve and is defined as the ratio between the "total adsorbed at breakthrough" and the "total adsorbed at saturation (equilibrium)." From Fig. 4.15, this ratio corresponds to the integral of $1 - c/c_0$ from $t = 0$ to $t = t_b$ (shaded gray) over the integral of $1 - c/c_0$ from $t = 0$ to $t = t_e$ (shaded gray + pink). A quicker approach is to approximate the integral of the adsorbed amount at breakthrough as a rectangle and the integral of the adsorbed amount at saturation as the sum of the rectangle and triangle. Using this simplistic approach, the total adsorbed at saturation may be approximated as:

$$\begin{bmatrix} \text{Total adsorbed} \\ \text{at saturation} \end{bmatrix} = Qy_{CO_2}\, t_b + \frac{1}{2} Qy_{CO_2}\, (t_e - t_b) \tag{4.53}$$

such that Q is the feed volumetric flow rate, y_{CO_2} is the mole fraction of CO_2 in the feed, t_b is the breakthrough time, and t_e is the time it takes for the sorbent bed to reach

[1] The Freundlich isotherm differs from Langmuir in that at high p_{CO_2}, the sorbent loading continues to increase, while for the Langmuir case the loading approaches monolayer coverage; $W = k\left(\frac{p}{p_0}\right)^{1/n}$, such that k and n are fitting parameters.

saturation (equilibrium). Notice the first term on the right-hand side of Eq. (4.53) is the total CO_2 adsorbed at breakthrough (shaded gray in Fig. 4.15) and the second term on the right-hand side corresponds to the pink shaded region of Fig. 4.15. Therefore, the fraction of the bed saturated, θ, is:

$$\theta = \frac{t_b}{t_b + \frac{1}{2}(t_e - t_b)} = \frac{2t_b}{t_b + t_e} \tag{4.54}$$

This approach assumes that when scaling up from the bench-scale to full-scale, that the length of the unused zone of the bed will not change and is constant, regardless of the actual bed length. The *length of the unused bed* (*LUB*) is found directly from the fraction of the bed saturated by, $LUB = l(1 - \theta)$, such that l is the actual length of the bed. Although a longer bed implies that a greater percentage of the bed will be used since *LUB* is constant, a longer bed also results in a greater pressure drop. Therefore, these parameters have to be carefully optimized for minimizing cost.

It is important to note that this approach can be dangerous depending upon the heat generated as a result of adsorption. For instance, a large bed operates nearly adiabatically, while a small-diameter laboratory column will be close to isothermal (unless it is well insulated). The assumption that the *LUB* measured in the laboratory column will be the same as in the larger full-scale bed may therefore result in the underdesign of the full-scale bed [16b].

Example 4.4 The following data [85] is for the adsorption of CO_2 from a CO_2/CH_4 mixture in fixed bed of Mg-MOF-74. The following parameters were used:

Bed length (depth) = 0.5 cm; Bed diameter = 0.04 cm Temperature = 25°C; Pressure = 15 psi, with no significant pressure drop; Entering volumetric flow rate, $Q = 10$ cm^3/min; Entering CO_2 concentration is 19.1% (by volume).

Time (min)	Mass fraction
0.13	0
1.98	0
3.02	0
3.99	0.002
4.59	0.053
5.11	0.097
5.72	0.141
6.28	0.174
6.88	0.191

With goal of scaling this process up 10,000 times, determine the amount of sorbent required for a bed depth of: (a) 0.5 cm and (b) 10 m.

Solution: Using the data available from the bench-scale experiment, the length of unused bed can be calculated from the fraction of bed saturated, θ using Eq. (4.54) yielding:

$$\theta = \frac{2(3.99\ \text{min})}{3.99\ \text{min} + 6.88\ \text{min}} = \frac{7.98}{10.87} = 0.73$$

and the *LUB* as

$$LUB = l(1 - \theta) = 0.5\ \text{cm}(1 - 0.73) = 0.135\ \text{cm}$$

(a) in the case of a 0.5-cm high bed (short and wide), the length of bed used for the experiment (at bench-scale) is 0.500 cm – 0.135 cm = 0.365 cm. The sorbent volume required for 10,000 times the scale of the experiment is:

$$V = 10{,}000[\pi(0.02\ \text{cm})^2(0.365\ \text{cm})] = 4.58\ \text{cm}^3$$

Taking into account *LUB*, the sorbent volume required for a 0.5-cm high bed is:

$$V = 10{,}000[\pi(0.02\ \text{cm})^2\ (0.5\ \text{cm})] = 6.28\ \text{cm}^3$$

which results in a bed with a diameter of approximately 4 cm at a height of 0.5 cm. In this case, the pressure drop remains the same as in the bench-scale experiment.

(b) in the case of a 10-m (1000 cm) high bed (tall and thin), *LUB* is still 0.135 cm, resulting in the fraction of bed used approximately 0.999, or 999 cm. The required volume of sorbent associated with the bed volume used is still 4.58 cm^3, which can be used to estimate the bed diameter by:

$$(999\ \text{cm})[\pi(r)^2] = 4.58\ \text{cm}^3 \ \rightarrow\ r = 0.038\ \text{cm},\ d = 0.076\ \text{cm}$$

The total sorbent required for a 10-m bed is:

$$V = (4.58\ \text{cm}^3)\left(\frac{1000\ \text{cm}}{999\ \text{cm}}\right) \sim 4.58\ \text{cm}^3$$

which is approximately 27% less sorbent than the 0.5-cm bed. However, to maintain the same bed velocity, overcoming a pressure drop approximately 2000 times larger in the taller column is required.

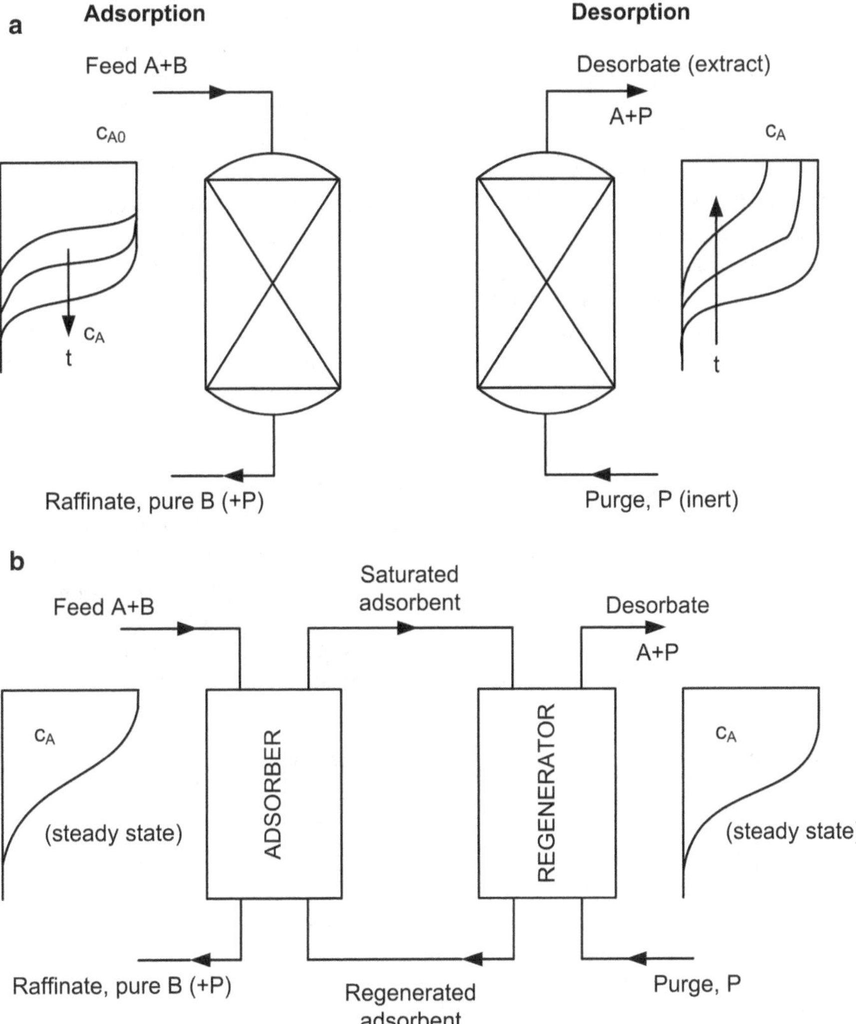

Fig. 4.16 Two basic modes of operation for an adsorption process; **a** fixed cyclic batch system and **b** moving continuous countercurrent system with sorbent recirculation, with concentration profiles indicated throughout the bed and A being the strongly adsorbing species. (Reprinted with permission of John Wiley & Sons, Inc. [3])

4.5 Adsorption Processes

There are two general types of systems associated with separation via adsorption on a large-scale, *i.e.*, fixed-bed and moving-bed adsorption processes. In a cyclic-batch process as shown in Fig. 4.16a, the sorbent bed is alternately saturated and regenerated in a cyclic fashion, while in a moving-bed adsorption process, *i.e.*, Fig. 4.16b there is continuous contact between the sorbent and the feed.

Fixed-bed adsorbers In fixed-bed adsorption, the feed gas is transported through one of the beds while the other bed is regenerated. The downward flow of gas in adsorption is preferred since upflow may result in particle fluidization and potential loss of the fine particles, as well as enhanced axial mixing (dispersion). The valves are switched so that the feed passes through the second bed, and regeneration occurs in the saturated bed when the concentration of adsorptive in the exiting gas reaches a certain point, or at a scheduled time. Regeneration can take place by lowering the bed pressure or increasing the bed temperature, or using either a hot inert gas or steam, which condenses in the bed raising the bed temperature providing sufficient energy for desorption. In many cases, water displaces CO_2 due to the stronger adsorption. At this point there is a mixture of CO_2 with either an inert gas or water, which will have to further be separated or simply condensed in the case of steam regeneration.

The bed size can be determined from the gas flow rate and cycle time. Traditionally, the cross-sectional area of the bed is usually designed to yield a superficial velocity through the bed of 0.15–0.45 m/s, resulting in a pressure drop of several inches of water per foot when using a sorbent of 4 × 10 or 6 × 16 mesh size [62]. The use of a longer bed may achieve addition separation, but could extend the cycle time significantly, which would require increased capital cost and lead to an increase in pressure drop as shown in Example 4.4. Bed heights of 0.3 m have been recommended for minimizing pressure drop, but would have the limitation of incomplete separation. The throughput of the process is governed by the product of the bed capacity and the cycle frequency; hence, increased bed capacity and faster cycle times reduce the required bed size.

Moving-Bed Adsorbers Due to the pressure drop associated with fixed beds and the scale of CO_2 capture that is required for any significant mitigation, moving beds offer the advantage of reduced pressure drop, but do suffer from sorbent attrition in addition to mechanical complexity associated with the equipment involved. Continuous countercurrent and cross-flow systems will be discussed in particular.

Within a continuous countercurrent system, the sorbent particles are circulated from an adsorbing section to a regeneration section as shown in Fig. 4.16b. The inefficiency associated with the mass-transfer zone is minimized since the sorbent leaves the adsorption section while it is essentially at equilibrium with the feed composition. The adsorption section is comprised solely of the MTZ, absent of the *LUB*, which is present in a fixed-bed system. Similarly, the countercurrent regeneration process is also more efficient. Also since the sorbent particles are moved from the adsorption to the regeneration section, the design of each section can be specific. For instance, the regeneration process may involve elevated temperatures and the adsorber may not need to be designed for such harsh conditions, which adds additional flexibility to the design of the system. Due to the sorbent attrition from moving the particles, there has been a movement toward the design and fabrication of attrition-resistant materials using graphite and other types of coating to increase their durability [69]. In general, countercurrent contact maximizes the mass-transfer driving force, which can lead to greater efficiency in terms of sorbent capacity, compared to a fixed batch process.

The three applications of cross-flow systems include panel beds, sorbent wheels, and rotating annular beds [17a]. In a cross-flow system, the sorbent is moved in a direction perpendicular to the fluid flow. In a panel bed, the used sorbent drops onto a "load-out" bin and fresh regenerated sorbent is added at the top with the gas flowing across the panel. The sorbent particles are placed in baskets representative of a collection of fixed beds to form a wheel that rotates around a horizontal axis. The wheel is continuously treating gas and this configuration is often used to treat ambient air due to the minimal pressure drop. Their limitations include reduced efficiency from short contact time and mechanical leakage at the seals. Rotary wheel adsorbers are widely used in desiccant cooling systems and in volatile organic carbon removal from air.

4.6 Adsorption Cycles

Details associated with the adsorption cycles for moving-bed adsorbers (continuous countercurrent systems) are available in the literature [16b] and only adsorption cycles associated with fixed-bed adsorption are discussed in this section. For a cyclic adsorption processes valve control is crucial, with the on/off valves responsible for switching flows among beds to simulate as close as possible a continuous process. It is crucial that valves can handle the times scales associated with the adsorption process. It is important to keep in mind that different CO_2 gas mixtures have varying associated capture parameters, and that a separation process that works optimally for postcombustion capture may not work as well for a precombustion application. Examples of parameters that might vary depending upon the conditions of the CO_2 gas mixture include cycle configuration and duration, temperature and pressure conditions, regeneration strategy, and bed dimensions. For instance, the CO_2 is at a significantly higher partial pressure in a fuel gas stream (*i.e.*, precombustion) than it is in a flue gas stream (*i.e.*, postcombustion). For an adsorption process to be successful at the scale of a traditional 500-MW power plant, the handling of large feed rates will be required. Recalling for instance that this sized power plant emits on average, 11,000 tons of CO_2 per day. It is important to recognize that the adsorption and regeneration cycles will have to match the application of CO_2 capture. Adsorption processes can operate with several different physical arrangements and cycles. Three basic cycles will be discussed, *i.e.*, temperature swing (TSA), displacement purge, and pressure swing adsorption (PSA), which also includes vacuum swing adsorption (VSA).

To assist in determining the appropriate cycle for regeneration, Keller et al. [70] have created a matrix to serve as a guide for the selection process as shown in Table 4.6.

Temperature-Swing Adsorption Figure 4.17 shows an example of a temperature-swing cycle, in which a feed stream containing CO_2 at a partial pressure of p_1 is passed through the adsorption bed at a temperature of T_1. The initial equilibrium loading is expressed as x_1 in units of mass of adsorbate per mass of sorbent. After

Table 4.6 Matrix for adsorption cycle selection [70]

Process condition	Temperature swing	Displacement purge	PSA
Adsorbate concentration in feed, <3%	Yes	Unlikely	Unlikely
Adsorbate concentration in feed, 3–10%	Yes	Yes	Yes
Adsorbate concentration in feed, >10%	No	Yes	Yes
High product purity required	Yes	Yes	Possible
Thermal regeneration required	Yes	No	No

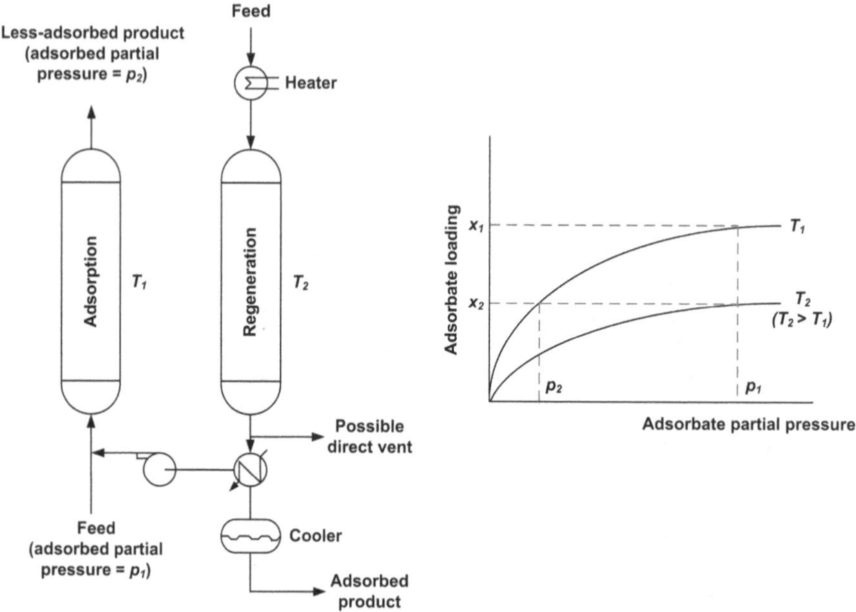

Fig. 4.17 Temperature-swing cycle [1]

equilibrium is reached between the sorbent and CO_2 in the gas phase, regeneration requires the bed temperature to be increased to T_2 (either directly with a hot purge gas or indirectly by heating the bed using an external heat exchanger) with the desorption process taking place as additional feed is passed through the bed and a new equilibrium loading, x_2 is established. In a two-bed system, the adsorption and desorption cycles (including the steps of heating and cooling) must be equal, which can make it difficult to full take advantage of the sorbent capacity. The use of multi-bed systems provides flexibility, but comes at an increase in capital cost. The net theoretical removal capacity (*e.g.*, *working capacity* or *delta loading*) of the bed can be determined from the difference between x_1 and x_2. In particular, for steam regeneration of carbon with steam at 130–150°C, regeneration is stopped soon after the temperature front reaches the top of the bed, with typical steam consumption of 0.2–0.4 kg steam per kg of carbon [62]. Within a TSA cycle, the loading at equilibrium can be expressed using the integrated form of the van't Hoff equation as:

$$H = H_0 e^{-\frac{\Delta H_{ad}}{RT}} \tag{4.55}$$

such that H is the dimensionless Henry's law adsorption equilibrium constant, defined as the ratio between the moles per unit volume of CO_2 in the fluid phase versus the adsorbed phase, ΔH_{ad} is the difference in the energy between the adsorbed phase and the corresponding gas phase, and H_0 is the dimensionless Henry's law constant of CO_2 for some initial temperature condition. From Eq. (4.55), it can be seen that a slight increase in the temperature can lead to a significant decrease in the adsorbed phase concentration. An advantage to this is that the desorbed CO_2 may be recovered at high concentration, but this comes at the cost of time delays associated with heating and cooling. Therefore, this method of regeneration may be unsuitable for systems requiring rapid cycling, such as power plants that generate emissions on the order of tens of thousands of tons of CO_2 per day.

Details associated with the mathematical models used to simulate the dynamic behavior of a cyclic adsorption process are available in the literature [16b], but some of the challenges associated with the accurate modeling of this complex process will be mentioned briefly. In real adsorption processes both adsorption and desorption are not typically carried out to completion, which means at the start of a new cycle, there will be some residual CO_2 that was not desorbed during regeneration. The challenge with modeling the real adsorption process is not knowing the initial sorbate distribution. In this case both adsorption and desorption processes must be modeled with an initial arbitrary sorbate distribution. These convergence calculations necessary to reach cyclic steady-state require numerical simulation and cannot be solved analytically.

Displacement-Purge In the displacement-purge cycle the displacement purge fluid in the regeneration step adsorbs nearly as strongly as the adsorbate so that desorption is favored by both change in partial pressure and competitive adsorption through the displacement of surface-bound CO_2. Typical cycle times are on the order of several minutes. In this process since the heat of adsorption of the displacement purge fluid is approximately equal to that of the adsorbate, the net heat generated or consumed is essentially negligible, maintaining nearly isothermal conditions throughout the process, which allows for higher sorbent loading compared to an inert-purge process. A disadvantage of this option is the requirement for product separation and final recovery of CO_2 from the displacement gas; however, in the case of a strongly adsorbing displacement gas, a sharp front may be achieved.

Pressure-Swing Adsorption Figure 4.18 represents a pressure-swing cycle in which the partial pressure of the adsorbate can be reduced by lowering the total pressure of the gas. The lower the total pressure during the regeneration step, the greater the working capacity is during the adsorption step. The time required to load, depressurize, regenerate, and repressurize a bed takes on the order of seconds to several minutes, making it potentially suitable for flue gas applications. Faster cycle times are most achievable with the use of parallel passage contactors, such as

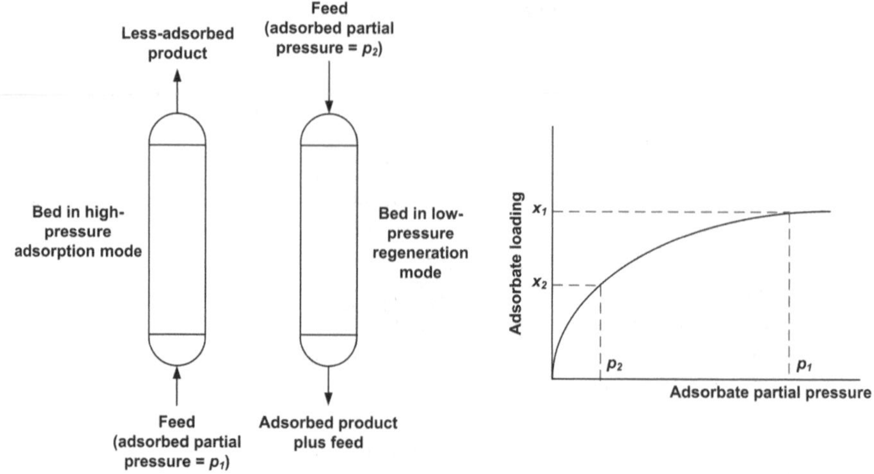

Fig. 4.18 Pressure-swing cycle [1]

sorbent-loaded monoliths. Therefore, to minimize thermal gradient issues, very short times make the pressure-swing cycle an option for bulk-gas separation processes. A disadvantage of this process for postcombustion CO_2 capture applications would be the required compression costs of handling large flow rates. For precombustion CO_2 capture applications, however, this process is advantageous since the feed is already at an elevated pressure. In the case of VSA, there are no compression costs, but rather, costs associated with achieving vacuum conditions.

4.7 Work Required for Separation

Pressure Drop Pressure drop in an adsorption process is important as it determines the cost of blowers or fans required to pass the gas through the sorbent material in the packed bed. Particle size and other physical characteristics, fluid velocity, and bed dimension all play a role in determining the pressure drop of a given system. Consider a fluid particle (*e.g.*, CO_2 entrained in air or a flue gas) forced by a pressure gradient to flow through channels of a packed bed. The fluid particle, depending upon the size of the channel may proceed by undergoing momentum exchanges with other fluid particles and the wall. Viscous flow, which can be described by Poiseuille's law [71], consists of a distribution of fluid velocities, with the fluid particles at the center of the channel at a maximum and the fluid particles adsorbing momentarily on a wall having a zero velocity component in the direction of flow. At room temperature, a gas molecule may be adsorbed on a surface for a short time, but no less than the required time for a single vibrational cycle, *e.g.*, 10^{-13} s. When the mean free path of the fluid particles is on the order of the channel diameter, slippage at the walls is possible. This can take place due to the collisions between fluid particles rebounding from the walls and the flowing fluid particles nearby in the center of the channel. In this case, there is no apparent decrease or distribution in the velocity of the fluid particles from the channel center to the walls (Knudsen diffusion). It is important to be reminded of these various flow processes when considering the appropriate relationships associated with pressure drop in a capillary or bundle of capillaries, representative of a packed bed of sorbent particles.

According to Darcy's law [72], the average velocity, u, of a fluid through a packed bed is directly proportional to the pressure gradient, Δp, across the bed and inversely proportional to the length, L, of the bed as:

$$\frac{\Delta p}{L} = \frac{u}{K} \tag{4.56}$$

such that K is a proportionality constant. From Poiseuille's law [71], which only applies in the laminar flow regime, it can be shown that the fluid viscosity, μ, can be expressed in terms of the volume, V, of a channel with radius, r and length, L in a time, t, under a pressure gradient such that:

$$\frac{\Delta p}{L} = \frac{8V\mu}{\pi t r^4} \tag{4.57}$$

Through rearrangement of Eq. (4.57) by substitution of $\pi r^2 l$ for v and u_0 for l/t, yields:

$$\frac{\Delta p}{L} = \frac{32\mu u_0}{d^2} \tag{4.58}$$

such that u_0 is the superficial velocity and d is the channel diameter. Kozeny [73] recognized the similarity between Eqs. (4.57) and (4.58) and derived the following expression for flow through a packed bed by assuming that the bed was comprised of a series of many small channels (or capillaries).

$$\frac{\Delta p}{L} = \frac{150 u_0 \mu}{\Phi^2 d^2} \frac{(1-\varepsilon)^2}{\varepsilon^3} \tag{4.59}$$

such that Φ is the sphericity, which is defined as the surface-volume ratio for a sphere of diameter, d. The diameter of the particle is assumed to also be the approximate diameter of the channels within the packed bed. Typical values of sphericity are available in Table 4.7.

Equation (4.59) is formally known as the Kozeny-Carman [73, 74] equation and is applicable for laminar, or low Re number flows. This expression may be generalized to the Hagen-Poiseuille equation to include tortuosity, τ as follows:

$$\frac{\Delta p}{L} = \frac{72\tau u_0 \mu}{\Phi^2 d^2} \frac{(1-\varepsilon)^2}{\varepsilon^3} \tag{4.60}$$

In the case of Eq. (4.59), the tortuosity is estimated at ~2.1. In general, these expressions indicate that flow is proportional to the pressure drop and inversely

Table 4.7 Sphericity of miscellaneous materials [15a]

Material	Sphericity	Material	Sphericity
Spheres, cubes, short cylinders	1.0	Ottoawa sand	0.95
Raschig rings		Rounded sand	0.83
$L=d_o$; $d_i=0.5d_o$	0.58 [62]	Coal dust	0.73
$L=d_o$; $d_i=0.75d_o$	0.33 [62]	Crushed glass	0.65
Berl saddles	0.3	Mica flakes	0.28

d_o and d_i correspond to the outer and inner diameters of the individual packing material, respectively

proportional to the fluid viscosity. This general statement is often referred to as Darcy's law, and is used to describe liquid flow through porous media. In the case of turbulent flow ($Re > 1000$) an empirical correlation for pressure drop, known as the Burke-Plummer [75] equation may be used:

$$\frac{\Delta p}{L} = \frac{1.75\rho {u_0}^2}{\Phi d}\frac{1-\varepsilon}{\varepsilon^3} \tag{4.61}$$

Under the assumption that viscous and kinetic losses are additive, the following Ergun [76] equation represents the entire range of flow rates:

$$\frac{\Delta p}{L} = \frac{150 u_0 \mu}{\Phi^2 d^2}\frac{(1-\varepsilon)^2}{\varepsilon^3} + \frac{1.75\rho {u_0}^2}{\Phi d}\frac{1-\varepsilon}{\varepsilon^3} \tag{4.62}$$

There have been a number of previous investigations associated with the pressure drop in flow through packed beds including Ergun [76], Chilton and Colburn [77], among others [78]. The collective data from these studies can be correlated in terms of a dimensionless friction factor, f, defined by:

$$f = \left(\frac{d_p}{L}\right)\frac{\Delta p}{\rho {u_0}^2} \tag{4.63}$$

such that L is the length of the packed bed, Δp is the pressure drop in units of N/m^2, ρ is the fluid density, and u_0, which is equivalent to εu is the superficial velocity of the fluid. The two commonly used correlations for the friction factor determined by Chilton-Colburn are [77]:

$$\begin{aligned} &Re < 40 \quad f = 805/Re \\ &Re > 40 \quad f = 38/Re^{0.15} \end{aligned} \tag{4.64}$$

and the simplified Ergun relation [76]:

$$f = \left(\frac{1-\varepsilon}{\varepsilon^3}\right)\left[\frac{150(1-\varepsilon)}{Re} + 1.75\right] \tag{4.65}$$

The Reynolds number within the correlations is based on the particle diameter, which is representative of a given channel width and the superficial velocity, and show reasonable agreement when the void space of the packed bed, ε is 0.35.

Movement of particles within a fixed bed should be avoided since it can lead to sorbent degradation. To avoid particle movement, a rule of thumb is for the allowable upflow fluid velocity, u_{max} should be less than 80% of the *minimum fluidization velocity*, u_{fl}:

$$u_{max} = 0.8u_{fl} \approx 6 \times 10^{-4} g \frac{(d_p)^2}{\mu}(\rho_p - \rho) \tag{4.66}$$

such that μ and ρ are the fluid viscosity and density, respectively, ρ_p is the sorbent particle density, and g is the acceleration due to gravity. For a typical sorbent with particle density of 1.1 g/cm^3 with air as the fluid at 298 K with viscosity of 1.84×10^{-4} Poise, yields a maximum allowable interstitial velocity for upflow through a packed bed of 1-mm diameter particles approximately 35 cm/s. For downflow, velocities of up to 1.8 times the minimum fluidization velocity are allowed.

Gas Blowing Work As previously described in Chap. 3, the power, P, required for moving a gas with a mass flow rate of $\dot{m}$, an average density, ρ, with a fan of efficiency, ε, over a pressure drop Δp, is:

$$P = \frac{dw_f}{dt} = \frac{\dot{m}\Delta p}{\rho\varepsilon} \tag{4.67}$$

Blowers are more sophisticated and expensive than fans and can handle greater pressure drops, *i.e.*, up to 300 kPa. The difference between the work calculated for a fan versus a blower is that the density cannot be considered constant over large pressure drops. For adiabatic and reversible compression of an ideal gas, the blowing power is:

$$P = \frac{dw_b}{dt} = \frac{\dot{m}RTk}{M(k-1)\varepsilon}\left[\left(\frac{p_2}{p_1}\right)^{(k-1)/k} - 1\right] \tag{4.68}$$

such that T is the gas temperature, M is the molecular weight, p_1 is the initial gas pressure, p_2, the final gas pressure, and k, the ratio of specific heats, c_p/c_v, which are available for selected gases in Appendix B. Typical fan and blower efficiencies range between 65 and 85%. To determine the work required to compress a gas from a pressure of p_1 to p_2 using either fan (Eq. 3.67) or blowing power Eq. (3.68), the power expression can be divided by the molar flow rate of the gas, which will result in the work required per mole of gas compressed. Depending on the magnitude of the pressure drop in the column, Eq. (3.67) or Eq. (3.68) may be used to determine the power required to overcome the pressure drop.

In addition to the work required with overcoming pressure drop, additional work is required for thermal or compression (or vacuum) work associated with the cycle, *i.e.*, TSA, PSA (VSA), or some combination thereof. A brief overview of various process cycles will be given, with additional details available in the literature [80–82]. For instance, Meunier et al. [79] have proposed an indirect TSA approach that includes an internal heat exchanger with indirect heating and cooling to maximize sorbent usage and increase the process productivity, which are limitations of a traditional TSA process that are directly heat-driven. Within their laboratory-scale experiments they are able to achieve volumetric productivity of 37 kg/m^3 h of CO_2 with a specific heat consumption of 6 MJ/kg CO_2, in addition to an adiabatic estimate of 4.5 MJ/kg CO_2.

Another approach investigated by Pugsley et al. [80] includes circulating fluidized bed pressure-temperature swing adsorption (CFB-PTSA), in which the circulating fluidized bed allows for gas-solid contact through the vertical transport of solid particles by a high velocity gas stream, with the gas-solid separation achieved in a cyclone. Their results show that the process enriches a gas stream containing 15% CO_2 to 90% with 70% capture. Additionally, Suzuki et al. [81] investigated the recovery of CO_2 from stack gas using a piston-driven ultra-rapid PSA approach. They found the enrichment to be significantly improved, but with low capture. They suggest that decreasing the pressure drop of the sorbent and cycle optimization are both required to improve the process efficiency.

Park et al. [82] have found that within a 2-stage PSA process, an enrichment of a gas mixture containing 10–15% CO_2 to 40–60% can be achieved in the first stage, with an enrichment up to 99% in a second stage. Although Park et al. investigate a number of different cycles in order to optimize the PSA process, only the Skarstom cycle is discussed, which involves the following steps: (1) feed pressurization, (2) gas adsorption, (3) adsorbed gas evacuation (depressurization), (4) blow-down and bed purge, and (5) product recovery. These steps are also illustrated in Fig. 4.18 in the discussion of the PSA cycle. In a two-column set-up as shown in Fig. 4.18, each of the steps is carried out, yet reversed in each column. For instance, while one column is adsorbing a gas, the other column is evacuating the gas. In a PSA process the majority of the power is consumed with the use of the blower to pressurize the feed and purge, and the use of a vacuum pump to depressurize the bed for desorption. Park et al. [82] define the P/F ratio as the amount of gas used in the purge step divided by the amount of feed introduced in the adsorption and feed pressurization steps, and they investigate the power optimization of PSA based upon this parameter. If on the other hand, a purge step was not included, the calculation of power (and work) associated with PSA would comprise of only the feed pressurization and column evacuation steps. Since both cases involve gas compression, Eq. (4.68) can be used to estimate the power required for a simplified PSA process without a purge step. The work per mole of CO_2 captured can be calculated by dividing the compression power by the molar flow rate of recovered (captured) CO_2.

Example 4.5 Given the following parameters associated with a packed-bed column for an adsorption process, determine:

(a) pressure as a function of packed-bed length for combustion flue gas and for DAC at $u_0 = 1.0\ \mathrm{m/s}$ and 2.0 m/s, and
(b) work per mole of CO_2 captured required to overcome the pressure drop for each scenario. For postcombustion capture, assume the column dimensions are $L = 38.5$ m and $D = 1$ m and for DAC, assume the column dimensions are $L = 2.8$ m and $D = 12$ m.

Parameters	Flue gas	Direct air capture (DAC)
Superficial velocity, u_0 (m/s)	3.0	1.0–2.0
Fluid viscosity, μ (Pa s)	0.000018	0.0000183
Fluid density, ρ (kg/m^3)	1	1.2
Fraction void, ε (dimensionless)	0.4	0.4
Sphericity, Φ (dimensionless)	1	1
Effective particle diameter, d_p (m)	0.01	0.01
Tortuosity, λ (dimensionless)	1	1
Initial CO_2 concentration	11.9 mol %	390 ppm
Outlet pressure, p_2 (Pa)	101325	101325

Solution: (a) Assume 90% CO_2 capture for combustion flue gas and 50% capture for DAC. Using the Ergun equation to calculate the pressure drop as a function of the length of packed bed yields:

$$\Delta p = \left[\frac{150 u_0 \mu (1-\varepsilon)^2}{\Phi^2 D_p^2 \varepsilon^3} + \frac{1.75 p u_0^2 (1-\varepsilon)}{\Phi D_p \varepsilon^3}\right] \cdot L$$

The pressure drop as a function of bed length for each scenario is:

$$\Delta p/L = 182.2\,\text{Pa/m for flue gas}$$

$$\Delta p/L = 61.8\,\text{Pa/m for DAC at } u_0 = 1.0\,\text{m/s}$$

$$\Delta p/L = 123.5\,\text{Pa/m for DAC at } u_0 = 2.0\,\text{m/s}$$

(b) The power required to overcome the pressure drop across the packed bed can be calculated. Assume the adiabatic and reversible compression of an ideal gas, the work associated with blower is determined by Eq. (3.68),

$$P = \frac{dw_b}{dt} = \frac{\dot{m} RTk}{M(k-1)\varepsilon}\left[\left(\frac{p_2}{p_1}\right)^{(k-1)/k} - 1\right]$$

This equation may be rewritten as:

$$P = \frac{P_1 Qk}{\varepsilon(k-1)}\left[\left(\frac{p_2}{p_1}\right)^{(k-1)/k} - 1\right]$$

such that Q is the volumetric flow rate of gas entering the adsorption column. Assume the blower efficiency, ε is 85% and the ratio of specific heat, k is 1.4,

$$P = \frac{3.5 p_1 Q}{0.85}\left[\left(\frac{p_2}{p_1}\right)^{0.285} - 1\right]$$

For postcombustion capture, assume the column dimensions are $L = 38.5$ m and $D = 1$ m. For DAC, assume the column dimensions are $L = 2.8$ m and $D = 12$ m. The volumetric flow rate of flue gas is 2.36 m^3/s and air is 113.1 m^3/s (at 1.0 m/s) and 226.2 m^3/s (at 2.0 m/s). The inlet pressure (p_1) can be determined from $p_1 = p_2 + \Delta p$ where the outlet pressure, (p_2) is 101.325 kPa for both cases and Δp is obtained from part (a). For postcombustion capture, $\Delta p = 6379$ Pa, so p_1 is 107703 Pa. For DAC at 1 m/s, p_1 is 101497 Pa and at 2 m/s, p_1 is 101671 Pa.

From this information, the power required in each situation is:

Power = 18.0 kW for flue gas
Power = 22.9 kW for DAC at $u_0 = 1.0$ m/s
Power = 91.9 kW for DAC at $u_0 = 2.0$ m/s

The work required per mol of CO_2 captured is determined from:

$$w_b = \frac{P}{\dot{n}_{CO_2,captured}}$$

The molar flow rate of CO_2 in flue gas and air can be calculated as:

Flue gas

$$\dot{n}_{CO_2} = \frac{0.119 p_1 Q}{RT} = \frac{(0.119)(107703\,\text{Pa})(2.36\,\text{m}^3/\text{s})}{\left(8.314\frac{\text{m}^3\,\text{Pa}}{\text{mol K}}\right)(298.15\,\text{K})} = 12.2\,\frac{\text{mol CO}_2}{\text{s}}$$

Air (1.0 m/s)

$$\dot{n}_{CO_2} = \frac{387 \rho Q}{10^6 M} = \frac{(390)\left(1200\frac{\text{g}}{\text{m}^3}\right)\left(113.1\frac{\text{m}^3}{\text{s}}\right)}{(10^6)\left(44.01\frac{\text{g}}{\text{mol}}\right)} = 1.19\,\frac{\text{mol CO}_2}{\text{s}}$$

Air (2.0 m/s)

$$\dot{n}_{CO_2} = \frac{387 \rho Q}{10^6 M} = \frac{(390)\left(1200\frac{\text{g}}{\text{m}^3}\right)\left(226.2\frac{\text{m}^3}{\text{s}}\right)}{(10^6)\left(44.01\frac{\text{g}}{\text{mol}}\right)} = 2.39\,\frac{\text{mol CO}_2}{\text{s}}$$

The work required per mol of CO_2 captured is:

Flue gas

$$w_b = \frac{Power}{\dot{n}_{CO_2,captured}} = \frac{18.0\,\text{kW}}{(0.9)\left(12.2\frac{\text{mol CO}_2}{\text{s}}\right)} = 1.63\,\frac{\text{kJ}}{\text{mol CO}_2}$$

Air (1.0 m/s)

$$w_b = \frac{Power}{\dot{n}_{CO_2,captured}} = \frac{22.9\,\text{kW}}{(0.5)\left(1.19\frac{\text{mol CO}_2}{\text{s}}\right)} = 38.5\,\frac{\text{kJ}}{\text{mol CO}_2}$$

Air (2.0 m/s)

$$w_b = \frac{Power}{\dot{n}_{CO_2,captured}} = \frac{91.9\,\text{kW}}{(0.5)\left(2.39\frac{\text{mol CO}_2}{\text{s}}\right)} = 76.9\,\frac{\text{kJ}}{\text{mol CO}_2}$$

4.8 Problems

Problem 4.1 The following experimental data for the equilibrium adsorption of pure N_2 on MOF Cu-BTC-MeOH at 298 K were obtained by Liu et al. [86]:

Pressure (bar)	Excess adsorption (wt%)
2.3	4.68
5.4	8.94
8.3	11.32
10.9	13.56
13.4	15.52
16.0	16.71
19.4	18.11
21.9	18.95
24.2	19.93
27.3	20.21
30.2	20.98
32.7	21.27
35.3	21.76
38.1	22.25
40.3	22.46

Fit the data provided to a Langmuir isotherm. The adsorption data reported in this example is the excess rather than the total. *Excess adsorption* is defined as the additional amount of fluid particles present as a consequence of adsorption, *i.e.*, the additional amount present over the gas phase at the same temperature, pressure, and volume.

Problem 4.2 Using data from Table 4.3 in the text, determine:

a. the volume fraction of pores in zeolite 13X filled with adsorbed water when its partial pressure is 4.6 mmHg at 25°C, and
b. whether the amount of water adsorbed is equivalent to more than a monolayer. At these conditions, the partial pressure is considerably below the water vapor pressure of 23.75 mmHg.

Problem 4.3 Fit the equilibrium data in the worked Example 4.1 to a Freundlich isotherm. The Freundlich isotherm is expressed as:

$$W = k\left(\frac{p}{p_0}\right)^{1/n}$$

such that W is the loading, p is the pressure, p_0 is the CO_2 vapor pressure, and k and n are fitting parameters. Compare this fit to that of the Langmuir isotherm and comment on the reason one isotherm fits better than another for this particular system. Use the vapor pressure parameters in Appendix J to determine the CO_2 vapor pressure at 25°C.

Problem 4.4 The following data was presented by Zheng et al. [87] for the adsorption of CO_2 from a CO_2/N_2 mixture in a fixed bed of ethylenediamine-modified SBA-15, silica-based sorbent with a diameter of 5.88 and height of 30 mm. The breakthrough experiment was performed by exposing the bed to a volumetric flow rate of 20 cm^3/min with 1.8×10^{-4} g/cm^3 of CO_2 at 25°C and 1 atm.

Time (min)	CO_2 concentration ($\times 10^{-4}$ g/cm^3)
0.39	0.07
0.75	0.10
0.80	0.60
0.85	1.16
0.90	1.54
1.60	1.75
3.32	1.80

With goal of scaling this process up 10,000 times, determine the amount of sorbent required for a bed depth of: a. 30 mm, b. 10 cm, and c. 10 m.

Problem 4.5 Consider a simplified PSA process, where purge is ignored, and the primary energy-intensive step is the compression of the feed gas. Assume the following scenario: a flue gas at 40°C with 90% CO_2 captured at 90% purity. Assuming the PSA involves adiabatic compression from 0.1 to 1 MPa, determine the:

a. theoretical minimum work per mol of captured CO_2,
b. real work required for compression, assuming $k = 1.4$ for flue gas, and
c. 2nd-Law efficiency of the process.

Problem 4.6 The physical characteristics of the packing material in a packed bed may impact the work required for separation. For a void fraction, ε of 0.25,

a. determine the pressure drop over a 2-m tall bed of spheres, berl saddles and granular activated carbon ($\Phi = 0.65$). Assume the fluid is air at 101 kPa and 25°C and the packing material has an average diameter of 0.01 m, and
b. assuming a blower is used with 85% efficiency, determine the work required to move the fluid through the bed.

References

1. Keller GE, Anderson RA, Yon CM (1987) In: Rousseau RW (ed) Handbook of separation process technology. Wiley, New York
2. Unit Operations of Chemical Engineering, 7th Ed., McCabe WL, Smith JC, Harriott P, Copyright (2005)
3. Ruthven DM (1997) Encyclopedia of separation technology, vol I. Adsorption, gas separation
4. Characterization of porous solids III, Jagiello J, Bandosz TJ, Putyera K, Schwarz JA Adsorption energy and structural heterogeneity of activated carbons
5. Springer Science + Business Media B.V., Springer and Kluwer Academic Pub, Characterization of porous solids and powders: surface area, pore size, and density, Lowell S, Shields JE, Thomas MA, Thommes M (2004) 39
6. Parra JB, Pis JJ, De Sousa JC, Pajares JA, Bansai RC (1996) Effect of coal preoxidation on the development of microporosity in activated carbons. Carbon 34(6):783–787
7. Sing KSW, Everett DG, Haul RAW, Moscou L, Pierotti RA, Rouquerol J, Siemieniewska T (2008) In: Ertl H et al (eds) Reporting physisorption data for gas/solid systems. Handbook of heterogeneous catalysis, pp 1217–1230. Copyright Wiley-VCH Verlag GmbH & Co. KGaA
8. Ind Eng Chem Res, Lively RP, Chance RR, Kelley BT, Deckman HW, Drese JH, Jones CW, Koros WJ, Hollow fiber adsorbents for CO_2 removal from flue gas (Copyright 2009). American Chemical Society
9. Christopher E. Wilmer, Northwestern University (2011)
10. Yang RT (2003) Adsorbents: fundamentals and applications. Wiley, Hoboken, p 415
11. Medek J (1977) Possibility of micropore analysis of coal and coke from the carbon dioxide isotherm. Fuel 56(2):131–133
12. Shen D, Bülow M, Siperstein F, Engelhard M, Myers AL (2000) Comparison of experimental techniques for measuring isosteric heat of adsorption. Adsorption 6(4):275–286
13. Yang Q, Zhong C, Chen JF (2008) Computational study of CO_2 storage in metal-organic frameworks. J Phys Chem C 112(5):1562–1569
14. (a) Furukawa H, Ko N, Go YB, Aratani N, Choi SB, Choi E, Yazaydin AÖ, Snurr RQ, O'Keeffe M, Kim J (2010) Ultrahigh porosity in metal-organic frameworks. Science 329(5990):424; (b) Farha OK, Yazayd, AÖ, Eryazici I, Malliakas CD, Hauser BG, Kanatzidis MG, Nguyen SBT, Snurr RQ, Hupp JT (2010) De novo synthesis of a metal-organic framework material featuring ultrahigh surface area and gas storage capacities. Nat Chem 2(11):944–948
15. (a) Green DW (2008) Perry's chemical engineers' handbook. McGraw-Hill, New York; (b) Hagen J (2006) Industrial catalysis: a practical approach, 2nd edn. Wiley-VCH Verlag GmbH &Co, Weinheim, p 525; (c) Rubel AM, Stencel JM (1996) Effect of pressure on NO_x adsorption by activated carbons. Energy Fuels 10(3):704–708; (d) Moon SI, Extrand CW (2011) Hydrogen chloride and ammonia permeation resistance of tetrafluoroethylene-perfluoroalkoxy copolymers. Ind Eng Chem Res 50(5), pp 2905–2909; (e) Baker RW (2004) Membrane technology and applications, 2nd edn. Wiley, Chichester
16. (a) Lide DR (2008) CRC handbook of chemistry and physics. CRC Press, Boca Raton, p 2736; (b) Ruthven DM (1984) Principles of adsorption and adsorption processes. Wiley, New York, p 433
17. (a) Ruthven DM (1997) Encyclopedia of separation technology. Wiley, New York; (b) Buckingham AD (1959) Molecular quadrupole moments. Q Rev Chem Soc 13(3):183–214; (c) Stogryn DE, Stogryn AP (1966) Molecular multipole moments. Mol Phys 11(4):371–393; (d) Prausnitz JM, Lichtenthaler RN, de Azevedo EG (1986) Molecular thermodynamics of fluid-phase equilibria. Prentice-Hall, Upper Saddle River
18. Sing KSW, Everett DH, Haul RAW, Moscou L, Pierotti RA, Rouquerol J, Siemieniewska T (1985) Reporting physisorption data for gas/solid systems, with special reference to the determination of surface area and porosity (recommendations 1984). Pure Appl Chem 57(4):603–619
19. Thorny A, Duval X (1994) Stepwise isotherms and phase transitions in physisorbed films. Surf Sci 299:415–425

20. Polley MH, Schaeffer WD, Smith WR (1953) Development of stepwise isotherms on carbon black surfaces. J Phys Chem US 57(4):469–471
21. Greenhalgh E, Redman E (1967) Stepped isotherms on carbons. J Phys Chem US 71(4):1151–1152
22. Langmuir I (1918) The adsorption of gases on plane surfaces of glass, mica and platinum. J Am Chem Soc 40(9):361–1403
23. Brunauer S, Emmett PH, Teller E (1938) Adsorption of gases in multimolecular layers. J Am Chem Soc 60(2):309–319
24. De Boer JH (1968) Dynamical character of adsorption, 2nd edn. Oxford University Press, London, p 256
25. Lowell S, Shields JE, Thomas MA, Thommes M (2004) Characterization of porous solids and powders: surface area, pore size, and density. Kluwer, Dordrecht, p 347
26. Thiele EW (1939) Relation between catalytic activity and size of particle. Ind Eng Chem 31(7):916–920
27. Juttner F (1909) Reaktionskinetik und Diffusion. Z Elektrochem Angew P 15 6:169–170
28. (a) Rowsell JLC, Spencer EC, Eckert J, Howard JAK, Yaghi OM (2005) Gas adsorption sites in a large-pore metal-organic framework. Science 309(5739):1350; (b) Collins DJ, Zhou HC (2007) Hydrogen storage in metalorganic frameworks. J Mater Chem 17(30):3154–3160; (c) D'alessandro DM, Smit B, Long JR (2010) Carbon dioxide capture: prospects for new materials. Angew Chem Int Ed 49:6058–6082
29. (a) Deng S (2006) In: Lee S (ed) Encyclopedia of chemical processing, vol 5. Taylor & Francis Group, New York 2006; (b) Tan X, Liu S, Li K (2001) Preparation and characterization of inorganic hollow fiber membranes. J Membr Sci 188(1): 87–95
30. The Techno Source. http://www.thetechnosource.net/adsorbents-dessicants.html
31. (a) Kim S, Ida J, Guliants VV, Lin JYS (2005) Tailoring pore properties of MCM-48 silica for selective adsorption of CO_2. J Phys Chem B 109(13):6287–6293; (b) Sjostrom S, Krutka H (2010) Evaluation of solid sorbents as a retrofit technology for CO_2 capture. Fuel 89(6):1298–1306; (c) Sjostrom S, Krutka H, Starns T, Campbell T (2011) Pilot test results of post-combustion CO_2 capture using solid sorbents. Energy Proc 4:1584–1592
32. (a) McKenzie AL, Fishel CT, Davis RJ (1992) Investigation of the surface structure and basic properties of calcined hydrotalcites. J Catal 138(2):547–561; (b) Occelli ML, Olivier J, Auroux A, Kalwei M, Eckert H (2003) Basicity and porosity of a calcined hydrotalcite-type material from nitrogen porosimetry and adsorption microcalorimetry methods. Chem Mater 15(22):4231–4238; (c) Choi S, Drese JH, Jones CW (2009) Adsorbent materials for carbon dioxide capture from large anthropogenic point sources. ChemSusChem 2(9):796–854
33. (a) Lapkin A, Bozkaya B, Mays T, Borello L, Edler K, Crittenden B (2003) Preparation and characterisation of chemisorbents based on heteropolyacids supported on synthetic mesoporous carbons and silica. Catal Today 81(4):611–621; (b) Wang S, Yan S, Ma X, Gong J (2011) Recent advances in capture of carbon dioxide using alkali-metal-based oxides. Energy Environ Sci 4: 3805–3819
34. Parra JB, Pis JJ, De Sousa JC, Pajares JA, Bansal RC (1996) Effect of coal preoxidation on the development of microporosity in activated carbons. Carbon 34(6):783–787
35. Lewis WK, Gilliland ER, Chertow B, Cadogan WP (1950) Pure gas isotherms. Ind Eng Chem 42(7):1326–1332
36. (a) Nandi SP, Walker Jr PL (1975) Carbon molecular sieves for the concentration of oxygen from air. Fuel 54(3):169–178; (b) Koresh J, Soffer A (1980) Study of molecular sieve carbons. Part 1—Pore structure, gradual pore opening and mechanism of molecular sieving. J Chem Soc Farad Trans 1 76:2457–2471
37. Gan H, Nandi SP, Walker Jr PL (1972) Nature of the porosity in American coals. Fuel 51(4):272–277
38. (a) Jüntgen H, Seewald H (1975) Charakterisierung der Porenstruktur mikroporser Adsorbentien aus Kohlenstoff Ber Bunsenges. Phys Chem 79(9):734–738; (b) Moore SV, Trimm DL (1977) The preparation of carbon molecular sieves by pore blocking. Carbon 15(3):177–180
39. Juntgen H (1977) New applications for carbonaceous adsorbents. Carbon 15(5):273–283

40. Walton KS, Abney MB, Douglas LeVan M (2006) CO_2 adsorption in Y and X zeolites modified by alkali metal cation exchange. Micropor Mesopor Mater 91(1–3):78–84
41. Britt D, Furukawa H, Wang B, Glover TG, Yaghi OM (2009) Highly efficient separation of carbon dioxide by a metal-organic framework replete with open metal sites. Proc Natl Acad Sci U S A 106(49):20637–20640
42. Banerjee R, Furukawa H, Britt D, Knobler C, OíKeeffe M, Yaghi OM (2009) Control of pore size and functionality in isoreticular zeolitic imidazolate frameworks and their carbon dioxide selective capture properties. J Am Chem Soc 131(11):3875–3877
43. (a) Bae YS, Snurr RQ (2011) Development and evaluation of porous materials for carbon dioxide separation and capture. Angew Chem Int Edit 50:11586–11596; (b) Davis ME (2002) Ordered porous materials for emerging applications. Nature 417(6891):813–821; (c) James SL (2003) Metal-organic frameworks. Chem Soc Rev 32(5):276–288; (d) Rosseinsky MJ (2004) Recent developments in metal-organic framework chemistry: design, discovery, permanent porosity and flexibility: metal-organic open frameworks. Micropor Mesopor Mater 73(1–2):15–30; (e) Rowsell JLC, Yaghi OM (2004) Metal-organic frameworks: a new class of porous materials. Micropor Mesopor Mater 73(1–2):3–14; (f) Mueller U, Schubert M, Teich F, Puetter H, Schierle-Arndt K, Pastre J (2006) Metal-organic frameworks-prospective industrial applications. J Mater Chem 16(7):626–636; (g) Keskin S, Liu J, Rankin RB, Johnson JK, Sholl DS (2008) Progress, opportunities, and challenges for applying atomically detailed modeling to molecular adsorption and transport in metal-organic framework materials. Ind Eng Chem Res 48(5):2355–2371
44. Lively RP, Chance RR, Kelley BT, Deckman HW, Drese JH, Jones CW, Koros WJ (2009) Hollow fiber adsorbents for CO_2 removal from flue gas. Ind Eng Chem Res 48(15):7314–7324
45. Dubinin MM, Radushkevich LV (1966) Evaluation of microporous materials with a new isotherm. Dokl Akad Nauk SSSR 55:331–347
46. Dubinin MM, Asthakov VA (1970) Prediction of gas-phase adsorption isotherms. Adv Chem Ser 102:69–81
47. Horvath G, Kawazoe K (1983) Method for the calculation of effective pore size distribution in molecular sieve carbon. J Chem Eng Jpn 16(6):470–475
48. Thommes M (2010) Phsical adsorption characterization of nanoporous materials. Chem Ing Tech 82:1056–1073
49. (a) Ravikovitch PI, Vishnyakov A, Russo R, Neimark AV (2000) Unified approach to pore size characterization of microporous carbonaceous materials from N_2, Ar, and CO_2 adsorption isotherms. Langmuir 16(5):2311–2320; (b) Cazorla-Amoros D, Alcaniz-Monge J, De la Casa-Lillo MA, Linares-Solano A (1998) CO_2 as an adsorptive to characterize carbon molecular sieves and activated carbons. Langmuir 14(16):4589–4596; (c) Cazorla-Amorüs D, Alcaòiz-Monge J, Linares-Solano A (1996) Characterization of activated carbon fibers by CO_2 adsorption. Langmuir 12(11):2820–2824
50. Kuwabara H, Suzuki T, Kaneko K (1991) Ultramicropores in microporous carbon fibres evidenced by helium adsorption at 4.2 K. J Chem Soc Farad T 87(12):1915–1916
51. Manovic V, Anthony EJ (2009) Screening of binders for pelletization of CaO-based sorbents for CO2 capture. Energy Fuels 23(10):4797—4804
52. Yue MB, Sun LB, Cao Y, Wang ZJ, Wang Y, Yu Q, Zhu JH (2008) Promoting the CO_2 adsorption in the amine-containing SBA-15 by hydroxyl group. Micropor Mesopor Mater 114(1–3):74–81
53. Xu X, Song C, Andresen JM, Miller BG, Scaroni AW (2002) Novel polyethylenimine-modified mesoporous molecular sieve of MCM-41 type as high-capacity adsorbent for CO_2 capture. Energy Fuels 16(6):1463–1469
54. (a) Scott DS, Dullien FAL (1962) Diffusion of ideal gases in capillaries and porous solids. Am Inst Chem Eng 8(1):113–117; (b) Evans III RB, Watson GM, Mason EA (1961) Gaseous diffusion in porous media at uniform pressure. J Chem Phys 35:2076; (c) Rothfeld LB (1963) Gaseous counter diffusion in catalyst pellets. Am Inst Chem Eng 9(1):19–24
55. Maxwell JC (1860) Illustrations of the dynamical theory of gases. Philos Mag 19(1860):19–32
56. (a) Mason EA, Evans III RB, Watson GM (1963) Gaseous diffusion in porous media. III. Thermal transpiration. J Chem Phys 38:1808; (b) Jackson R (1977) Transport in porous catalysts.

Elsevier, Amsterdam, p 197; (c) Cunningham RE, Williams RJJ (1980) Diffusion in gases and porous media. Plenum Press, New York

57. Kärger J, Ruthven DM (1992) Diffusion in zeolites and other microporous solids. Wiley, New York, p 605
58. Carman PC, Raal FA (1951) Diffusion and flow of gases and vapours through micropores. III. Surface diffusion coefficients and activation energies. Proc R Soc Lond A Mater 209(1096):38
59. Weisz PB (1975) Diffusion transport in chemical systems—key phenomena and criteria. Ber Bunsenges Phys Chem 79(9):798–806
60. (a) Wakao N, Funazkri T (1978) Effect of fluid dispersion coefficients on particle-to-fluid mass-transfer coefficients in packed beds correlation of sherwood numbers. Chem Eng Sci 33(10):1375–1384; (b) Wakao N, Kaguei S (1982) Heat and mass-transfer in packed beds. Gordon and Breach, Science Publishers, Inc., New York, p 365
61. Glueckauf E (1955) Theory of chromatography. Part 10. Formulae for diffusion into spheres and their application to chromatography. Trans Faraday Soc 51:1540–1551
62. McCabe WL, Smith JC, Harriott P (2005) Unit operations of chemical engineering, 7th edn. McGraw-Hill, New York
63. Luss D (1986) In: Carberry J, Varma A (eds) Chemical reaction and reactor engineering, vol 26. Chemical Industries, New York, p 239
64. Carberry JJ, Kulkarni AA (1973) The non-isothermal catalytic effectiveness factor for monolith supported catalysts. J Catal 31(1):41–50
65. (a) Carberry JJ (1975) On the relative importance of external-internal temperature gradients in heterogeneous catalysis. Ind Eng Chem Fund 14(2):129–131; (b) Dullien FAL (1975) New network permeability model of porous media. AIChE J 21(2):299–307
66. Langer G, Roethe A, Roethe KP, Gelbin D (1978) Heat and mass-transfer in packed beds–III. Axial mass dispersion. Int J Heat Mass Trans 21(6):751–759
67. Hall KR, Eagleton LC, Acrivos A, Vermeulen T (1966) Pore- and solid-diffusion kinetics in fixed-bed adsorption under constant-pattern conditions. Ind Eng Chem Fund 5(2):212–223
68. Collins JJ (1967) The LUB/equilibrium section concept for fixed-bed adsorption. In Chemical engineering progress symposium, Science Press, pp 31–35
69. Acharya A, BeVier WE (1985) Attrition resistant molecular sieve. Union Carbide Corporation, Danbury
70. Keller GEI, Anderson RA, Yon CM (1987) In: Rousseau RW (ed) Handbook of separation process technology. Wiley, New York, pp 644–696
71. Poiseuille JLM (1846) Recherches experimentales aur le mouvements les liquides dans les tubes de tres petits diametres. Inst France Acad Sci 9:433–545
72. Darcy H (1856) Les fontaines publiques de la ville de Dijon. Victor Dalmont, Paris, p 674
73. Kozeny J (1927) Über kapillare Leitung des Wassers im Boden. In: Sitzungsberichte der Wiener Akademie der Wissenschaften, Vienna, 1927, vol 139(Kl.abt.IIa), pp 271–306
74. Carman PC (1939) Permeability of saturated sands, soils and clays. J Agric Sci 29(2):262–273
75. Burke SP, Plummer WB (1928) Gas flow through packed columns 1. Ind Eng Chem 20(11):1196–1200
76. Ergun S (1952) Fluid flow through packed columns. Chem Eng Prog 48:89
77. Chilton TH, Colburn AP (1931) II-Pressure drop in packed tubes. Ind Eng Chem 23(8):913–919
78. (a) Furnas CC (1929) The flow of gases through beds of broken solids. Bureau of Mines, Washington, DC, p 164; (b) Leva M (1949) Fluid flow through packed beds. Chem Eng 56(5):115
79. (a) Clausse M, Bonjour J, Meunier F (2003) Influence of the presence of CO_2 in the feed of an indirect heating TSA process for VOC removal. Adsorption 9(1):77–85; (b) Bonjour J, Chalfen JB, Meunier F (2002) Temperature swing adsorption process with indirect cooling and heating. Ind Eng Chem 41(23):5802–5811; (c) Merel J, Clausse M, Meunier F (2008) Experimental investigation on CO_2 post-combustion capture by indirect thermal swing adsorption using 13X and 5 A zeolites. Ind Eng Chem 47(1):209–215
80. Pugsley T, Berruti F, Chakma A (1994) Computer simulation of a novel circulating fluidized bed pressure-temperature swing adsorber for recovering carbon dioxide from flue gases. Chem Eng Sci 49(24):4465–4481

81. Suzuki T, Sakoda A, Suzuki M, Izumi J (1997) Recovery of carbon dioxide from stack gas by piston-driven ultra-rapid PSA. J Chem Eng Jpn 30(6):1026–1033
82. Park JH, Beum HT, Kim JN, Cho SH (2002) Numerical analysis on the power consumption of the PSA process for recovering CO_2 from flue gas. Ind Eng Chem 41(16):4122–4131
83. Liang Z, Marshall M, Chaffee AL (2009) CO_2 adsorption-based separation by metal organic framework (Cu-BTC) versus zeolite (13X). Energ Fuel 23(5):2785–2789
84. Ruthven DM (1984) Principles of adsorption and adsorption processes. Wiley, NewYork
85. Britt D, Furukawa H, Wang B, Glover TG, Yaghi OM (2009) Highly efficient separation of carbon dioxide by a metal-organic framework replete with open metal sites. Proc Natl Acad Sci U S A 106(49):20637–20640
86. Liu J, Culp JT, Natesakhawat S, Bockrath BC, Zande B, Sankar SG, Garberoglio G, Johnson JK (2007) Experimental and theoretical studies of gas adsorption in Cu_3 $(BTC)_2$: an effective activation procedure. J Phys Chem C 111(26):9305–9313
87. Zheng F, Tran DN, Busche BJ, Fryxell GE, Addleman RS, Zemanian TS, Aardahl CL (2005) Ethylenediamine-modified SBA-15 as regenerable CO_2 sorbent. Ind Eng Chem Res 44(9):3099–3105

Chapter 5
Membrane Technology

Membrane separation processes have many advantages over absorption and adsorption processes, some of which include the following: no regeneration, ease of integration into a power plant, process continuity, space efficiency, and absence of a phase change, which can lead to increases in efficiency. Membrane applications, however, require a sufficient driving force for effective separation of a more permeable species. In postcombustion capture of CO_2 for a traditional coal-fired or natural gas-fired power plant this is a challenge due to the somewhat low concentration of CO_2 in the flue gases of these processes. This is in the case that CO_2 is the selective component for separation from the gas mixture. For membrane technology to be applicable for these somewhat dilute systems, either the CO_2 concentration in the flue gas would have to be increased or the selective component would have to be the dominant species (*i.e.*, N_2) in the gas mixture.

Membrane technology requires durable, high mass-transfer flux, defect-free membrane fabrication into compact cost-effective modules with high surface area per unit volume to be competitive with absorption and adsorption technologies. A semipermeable membrane allows for the selective transport and subsequent separation of one or more components of a gas mixture. In terms of carbon capture, the membrane may be selective to CO_2 or the other gas species (*e.g.*, CH_4, N_2, O_2, H_2, etc.) within the mixture, acting as an indirect CO_2 separation process. In general, sharp separations are not achievable using semipermeable membranes, although defect-free dense metallic membranes and dense oxides (*e.g.*, perovskites and related structures) are an exception for atomic hydrogen and anionic oxygen (*i.e.*, O_2^-), respectively. Dense metallic membranes are also being realized for selective oxygen and nitrogen separation processes [10].

A schematic outlining the details of the chapter is shown in Fig. 5.1. As discussed in Sect. 1.6 of Chap. 1, the costs can be divided into non-technical and technical. The factors that influence the cost of the technology associated with membrane separation processes are discussed in this chapter and hence, the compression of CO_2 required for transport is not included in the schematic. It should be noted, however, that it may be possible to incorporate compression energy from the membrane separation process into the requirement associated with CO_2 transport, provided the CO_2 is captured on the high-pressure side of the membrane. This concept is discussed in

J. Wilcox, *Carbon Capture*,
DOI 10.1007/978-1-4614-2215-0_5, © Springer Science+Business Media, LLC 2012

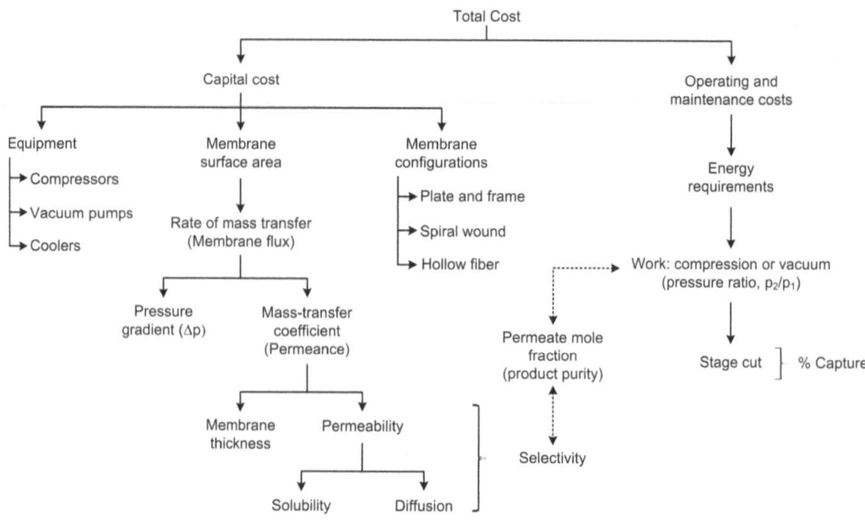

Fig. 5.1 Schematic outlining the components that comprise the cost of the CO_2 membrane separation process

more detail throughout the chapter. Figure 5.1 is a simplistic view of the relationship between cost, equipment, material, and process parameters to emphasize that these aspects are not disconnected, but rather, highly dependent and intertwined in a complex fashion. This schematic is not meant to be inclusive of all connections and components of membrane separation processes, but rather a general overview of the primary constituents. The capital cost of membrane separation processes is dominated by equipment costs (*i.e.*, compressors, vacuum pumps, blowers, etc.) as shown in Fig. 5.1. Capital costs also include the required membrane surface area, which is ultimately determined from the membrane flux and permeance (flux/driving force) assuming that an acceptable selectivity can be achieved. Carbon capture from a power plant will require membrane surface areas that can handle tens of thousands of tons of CO_2 per day. The Ashkelon Seawater Reverse Osmosis plant in Israel has a total production capacity of 108 million m^3/year (*e.g.*, 330,000 m^3/day) and is the largest scale membrane technology application in existence. As a comparison, a 500-MW power plant generates approximately 11,000 tons of CO_2 per day. Assuming a membrane assembly is capable of capturing 90% CO_2 and the operating conditions at the flue gas exit are 350 K and 1 atm, the membrane must process on the order of 6,000,000 m^3/day of CO_2, which is nearly a factor of 20 greater than the largest existing membrane plant. Membrane separation processes also play a role in CO_2 capture from natural gas purification and in air separation for oxyfuel combustion and integrated gasification combined cycle (IGCC) plants. Air separation is discussed in more detail in Chap. 6. In addition, for an IGCC process, in which both the CO_2 and H_2 concentrations in the fuel gas are significant, either selective-CO_2 or -H_2 membrane separation processes may be suitable.

5.1 Common Membrane Materials

Membranes for CO_2-selective separation are commonly synthesized from natural or synthetic polymers including wool, rubber, and cellulose or polyimide, polysulfone, among others. Polymer membranes can be glassy or rubbery, and fabricated into flat asymmetric or thin-composite sheets, tubules, or hollow fibers. Polymer membranes are limited in that they generally operate below approximately 150°C. Common membrane housings or *membrane modules*, consist of plate-and-frame, spiral-wound, hollow fiber, tubular, and monolith. Depending upon the environment of CO_2 separation, *i.e.*, flue gas, air, natural gas, etc., CO_2 may also be indirectly separated based upon a membrane selectivity toward the dominant chemical (*i.e.*, highest concentration) species in a given gas mixture. For instance, in the case of flue gas and air, the dominant species is N_2, rather than CO_2; therefore, the development of a N_2-selective membrane (*i.e.*, similar to H_2-selective metallic membranes) for separation may be of interest since the partial pressure driving force would be greater in this case. Similarly, this is the case for the use of H_2-selective membranes for the separation of CO_2 from fuel or synthesis gas of a gasification process (precombustion capture). Additionally important is the desired product and pressure. For instance, in natural gas purification or CO_2 capture, the desired products are natural gas and CO_2, respectively. Since both of these gases are then compressed for pipeline transport, they should remain on the high-pressure side of the membrane to avoid the high cost of recompression. Hence, a N_2-selective membrane for postcombustion CO_2 capture may be favored as is a H_2-selective membrane for precombustion CO_2 capture. For natural gas purification, since natural gas is the desired product, a CO_2-selective membrane would be favored.

In general, in the case of membrane-based separations, exploitation of the differences between the chemical, electronic, or physical properties of the species involved in the separation is required. For instance, a membrane may serve to separate based upon size, solubility, rate of diffusion or some combination of these properties.

5.2 Membrane Transport Mechanisms

The two primary options for gas separation using membranes are via porous or solution-diffusion mechanisms as shown in Fig. 5.2. If a membrane's pores are large enough for convective flow, separation will not take place. There are several modes of separation possible in porous membranes, *i.e.*, *Knudsen diffusion*, surface diffusion, and *molecular sieving*. In the case of diffusion-based separation it is difficult to achieve sufficient separation, although Knudsen diffusion in some cases may be enhanced by surface diffusion, in which molecules will adsorb to and diffuse across the pore surface. It should be noted that commercial application of a Knudsen diffusion-based technology might be limited since it is most suitable for systems of large molecular weight ratios. For instance, large-pore zeolite-based membranes operating through a Knudsen diffusion mechanism were developed in the 1940s by

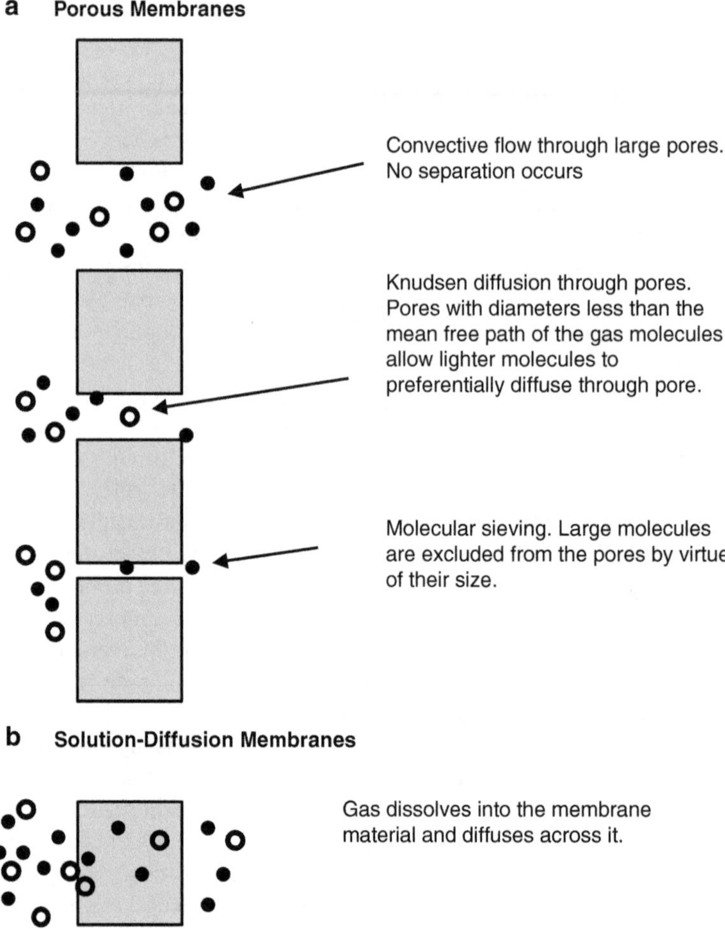

Fig. 5.2 Membrane mechanisms for gas separation for **a** porous versus **b** solution-diffusion membranes. (Reprinted with permission of Wiley, Inc. [7])

the Manhattan Project to enrich uranium by separating uranium isotopes as gaseous UF_6. If pores are small enough so that larger molecules are unable to pass through, molecular sieving can be used to separate molecules of different sizes. Additionally, surface diffusion can assist in separation. In cases in which one component has preferential adsorption on the surface, high selectivity is possible.

5.3 Nonporous Membranes

Gas transport through dense and nonporous membranes takes place by a solution-diffusion mechanism as illustrated in Fig. 5.2b. The membrane material can be either polymer-based for selective transport of CO_2 or methane, or metal-based for

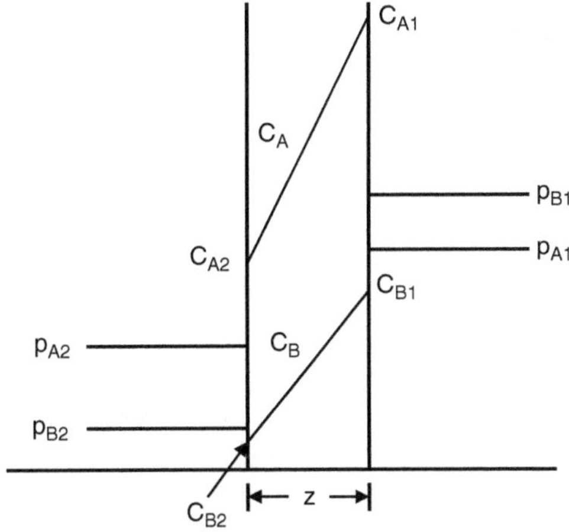

Fig. 5.3 Concentration and pressure gradients in a dense nonporous membrane. (With permission from McGraw-Hill Companies, Inc. [12])

selective transport of H_2 or perhaps O_2, or N_2. The selective nature of the membrane is dependent, in part, upon the solubility of a gas within a given membrane material. The gas will dissolve on the high-pressure side of the membrane, diffuse through the dense phase, and desorb or evaporate at the low-pressure side. The mass-transfer rate depends on the concentration gradient within the membrane, which is directly proportional to the pressure gradient[1] across the membrane, the membrane's solubility, diffusivity and thickness. An example of possible pressure and corresponding concentration gradients for binary mixtures is provided in Fig. 5.3.

Henry's law is assumed to be applicable for each gas species, with equilibrium taking place at the interface. Traditionally, there will exist some extent of gas-film resistance (commonly called *concentration polarization*), but with the partial pressures at the interface equal to those in the bulk (*e.g.*, notice that the pressure profiles are horizontal from the interface to bulk regions in Fig. 5.3), this example neglects the resistance in the gas film. Since membranes may be designed to selectively separate CO_2 from gas mixtures directly or indirectly, the expression for flux is generalized. The flux for gas A, J_A can be represented as:

$$J_A = -D_A\left(\frac{dc_A}{dz}\right) = D_A\left(\frac{c_{A1} - c_{A2}}{z}\right) \tag{5.1}$$

such that D_A is the diffusivity of gas A, c_{A1} and c_{A2} are the concentrations of gas A at the high-and low-pressure sides, respectively and z is the thickness of the membrane. Notice that the membrane thickness, z, represents the length-scale over which the mass transfer is taking place and was termed δ in the absorption and adsorption separation processes described in Chaps. 3 and 4, respectively. In general, the high-pressure side of the membrane is termed the *feed* with the low-pressure side termed

[1] For nonideal gases, the driving force is proportional to the fugacity difference across the membrane for a given gas species permeating the membrane.

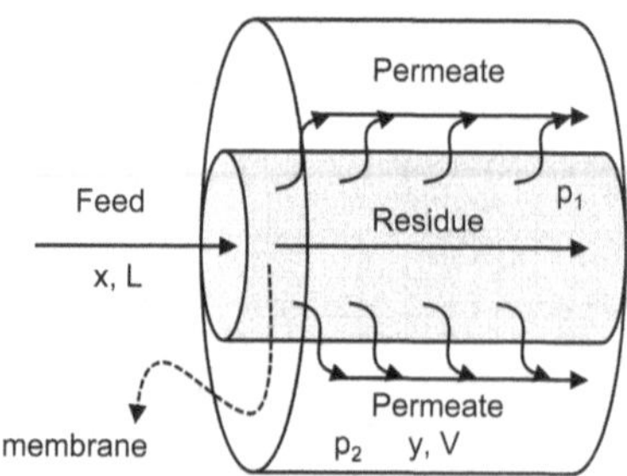

Fig. 5.4 Schematic of simplified membrane gas separation process

the *permeate*. The *residue* (or *retentate*) is the remainder of the feed stream after exiting the membrane system. A simplistic schematic of a membrane separation process illustrating each of these gas streams is shown in Fig. 5.4. Similar to the notation used in an absorption process, the feed stream composition and flow rate are represented by the variables x and L, while the permeate stream composition and flow rate are represented by the variables y and V, respectively. Throughout the worked problems of the chapter, the molar flow rate is represented by i, such that i is equal to F, R, or P, to designate feed, residue, or permeate streams, respectively. The total feed and permeate pressures in this case are p_1 and p_2, respectively.

The concentrations of a given gas mixture comprised of components A and B are related to the partial pressures in the following way through the solubility coefficient, S, which has units representative of the reciprocal of the Henry's law coefficient, mol/cm^{3}atm:

$$c_A = p_A S_A \quad \text{and} \quad c_B = p_B S_B \tag{5.2}$$

It is important to note that the units of Henry's law may also be in terms of pressure if the mole fraction is considered in place of the concentration to represent the composition of the gas stream. Substitution of Eq. (5.2) into Eq. (5.1) yields,

$$J_A = \frac{D_A S_A (p_{A1} - p_{A2})}{z} \tag{5.3}$$

such that the product of the diffusivity and the solubility of component A is equivalent to its permeability. Note that the ratio of $D_A S_A / z$ in Eq. (5.3) is equivalent to the mass-transfer coefficient.

5.3.1 *Permeability and Selectivity*

The *permeability*, P_A, is defined as the product of the diffusivity and solubility, $D_A S_A$, for a gas A in a given gas mixture or more simply, the membrane's ability to permeate gas, and is often expressed in units of Barrer[2] defined by:

$$1 \text{ Barrer} = 10^{-10} \frac{\text{cm}^3(\text{STP}) \cdot \text{cm}}{\text{cm}^2 \cdot \text{s} \cdot \text{cmHg}} \tag{5.4}$$

[2] Multiply by 3.348×10^{-19} to convert from Barrer to (kmol m)/(m^{2}sPa).

Table 5.1 Gas permeabilities [12] [Barrer] of common membrane materials at 25–30°C

Polymers	H_2	He	CH_4	N_2	O_2	CO_2
Silicone (PDMS)	940	560	1370	440	930	4600
Natural rubber	49	30	29	8.7	24	134
Polysulfone	14	13	0.27	0.25	1.4	5.6
Polycarbonate	–	14	0.28	0.26	1.5	6.5
Polyimide	2.3	–	0.007	0.018	0.13	0.41

The Barrer is named after the gas permeability pioneer, Richard Maling Barrer. Since the volume flux is easier to measure than the mass flux in gas permeability experiments, this unit is most frequently used [11]. Example gas permeabilities for common membrane materials are listed in Table 5.1. In general, the solubility is thermodynamic in nature and is affected by gas-membrane interactions, while the diffusivity is a kinetic parameter and depends upon the gas dynamics within the bulk of the membrane. Typically, the diffusivity is calculated from measurements of the permeability and solubility, and in this way is an average between the feed and permeate concentrations.

The general gas flux expressed in terms of *permeance*, $\bar{P}_A$ is:

$$J_A = \frac{D_A S_A (p_{A1} - p_{A2})}{z} = \bar{P}_A (p_{A1} - p_{A2}) \tag{5.5}$$

Typical units of permeance or "pressure-normalized flux" are standard $\text{ft}^3/\text{ft}^2 \cdot \text{h} \cdot \text{atm}$, $\text{L(STP)/m}^2 \cdot \text{h} \cdot \text{atm}$, or $10^{-6}\,\text{cm}^3\text{(STP)/cm}^2\text{s} \cdot \text{cmHg}$, which are also known as gas processing units (GPU). Note that GPU is the ratio between the volumetric gas flow rate and the product of the membrane area and partial pressure driving force.

Example 5.1 A ZrO_2 microporous membrane [59] with a thickness of 150 nm and a pore size of 7 nm has been developed for separating H_2 and CO_2 at a temperature of 600°C. Experimental data indicates the gas permeance, $\bar{P}_A$, of H_2 and CO_2, respectively, are 30.0 GPU and 5.4 GPU. The transmembrane partial pressure, p_{1,H_2} is 1 bar for H_2 and 1.5 bar for CO_2. Determine the transmembrane fluxes in kmol/m^{2}s. Assume vacuum conditions on the permeate side of the membrane.

Solution: From Eq. (5.5), the flux for component H_2 is:

$$J_{H_2} = \bar{P}_{H_2}(p_{1,H_2} - p_{2,H_2})$$

Therefore, in the case of H_2,

$$J_{H_2} = (30\ \text{GPU})(1\ \text{bar})$$

$$= \left(30 \frac{\text{ft}^3}{\text{ft}^2\text{h atm}}\right)\left(\frac{0.305\ \text{m}}{1\ \text{ft}}\right)\left(\frac{1\ \text{h}}{3600\ \text{s}}\right)\left(\frac{0.987\ \text{atm}}{1\ \text{bar}}\right) = 0.0025\ \text{m}^3/\text{m}^2\text{s}$$

Assuming an ideal gas, at a total pressure of 2.5 bar and temperature of 600°C, 1 m^3 of gas will contain 34.4 moles. Thus, the H_2 flux across the membrane is 8.61×10^{-5} kmol/m^2s. Following the same steps, the CO_2 flux is 4.5×10^{-4} m^3/m^2s, equivalent to 1.55×10^{-5} kmol/m^2s.

The *separation factor* (selectivity or permeability selectivity), α, is often expressed as the ratio of permeabilities for a binary mixture as:

$$\alpha = \frac{P_A}{P_B} = \left(\frac{D_A}{D_B}\right)\left(\frac{S_A}{S_B}\right) \tag{5.6}$$

in which the ratio D_A/D_B is often referred to as the "mobility selectivity" and the ratio S_A/S_B is referred to as the "solubility selectivity," with gas A in each case being the more permeable species. From Eq. (5.6) it is that a high selectivity may be achieved by having either a high diffusivity ratio or a high solubility ratio. Gas diffusivity within a membrane can depend upon the size, shape and charge properties of the molecular species. A reasonable rule of thumb is that a selectivity of 10 or greater[3] is required for the effective separation of a gas from a mixture. Example selectivities for a variety of CO_2 gas mixtures are listed in Table 5.2.

Example 5.2 Natural gas often will contain CO_2 and can be purified by gas permeation through a variety of polymeric membranes. A natural gas well produces 1000 m^3 (at 1 atm and 20°C) per day that must be treated. Natural gas, with a composition of 85 mol% CH_4 and 15 mol% CO_2 is sent to a membrane separator at 20°C and 80 psia. A CO_2-selective membrane with a 10 m^2 surface area and a thickness of 1.0 μm is used to achieve 98 mol% purity of CH_4 in the residue stream and 98 mol% CO_2 in the permeate stream. Determine the selectivity (α_{CO_2,CH_4}) of the membrane. Assume vacuum conditions on the permeate side.

Solution: At STP, 1000 m^3 per day is equivalent to 0.48 mol/s, assuming an ideal gas.

First, perform a mass balance around the membrane, such that $\dot{n}$ is the molar flow rate in units of mol/s. Recall the notation, such that *x* represents the concentration of permeable species in the feed and residue streams, while *y* represents the concentration of permeable species in the permeate stream. The subscripts *F*, *R*, and *P* refer to the feed, residue, and permeate, respectively. For CO_2:

$$x_{F,CO_2}\dot{n}_F = y_{P,CO_2}\dot{n}_P + x_{R,CO_2}\dot{n}_R$$

[3] For natural gas purification, the ideal separation factor for CO_2/CH_4 is ~20, and reduced to ~15 at high feed pressure; for N_2 production from air, the ideal separation factor is ~6–8, with the residue stream (N_2) as the product.

Table 5.2 Selected selectivities of gas mixtures with various membrane materials

Material	Gas mixture	Selectivity
Silicone rubber [3]	CO_2/H_2	4.9
	CO_2/O_2	5.0
	CO_2/CH_4	3.4
	O_2/N_2	2.1
Kapton (aromatic polyether diimide) [3]	CO_2/H_2	0.18
	CO_2/O_2	3.1
Ethyl cellulose [13]	CO_2/O_2	6.1
	CO_2/CH_4	11
	O_2/N_2	3.6
Poly (methyl methacrylate) [13]	CO_2/O_2	4.4
	CO_2/CH_4	119
	O_2/N_2	7
PMDA-4,4′-ODA polyimide [13]	CO_2/O_2	4.4
	CO_2/CH_4	46
	O_2/N_2	6.1
Polysulfone [14]	CO_2/O_2	4.1
	CO_2/CH_4	23.3
	O_2/N_2	6.0
Polydimethylsiloxane (PDMS) [15]	CO_2/O_2	5.8
	CO_2/CH_4	3.2
	O_2/N_2	2.2
Cellulose acetate [15]	CO_2/O_2	5.8
	CO_2/CH_4	31.7
	O_2/N_2	5.5

For CH_4:

$$x_{F,CH_4}\dot{n}_F = y_{P,CH_4}\dot{n}_P + x_{R,CH_4}\dot{n}_R$$

Using the desired outlet purities, $\dot{n}_p = 0.065$ mol/s.

The flux of CO_2 across the membrane can be calculated as:

$$J_{CO_2} = \frac{y_{P,CO_2}\dot{n}_P}{A} = \frac{(0.98)(0.065)}{100} = 5.3 \times 10^{-4}\frac{\text{mol}}{\text{s m}^2}$$

The CO_2 permeance is calculated using Eq. (5.5)

$$J_{CO_2} = \bar{P}_{CO_2}(p_{1,CO_2} - p_{2,CO_2}) = \bar{P}_{CO_2}(x_{F,CO_2}p_1 - y_{P,CO_2}p_2)$$

such that $p_2 = 0$ psia and $p_1 = 80$ psia, resulting in:

$$\bar{P}_{CO_2} = \frac{J_{CO_2}}{x_{F,CO_2}p_F} = \frac{5.3 \times 10^{-4}}{(0.15)(80)} = 2.47 \times 10^8\frac{\text{Barrer}}{\text{m}}$$

with a membrane thickness of 1 μm, $P_{CO_2} = 247$ Barrer.

Following these steps to determine the permeability of CH_4, $P_{CH_4} = 0.90$ Barrer.

The selectivity is then:

$$\alpha_{CO_2,CH_4} = \frac{P_{CO_2}}{P_{CH_4}} = \frac{247}{0.90} = 275.9$$

This membrane is near the Robeson limit according to Fig. 5.6c.

In general, the permeability of a gas increases with temperature since the increase in diffusivity is substantially higher than the decrease in solubility. However, an increase in temperature may decrease the selectivity of the membrane, leading to a required balance between high flux versus selectivity to determine the optimal operating temperature.

The fundamental parameters that characterize membrane separation performance are the permeance and the selectivity. Membranes with both a high permeability and selectivity are desired since higher permeabilities decrease the required membrane area, thereby decreasing the capital cost of the membrane assembly and a high selectivity leads to an increase in product purity. It is well-recognized that there is a trade-off between permeability and selectivity. In 1991, Lloyd M. Robeson [16] quantified this by plotting the permeability versus selectivity for a large number of available data including a variety of membrane materials for various gas pairs. The original Robeson plot is pictured in Fig. 5.5. The Robeson plot shows the

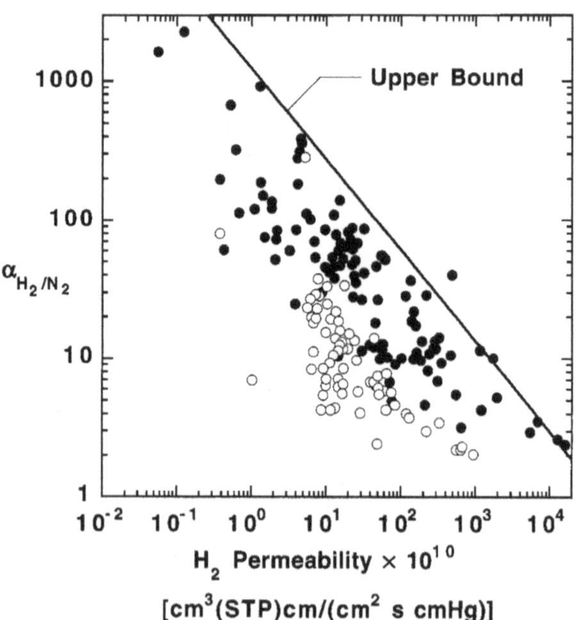

Fig. 5.5 Relationship between H_2 permeability and H_2/N_2 selectivity for rubber (*open circles*) and glassy (*closed circles*) polymers and the empirical upper-bound relationship. (With permission from Elsevier [9])

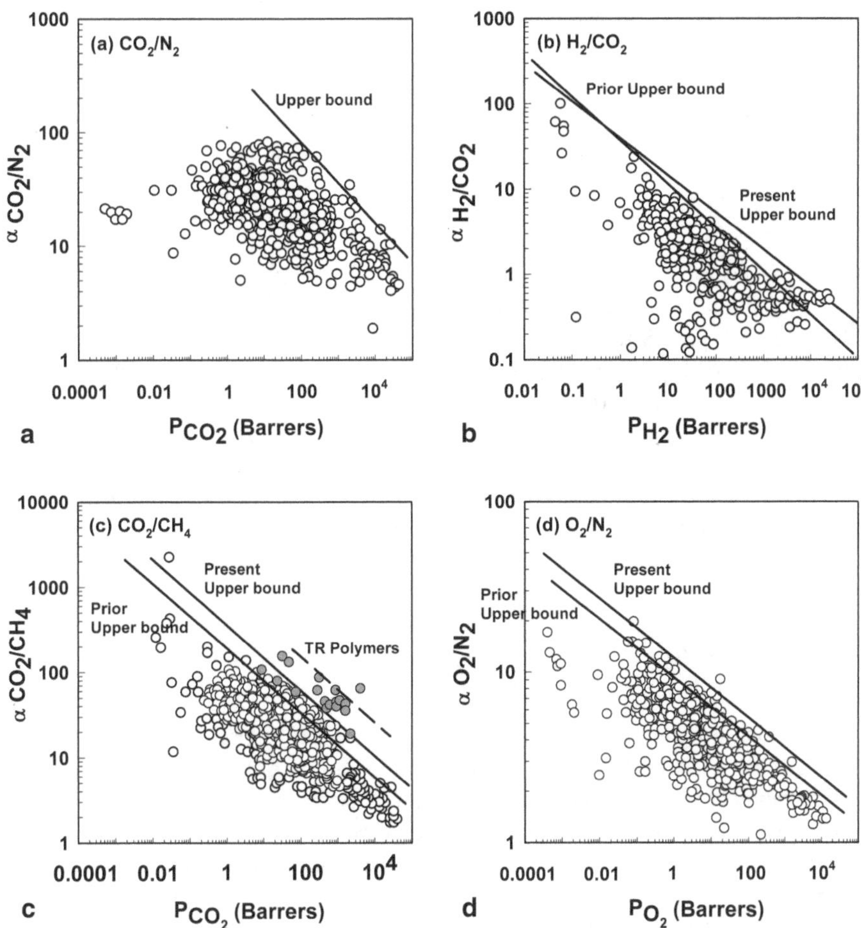

Fig. 5.6 Robeson plots of common CO_2 gas pairs **a** CO_2/N_2, **b** CO_2/H_2, **c** CO_2/CH_4, and **d** O_2/N_2. (With permission from Elsevier [2])

relationship between the permeability of H_2 and the H_2/N_2 selectivity. Common gas pairs for CO_2 separation include CO_2/N_2, CO_2/H_2, CO_2/CH_4, and O_2/N_2, and the Robeson plots [17] of each are shown in Fig. 5.6.

In 2008, Robeson revisited his 1991 study and found only several cases in which the upper bound shifted, with the shift only slight. In the case of CO_2/N_2 gas mixtures, it has been shown that polymers containing poly(ethylene oxide) units have unique separation properties [18] and that membranes comprised of poly(trimethylgermylpropyne) and poly(trimethylsilylpropyne) provide upper-bound properties with high permeabilities [19]. In particular, poly(ethylene oxide)-based polymers have resulted in increased solubility due to their polar ether segments. Unlike their nonpolar analogs, these polar polymers show significantly higher CO_2/H_2 selectivities, indicating the importance of the polymer structure [12].

Table 5.3 Ideal liquid and polymer gas solubilities [11] at 35°C

Gas	Vapor pressure [atm]	Ideal liquid solubility at 1 atm (mole fraction)	Ideal polymer solubility [10^{-3} cm^3(STP)/cm^3cmHg]
N_2	1400	0.0007	2.6
O_2	700	0.0014	4.8
CH_4	366	0.0027	18.4
CO_2	79.5	0.0126	29.5

The H_2/CO_2 upper-bound relationship has shifted slightly due to a number of poly(trimethylsilylpropyne)-based membranes that have been developed [20]. The CO_2/CH_4 gas pair is the second most studied after N_2/O_2. The *ladder polymers* (PIM-1 and PIM-7) [21] demonstrate reasonable separation properties, and in particular, thermally rearranged (TR) ladder-like polymer variants comprised of benzoxazole-phenylene or benzothiazole-phenylene structures [22] have exhibited superior CO_2/CH_4 separation performance as is demonstrated in Fig. 5.6c. In the cases of the CO_2/CH_4 upper-bound shifts, there is indication that the membranes may be performing by a transition transport mechanism, bordering solution-diffusion and Knudsen diffusion mechanisms due to their microporous structure [17].

5.3.2 *Solubility*

Polymer Membranes In general, polymer membranes exhibit low gas solubilities if the gas has a low boiling point or critical temperature. Additionally, the similarity in the chemical structure between the gas and polymer will influence its solubility. For instance, gases with a high quadrupole moment such as CO_2 are more soluble in polymers containing a high concentration of polar groups. In particular, the solubility of water vapor in a polymer membrane will depend on the ability of the water to form hydrogen bonds within the material. Ideal liquid and polymer solubilities for common gases are listed in Table 5.3. The values in Table 5.3 serve only to demonstrate general trends, as specific values of solubilities of a gas in a polymer will depend on the gas interactions with a particular polymer.

To select an optimal material for a given separation it is important to understand the differences between ideal and real polymers. The dual-sorption model proposed by Barrer et al. [23] and later extended by Michaels et al. [24], Paul and Koros [58], and Koros et al. [25] is useful in determining how to tune polymer membranes for a given set of gas separation conditions. Before discussing this model it is important to first define the *free volume* of a polymer. Figure 5.7 demonstrates the relationship of polymer volume (cm^3/g) to temperature. At high temperatures the polymer is in a rubbery state, in which free volume exists between the polymer chains since chains do not pack perfectly. Additionally, at high temperatures, although this unoccupied space represents a small portion, *e.g.*, approximately 3% of the entire polymer, the space is sufficient to allow for the rotation of the backbone polymer segments. As the temperature decreases, so does the free volume since the backbone flexibility is compromised

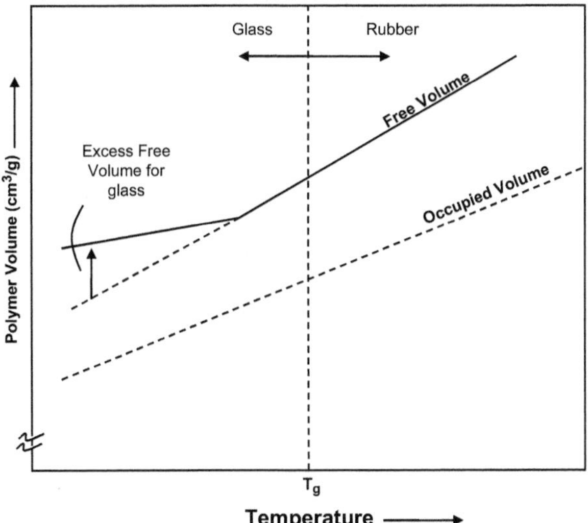

Fig. 5.7 The change in polymer volume as a function of temperature for a typical polymer. (Reprinted with permission of John Wiley & Sons, Inc. [4])

as internal energy and mobility decreases with temperature. Reducing the temperature further to the glass transition temperature, reduces the free volume to the extent at which the polymer chains are no longer free to rotate. As the temperature is reduced further, the occupied volume will continue to decrease, with the original free volume remaining essentially unchanged. Hence, a glassy polymer contains both the free volume contribution associated with the incomplete packing of the groups comprising the polymer chains, in addition to an *excess free volume* that remains frozen within the polymer matrix due to the inability of the polymer chains to rotate. In general, a rubbery polymer is considered an amorphous material above its softening or *glass transition temperature*, T_g, while a glassy polymer is an amorphous, nonequilibrium material below T_g [26]. In a rubbery polymer, sections of the backbone of the polymer can rotate freely allowing the polymer to be soft and elastic. On the other hand, in a glassy polymer, steric hindrance exists along the polymer backbone, which limits rotation of the polymer segments resulting in a more rigid and tougher polymer material.

Within the dual-sorption model, gas sorption in a polymer matrix involves two types of sites. The first type of site is associated with the free volume portion of the material that can be filled by dissolved gas, with the second type available only in glassy polymers. The second types of sites are those in which the gas dissolves and occupies space in the excess free volume space of the glassy polymer. In general, Henry's law applies to the sorption of gas in the free volume component of the polymer, where as Langmuir theory applies to gas occupation (adsorption) in the excess free volume that exists between the polymer chains. The equilibrium concentration of gas in a given polymer, c, at a given pressure, p, can be expressed as:

$$c = c_H + c_L = Hp + \frac{c_L' bp}{1 + bp} \tag{5.7}$$

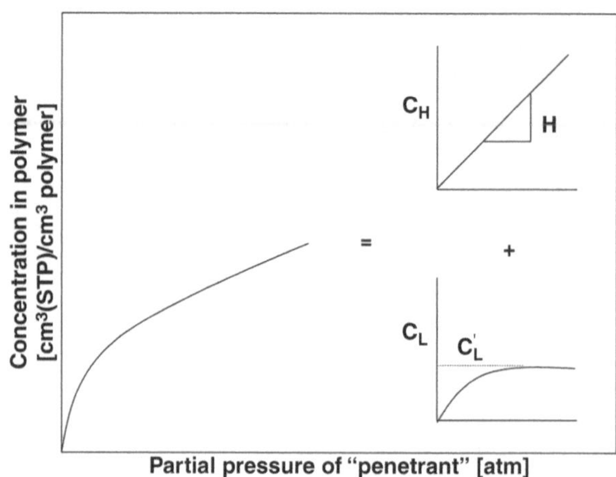

Fig. 5.8 Typical gas sorption isotherms for dual-mode model. (With kind permission from [6])

such that c_H and c_L are the Henry's Law- and Langmuir-mechanism concentrations, respectively, H is the Henry's law constant, c'_L and b are the Langmuir capacity constant and affinity constant, respectively. The two components that comprise the dual-mode model are depicted in Fig. 5.8. Various sorption models in glassy polymers will be discussed briefly, with thorough reviews on gas sorption in polymers available in the literature [27, 28].

Unique from sorption in polymers, the mechanism of gas solubility in metallic membranes is based upon atomic sorption of a given species (*i.e.*, hydrogen, oxygen, or nitrogen) within the crystal structure of the metal lattice versus the sorption of a molecular species within the framework of a given polymer.

Metallic Membranes Metallic membranes may be used to transport atomic hydrogen, oxygen, or nitrogen. The metallic surface of the dense membrane acts as a catalyst by dissociating a diatomic molecule into its atomic species, which then sorb into the bulk of the metal and subsequently diffuse through the metal by hopping through the metal's interstitial crystal sites. Hydrogen-selective membranes are the most common and are traditionally comprised of palladium, although alloying with silver, copper, and gold are also considered suitable and in some cases exhibit enhanced permeability [29]. Using H_2 as an example, its solubility c_H, within the bulk crystal structure of a given metallic membrane is $c_H = K_S p_{H_2}^{1/2}$, in which $p_{H_2}^{1/2}$ is the partial pressure of hydrogen in gas phase and K_S is Sieverts' constant represented by Sieverts' law [30] as:

$$K_S = exp\left(\beta\left[-\frac{E_d}{2} + \frac{h\nu_{H_2}}{4} - E_{abs} - \frac{3}{2}h\nu_H\right]\right) \times \frac{1}{\sqrt{\gamma}}\sqrt{1 - exp\left(-\beta h\frac{\nu_{H_2}}{2}\right)}\frac{1}{(1 - e^{-\beta h\nu_H})}, \quad (5.8)$$

and $\gamma = \left(\frac{2\pi mkT}{h^2}\right)^{3/2} \frac{4\pi^2 I(kT)^2}{h^2}$

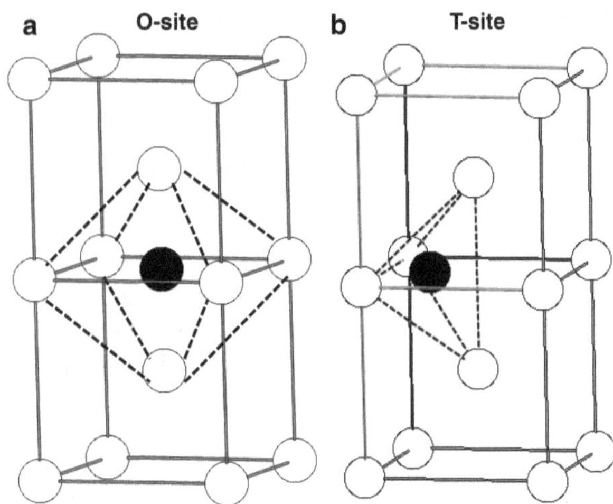

Fig. 5.9 **a** Octahedral (O-site) and **b** tetrahedral (T-site) sites in BCC metals

such that β is defined as $1/kT$, E_d is the dissociation energy of the diatomic gas, ν_{H_2} is the vibrational frequency of H_2 in the gas phase, ν_H is the vibrational frequency of atomic hydrogen within the crystal lattice, m is the mass of H_2, I is the moment of inertia of the molecular H_2 species, h is Planck's constant, and E_{abs} is the binding energy of the atom within the bulk of the crystal. Quantum chemical techniques may be used to determine the relative binding energies associated with the sorption sites within the crystal lattice [31]. Typical stable sorption sites include the octahedral and tetrahedral sites of the crystal lattice as shown in Fig. 5.9. Within palladium, which has an face-centered cubic (FCC) crystal structure, atomic hydrogen is most stable at the octahedral site and within vanadium, which has a body-centered cubic (BCC) crystal structure, hydrogen is more stable at the tetrahedral site [31a]. The binding strength of the atomic species at each site is directly linked to the solubility of the gas within the bulk crystal by Eq. (5.8). For additional information on H_2-selective membranes, a thorough review of palladium-based membranes for H_2 separation has been carried out by Paglieri [32].

5.3.3 Diffusion

Recall from Eq. (5.5) that to determine the flux of a gas through a nonporous membrane, knowledge of both the solubility and the diffusivity must be known. The diffusivity of a permeating molecule is a function of both the frequency of the movements in addition to the size of each movement [11]. It is interesting to consider the relative scales of atomic and molecular diffusivities in various environments since the order of magnitude of this parameter in part dictates the gas flux through a given membrane. Carbon atoms within the crystal lattice of diamond have a very small diffusivities with movements on the order of 1–2 Å. On the other hand, gas diffusivities

Table 5.4 Typical diffusivities [11] for various applications at 25°C

Material/mixture	Diffusivities (cm^2/s)
Oxygen in air (1 atm)	1.0×10^{-1}
Salt in water	1.5×10^{-5}
Oxygen in silicon rubber	1.0×10^{-5}
Sodium ions in sodium chloride crystals	1.0×10^{-20}
Aluminum ions in metallic copper	1.0×10^{-30}

are much larger since the gas molecules are in constant motion with each diffusive jump, on the order of 1000 Å or more. Table 5.4 provides some representative values of diffusivities for various applications including gas diffusion through solids, liquids, and within gas mixtures. Generally, gas diffusivities in liquids and polymers range between 10^{-5} to 10^{-10} cm^2/s.

More specifically, gas diffusion in polymers is dependent upon the polymer type. Gas diffusion in glassy polymers tends to be slower than in rubbery polymers due to limited thermal motion. If the temperature of a glassy polymer is increased high enough, it may acquire the thermal energy required to overcome the steric hindrance restricting the motion of the polymer backbone. In some cases, sufficiently strong interactions between polymer and dissolved gas can lead to polymer plasticization. Plasticization and performance degradation, resulting from loss of separation selectivity, plague current state-of-the-art polymeric membranes for CO_2 separation processes, especially at high CO_2 partial pressures, such is the case with many natural gas applications.

Rubbery Polymers As discussed previously, the sorption of gases in rubbery polymers may be described by Henry's law for scenarios in which the concentration of sorbed gas is low, and with the gas concentration c (cm^3 (STP)/cm^3 polymer) in the polymer described by, $c = H \cdot p$, such that H is the Henry's law coefficient, and p is the pressure. For cases in which the gas concentration in the rubbery polymer membrane is low and the gas diffusivity is consistent throughout, the permeability, P_A of a gas A is independent of the feed pressure and can be expressed as $P_A = H \cdot D$ [33]. In general, gas transport in rubbery polymers is quite similar to that of low molecular weight liquids. Table 5.5 provides a comparison of typical diffusivities for a variety of polymer membrane materials.

Glassy Polymers Since these materials have more restricted segmental motions, they provide enhanced "mobility selectivity" compared to their rubbery polymer counterparts [33, 34]. Due to their inherent restriction of structure size and shape they are often used as selective layers in membranes for gas separation processes. Glassy polymers are able to discriminate between molecular dimensions of 0.2–0.5 nm. Examples of permeability and corresponding selectivity for several gas mixtures of interest are presented in Table 5.6.

Transport in glassy polymers is traditionally dictated by the molecular size of the gas for gas pairs where diffusivity dominates transport behavior, and solubility differences among permeants are negligible, as is often the case for air separation (N_2/O_2) processes. Gas sorption and subsequent transport in glassy polymers does not follow

Table 5.5 Selected gas diffusivities [12] of common membrane materials ($D \times 10^9$ at 25°C, cm^2/s)

	O_2	N_2	CH_4	CO_2
Polymers				
Polyethylene terephthalate	3.6	1.4	0.17	0.54
Polyethylene ($\rho = 0.964\,\mathrm{g/cm^3}$)	170	93	57	124
Polyethylene ($\rho = 0.914\,\mathrm{g/cm^3}$)	460	320	193	372
Natural rubber	1580	1110	890	1110

Table 5.6 Permeabilities and selectivities of various gas pairs in silicone rubber (PDMS[a]) and polycarbonate (PC[b]) at 35°C

Polymer	P_{CO_2} (Barrer)	P_{CO_2}/P_{CH_4}	$P_{CO_2}/P_{C_2H_4}$	P_{CO_2} (Barrer)	P_{O_2}/P_{N_2}
PDMS [35]	4550	3.37	1.19	933	2.12
PC [25]	6.5	23.2	14.6[16]	1.48	5.12

[a]PDMS is poly (dimethyl siloxane) and values are for 100 psia
[b]PC is bisphenol-A-polycarbonate and values are for 147 psia

the same Henry's law behavior as in rubbery polymers. Below T_g the excess volume is hypothesized to be the result of trapped nonequilibrium chain conformations, frozen in quenched glasses, termed "frozen" due to the long times associated with the relaxation of the segmental motions in the glassy state [3]. The excess volume allows for higher gas occupation compared to liquids or rubbery polymers [36]. Based upon the previously discussed dual-mode model, the transport model representing the flux, J, in terms of a two-part contribution may be expressed as:

$$J = -D_{He}\frac{dc_H}{dx} - D_{La}\frac{dc_L}{dx} \tag{5.9}$$

such that D_{He} and D_{La} refer to the transport of the dissolved (based on Henry's law) and Langmuir-sorbed components, respectively. As expected, D_{He} is traditionally greater than D_{La} except in cases of a noncondensible gas such as hydrogen and helium. The corresponding permeability can be expressed as:

$$P = HD_{He}\left(1 + \frac{FK}{1+bp}\right) \tag{5.10}$$

such that $F = D_{La}/D_{He}$ and $K = c'_L b/H$ and are dimensionless parameters and p is the upstream driving pressure. In Eq. (5.10), naturally the first term describes the transport in the Henry's law regime, while the second relates to the Langmuir environment. Figure 5.10 shows the CO_2 permeability in a glassy polycarbonate polymer at 35°C [36], and as illustrated, increasing pressure leads to a decrease in permeability approaching the HD_{He} limit, as Langmuir sites become saturated at higher feed partial pressures.

Metallic Membranes Following catalytic dissociation at the metallic membrane surface, the atomic species are sorbed into the bulk crystal structure of the metal and

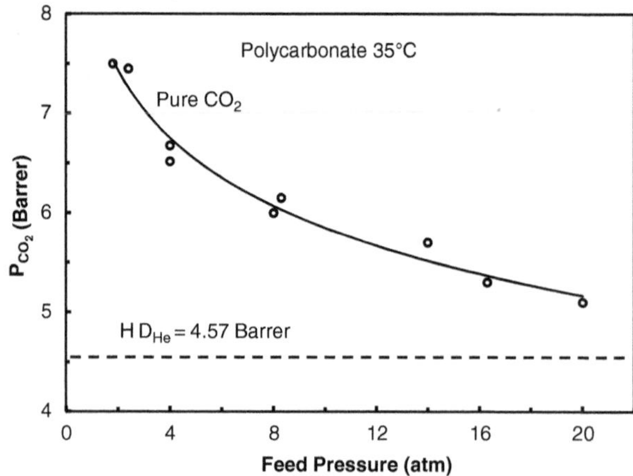

Fig. 5.10 Pressure dependence of CO_2 transport in glassy polycarbonate. (Reprinted with permission from [8])

diffuse through due to a partial pressure or concentration driving force. Hydrogen transport through metals has been studied extensively over the last 40 years [37]. The diffusion mechanism consists of the atomic species hopping from different interstitial sites within the crystal lattice, with the rate constant of each hop represented by an Arrhenius expression with an activation energy equivalent to the energy required to overcome the barrier associated with a given hop. Using quantum chemical methods combined with kinetic Monte Carlo, atomic diffusivities may be predicted [38]. Figure 5.11 illustrates an example pathway of an octahedral-to-tetrahedral hop in an FCC metal.

The concentration change from the gas phase to the bulk sorbed phase leads to a more complicated permeability relationship for gas transport in metallic membranes

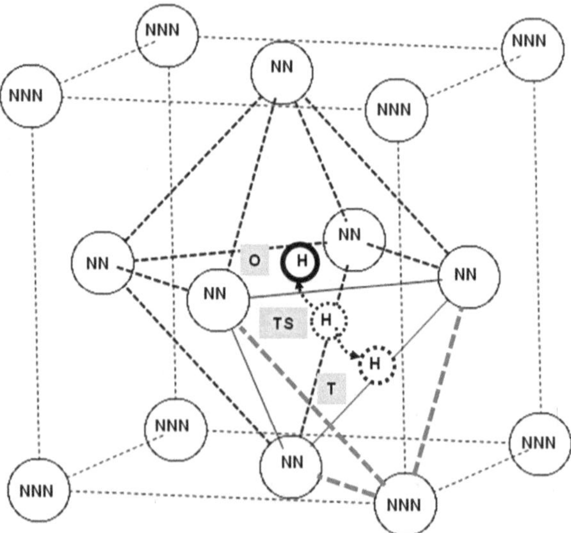

Fig. 5.11 Example of atomic hydrogen hopping between an octahedral (O) and tetrahedral (T) site via a transition structure, TS in an FCC metal, such that NN and NNN represent the nearest and next-nearest atomic (metal) neighbors, respectively

compared to polymer membranes. Since H_2 dissociates across the catalytic surface of the metal membrane into two atoms, that diffuse through the bulk crystal, the atomic concentration of hydrogen, c_H is related to the partial pressure of hydrogen, p_{H_2} in equilibrium with the metal as:

$$c_H = K_S\left(p_{H_2}^{1/2}\right) \tag{5.11}$$

such that K_S is Sieverts' constant, with the ½ exponent due to the dissociation of molecular hydrogen into two atoms. Assuming the permeate side is under vacuum, the flux of H_2 may be expressed as:

$$J_{H_2} = -D_H\left(\frac{K_S \Delta p_{H_2}^{1/2}}{2z}\right) \tag{5.12}$$

such that the "2" in the denominator is due to the molecular flux being half the atomic flux. This expression may also be expressed in terms of the H_2 permeability, P_{H_2} as:

$$J_{H_2} = -P_{H_2}\left(\frac{\Delta p_{H_2}^{1/2}}{z}\right) \tag{5.13}$$

Identical to polymer membranes, the permeance is dependent upon the membrane thickness as follows:

$$\text{permeance} \equiv \frac{J_{H_2}}{\Delta p_{H_2}^{1/2}} = \frac{P_{H_2}}{z} \tag{5.14}$$

Example H_2 permeabilities are exhibited in Fig. 5.12 and Table 5.7 for a variety of metals and metal alloys.

5.4 Facilitated-Transport Membranes

The previously described solution-diffusion mechanism in nonporous membranes is a passive transport process, where the permeating species travels by diffusion down a concentration gradient. Another class of nonporous membranes is facilitated, or carrier-assisted membranes, in which there exists an active transport mechanism that enhances the CO_2 flux through its reversible attachment to a carrier. In this type of membrane, CO_2 reacts with a "carrier" in the membrane to form a reversible complex that is generated at the feed side with the carrier regenerated at the permeate side, when CO_2 is released from the carrier-CO_2 complex. An example of facilitated transport of CO_2 in a supported liquid membrane is shown in Fig. 5.13. In this membrane, facilitated transport of CO_2 is taking place across a cellulose acetate membrane that contains an aqueous carbonate solution. After dissolution at the feed

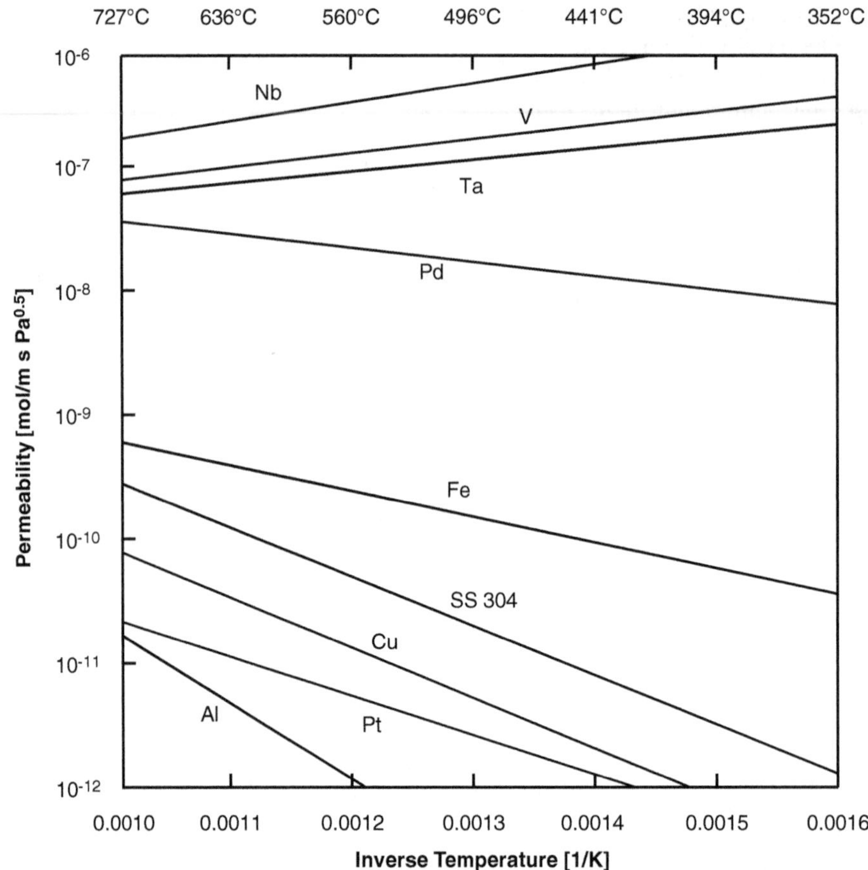

Fig. 5.12 Hydrogen permeabilities of selected metals. (With permission from Elsevier [2])

Table 5.7 Gas permeabilities of common metal membranes at 500°C

Material[a]	Permeability ($P \times 10^8$ mol/m s $Pa^{0.5}$)
$Pd_{80}Cu_{20}$ [39]	1.10
$Pd_{60}Cu^{b}_{40}$ [40]	1.28
$Pd_{95}Ru_5$ [41]	1.42
Pd [42]	2.24
$Pd_{74}Ag_{26}$ [41]	4.17
Ta [42]	13.5
$V_{95}W_5$ [44]	16.2
V [45]	19.1
$Nb_{95}Ru_5$ [46]	25.4
$Nb_{95}W_5$ [46]	34.8
Nb [43]	36.6

[a] Alloys are based upon wt%

[b] $Pd_{60}Cu_{40}$ would be higher than pure Pd at $T \leq 450$°C due to the BCC to FCC phase transition

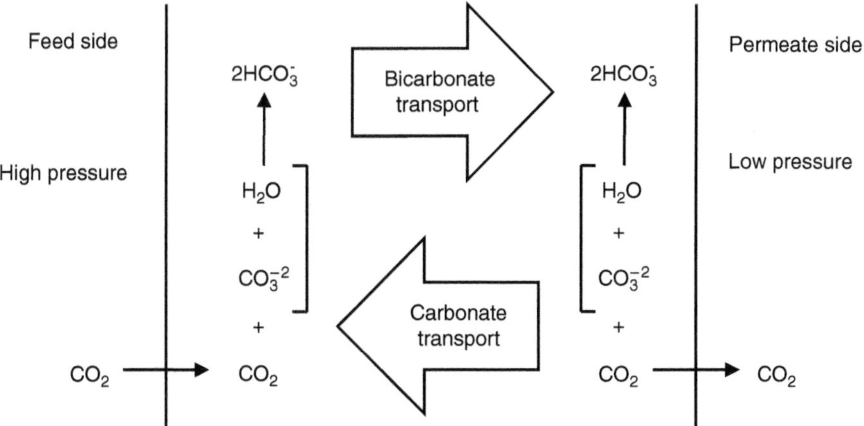

Fig. 5.13 Facilitated transport of CO_2 in a carbonate-containing supported liquid membrane. (This figure was published in Carbon Capture and Storage [5])

side, CO_2 reacts with the carbonate anion forming a bicarbonate ion via the following reaction:

$$CO_2 + H_2O + CO_3^{2-} \leftrightarrow 2HCO_3^- + heat \tag{5.15}$$

The bicarbonate species is transported from the high-pressure feed to the low-pressure permeate side of the membrane as shown in Fig. 5.13. The reaction is reversed with CO_2 being released at the permeate side of the membrane.

In practice, the examples previously described require that diffusion of the carrier complex across the membrane suitably high. Analogy is often drawn to the natural system, where O_2 is carried from the lungs to tissues throughout the body as a carrier complex with hemoglobin. However, this system has the added component of pumping by action of the heart, so that the complexation and subsequent transport of O_2 is not limited by its diffusion rate.

The carrier of CO_2 may also be fixed in place within the membrane. For instance, amine functional groups may be anchored to the polymer backbone of a membrane with the amine reacting directly with a proton from water allowing for the formation of bicarbonate anion HCO_3^-. In this system, the bicarbonate species travels from the high-pressure feed to the low-pressure permeate side of the membrane subsequently complexing with a protonated amine group to form H_2O and CO_2 at the permeate side as shown in Fig. 5.14. Additional information regarding facilitated transport membranes is available in the literature [11, 47].

5.5 Porous Membranes

In a porous membrane, the difference between the rates of which the molecular species diffuse through the pores dictates the extent of separation. The term "porous membranes" encompasses a range of membranes, each having a given nominal pore

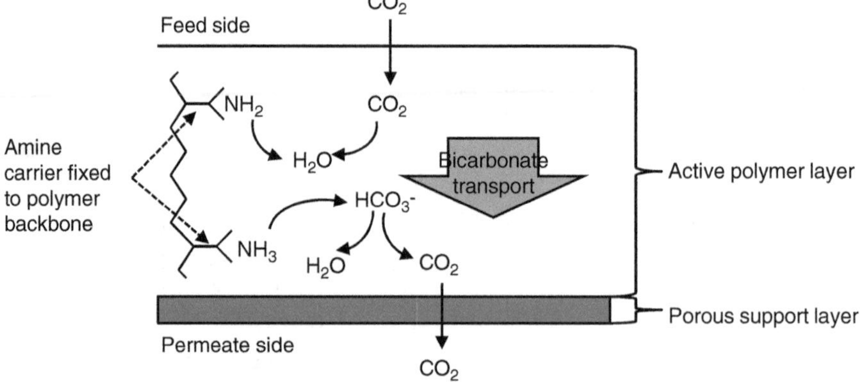

Fig. 5.14 Amine-facilitated transport of CO_2 in an aqueous contained liquid membrane. (This figure was published in Carbon Capture and Storage [5])

Table 5.8 Porous membrane classifications [11]

Pore type	Pore size range	Filtration type
Macroporous	>50 nm (500 Å)	Microfiltration
Mesoporous	2–50 nm (20–500 Å)	Ultrafiltration
Microporous	1–2 nm (10–20 Å)	Nanofiltration
Nanoporous	<1 nm (10 Å)	Molecular sieving

size or average pore size in its distribution of pore sizes. Only those porous membranes having the smallest pore sizes will be considered here, as they have relevance to applications involving gas separation processes and more specifically, carbon capture.

Consider a gas mixture of CO_2 and N_2 or CO_2 and methane. All of these gases will easily fit through pores greater than their molecular diameters and since their molecular diameters are so similar in size (see Chap. 4, Table 4.1), molecular size may not be an adequate parameter to ensure significant separation. If, however, each molecule diffuses through the porous membrane at differing rates, this process may be applied for separation in a batch-like set-up. A flow set-up would be difficult since this would allow the slower gas to compete with the faster one thereby limiting the extent of separation. The porous membrane classification as show in Table 5.8 is quite similar to the pore size classification in sorbents as previously discussed in Chap. 4.

In membranes with meso and/or micropores, the mean free path of the gas phase is much larger than the pore size, with Knudsen and surface diffusion mechanisms dominating in meso and micropores, respectively. As discussed previously in Chap. 4, Knudsen diffusion of a gas A in cylindrical pores can be represented by:

$$D_A = 9700r\left(\frac{T}{M_A}\right)^{0.5} \tag{5.16}$$

such that D_A is the diffusivity of gas A in units of cm^2/sec, r is the pore radius in cm, T is the temperature in K, and M_A is the molecular weight of gas A. Therefore, at a given temperature and pore size, the diffusion of N_2 in a mixture of CO_2 and N_2 is 1.25 times higher than that of CO_2. Similarly, the diffusion of methane in a mixture of CO_2 and CH_4 is 1.65 times higher than that of CO_2. The flux per unit surface area is dependent upon the effective diffusivity D_e, which is lower than the diffusivity in the pore by a factor of ε/τ, such that ε is the porosity and τ is the tortuosity. The tortuosity factor attempts to account for the longer distance traveled in a complex three-dimensional pore network and usually ranges between 2 to 6 [48]. The flux of each gas is directly proportional to the concentration gradient, which is linear for a uniform membrane structure with non-interacting gases. For ideal gases, the flux of gas A, J_A can be expressed as:

$$J_A = D_e \left(\frac{\Delta c_A}{\Delta z}\right) = D_e \left(\frac{\Delta p_A/RT}{\Delta z}\right) \tag{5.17}$$

such that c_A is the concentration of A, p_A is the partial pressure of A and z is the membrane thickness.

Figure 5.15 shows profiles of the pressure gradients of gases A and B through a porous membrane. It is assumed that for this particular case, gas A has a diffusivity two times that of gas B. The total pressure on the feed side of the membrane 2.8 atm such that the $p_{A1} = p_{B1}$ and the total pressure on the permeate side is 1.0 such that $p_{A2} = 0.6$ atm and $p_{B2} = 0.4$ atm. The permeate side has 60% A, which is slightly enriched over the feed side, in which it comprises 50% of the feed. The gradient for gas A is less than that of B, *e.g.*, $\Delta p_A = 1.4 - 0.6 = 0.8$ and $\Delta p_B = 1.4 - 0.4 = 1.0$, which is due to the slight enrichment of A on the permeate side. Therefore, the flux of A is only $2 \times (0.8/1.0) = 1.6$ times that of B. It would be possible to additionally enrich the permeate side with gas A if the feed pressure were higher or the permeate side was less than atmospheric pressure, *i.e.*, if the membrane were operated at a higher pressure ratio.

Molecular sieves are a class of porous membranes, in which the pore diameter approaches that of the molecules attempting to diffuse through the porous membrane. In the case of molecular sieving, the selectivity is a function of molecular size, with smaller molecular species having higher rates of diffusion. Of course this is also

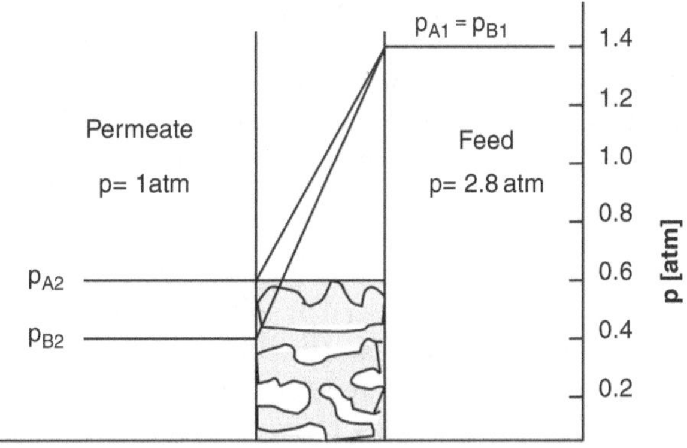

Fig. 5.15 Pressure gradient profile of gases A and B in a porous membrane. (With permission from McGraw-Hill Companies, Inc. [12])

Table 5.9 Selectivity and permeance [48] of selected zeolite ion-exchange membranes for CO_2-N_2 mixtures in the temperature range of 35–40°C

Material	CO_2-N_2 selectivity	CO_2 permeance (mol/m^{2}sPa)
Li (20%) Y	10	7×10^{-7}
LiY	4	2×10^{-6}
K (62%) Y	39	5×10^{-7}
KY	30	1.4×10^{-6}

ultimately dependent on the gas-pore surface interactions and extents of adsorption and surface diffusion. Transport in molecular sieve membranes involves a combination of pore diffusion and surface diffusion. Molecular sieving materials include zeolites, MOFs, activated carbon, and composite silica alumina. Examples of CO_2 selectivities and corresponding permeances for zeolite ion-exchange membranes in the 35–40°C temperature range are presented in Table 5.9.

In general, a major disadvantage of porous membranes for gas separation is the difficulty associated with the fabrication of membranes with a narrow pore size distribution. Another drawback is the potential of water and other vapors condensing in the small pores leading to pore blocking and inhibiting gas transport through the membrane. Research on improving porous membranes having suitably small sized pores for gas separation is a current field of research, from materials including ceramics [49], carbon molecular sieves [50], polyphosphazenes [51], and metals [52].

5.6 Membrane Architecture

From Eq. (5.1), it is clear that the flux through a nonporous dense membrane is inversely proportional to the membrane thickness, thereby leading to the requirement of a thin membrane for optimal performance. Due to the driving force requirements of membranes, gas separation processes usually operate with partial pressure differences up to 20 atm. Thin membranes require porous supports durable enough to withstand these pressure differences, while minimizing resistance to gas flow. Support materials are often made from porous ceramics, metals, or polymers, and should have approximately 50% porosity with the pore size comparable to the thin membrane film covering the support. A membrane consisting of more than one layer of material type is termed an asymmetric membrane. The concentration gradients throughout these membranes are complex since the mechanism of transport through the porous support is different than through the dense selective film known as the *skin*. In particular, for membranes exhibiting high flux, mass-transfer resistance may play a significant role at the interface of both the feed and permeate sides of the membrane. Figure 5.16 shows the pressure and concentration gradients of an asymmetric membrane in which the permeability of gas A is much greater than that of B, and with the flux of A several times that of B. From this figure it can be seen that the large decrease in c_A indicates that the membrane skin is primarily responsible for the mass-transfer resistance. It is also important to note that the concentration gradient for gas B in the boundary layer of the skin is negative, and that gas B is carried against its concentration gradient by the total flow, which mainly consists of gas A.

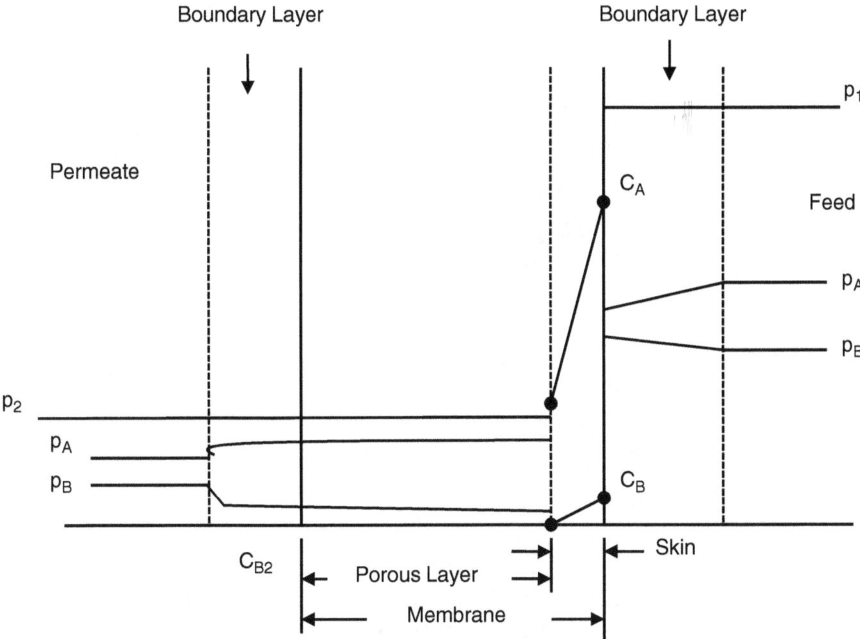

Fig. 5.16 Pressure and concentration gradients for an asymmetric membrane with boundary-layer resistance. (With permission from McGraw-Hill Companies, Inc. [12])

In the example of Fig. 5.16 adapted from McCabe et al. [12] it is assumed that the gases are in equilibrium with those in the dense membrane on both the feed and permeate sides. In this case, the bulk permeate is approximately 70% A and the gas exiting the skin layer composed of approximately 90% A. The gas composition in the pores neighboring the skin is generally not equal to the bulk composition of either the feed or the permeate. Additionally, the local gas composition nearest the membrane surface is dependent on the flow arrangement within the membrane separator especially in situations where the permeability of one component is very high relative to other components in the feed stream (concentration polarization). Mass-transfer limitations, such as the effects of concentration polarization normally do not become significant, unless membrane selectivities (or permeability ratios) for components exceed values of several hundreds to thousands.

5.7 Product Purity and Yield

The feed and permeate compositions depend primarily on the pressure difference across the membrane, feed composition and flow rate, the permeability of the various species, the total membrane surface area, and the flow arrangement. The fraction

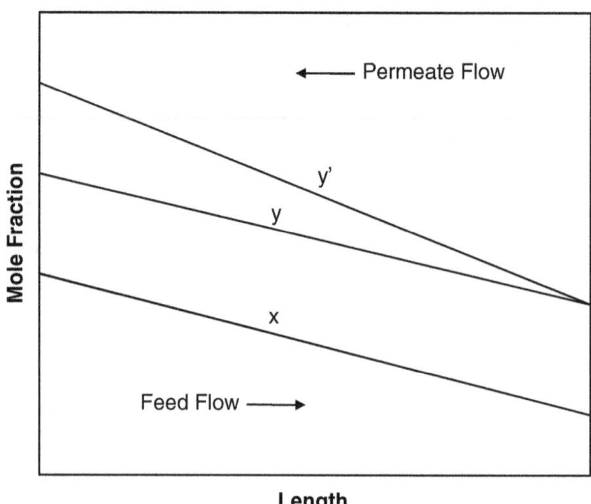

Fig. 5.17 Local and average permeate compositions and residue composition for counter flow separation. (With permission from McGraw-Hill Companies, Inc. [12])

of feed that is recovered as permeate is termed the *stage cut* (recovery). More specifically, the stage cut is defined as the permeate flow rate divided by the feed flow rate. For simplicity, consider the example of the separation of a binary mixture assuming negligible resistances in the porous support and boundary layers, along with negligible frictional pressure drops on the feed and permeate sides. For further simplification, assume the permeate stream flows countercurrent to the feed stream. The mole fraction of the more permeable species is represented by x in the feed stream and y in the permeate stream. For a CO_2-selective membrane, this would correspond to the feed-side and permeate-side CO_2 concentrations. At the exit of the membrane, the average gas composition y, at some axial position is equal to the local composition of the gas y^* leaving the membrane surface at the same axial position. As the permeate flows along the length of the membrane y^* increases since x is decreasing in the opposite direction as illustrated in Fig. 5.17. The average composition y increases, but not as rapidly as the surface composition at the same axial position. Next, the relationship between selectivity, pressure ratio, feed composition, and permeate mole fraction is derived. This derivation has been adapted from McCabe et al. [12] and the reader is encouraged to explore the original work for further details.

The relationship between x and y^* is based upon the relative permeabilities of the components in the gas mixtures and on the partial pressure differences. The relative flux of components A and B can be represented in terms of the permeances, $\bar{P}_A$ and $\bar{P}_B$ and the total feed and permeate pressures, p_1 and p_2, respectively.

$$J_A = \bar{P}_A(p_1 x - p_2 y) \tag{5.18}$$

$$J_B = \bar{P}_B[p_1(1 - x) - p_2(1 - y)] \tag{5.19}$$

To eliminate p_2, the ratio of total pressures, $r = p_2/p_1$, is used and rewriting Eqs. (5.18) and (5.19) accordingly yields:

$$J_A = \bar{P}_A p_1(x - ry) \tag{5.20}$$

$$J_B = \bar{P}_B p_1[1 - x - r(1 - y)] \tag{5.21}$$

The permeate concentration at the membrane surface is dependent upon the flux ratio at the given axial position, such that P_A and P_B are the permeabilities of components A and B and y* is:

$$y^* = \frac{J_A}{J_A + J_B} = \frac{P_A p_1(x - ry)}{P_A p_1(x - ry) + P_B p_1[1 - x - r(1 - y)]} \tag{5.22}$$

and substituting the selectivity, $\alpha = P_A/P_B$, yields:

$$y^* = \frac{x - ry}{x - ry + (1 - x - r + ry)/\alpha} \tag{5.23}$$

At the feed gas inlet of the membrane, in which the average composition is equal to the surface composition at a given axial position, *i.e.*, $y = y^*$, Eq. (5.23) can be rewritten as:

$$(\alpha - 1)(y^*)^2 + \left[1 - \alpha - \frac{1}{r} - \frac{x(\alpha - 1)}{r}\right] y^* + \frac{\alpha x}{r} = 0 \tag{5.24}$$

Equation (5.24) may be used to determine the local permeate composition dependence on the pressure ratio, r, selectivity, α, and feed composition of the more permeable gas, x. A decrease in the total pressure ratio, r will lead to an increase in y^*; however, as the limit of r goes to zero, there will be an associated limit of y^*, such that,

$$y^* = \frac{\alpha x}{1 + (\alpha - 1)x} \tag{5.25}$$

In the case of $r = 1.0$ for a binary mixture, no separation takes place since there is no driving force for diffusion to take place. If a third component is added to the mixture, which would be the case if one introduced a sweep gas on the permeate side of the membrane, the partial pressures of gases A and B would be lowered, and separation would take place even if $r = 1.0$. The separation of gas A from a mixture of A and B may improve with increasing membrane selectivity, but the partial pressure of A in the permeate cannot exceed that of the feed. The maximum value of y^* is determined by equating the partial pressures on either side of the membrane such that:

$$p_1 x \geq p_2 y^* \tag{5.26}$$

$$y^*_{max} = \frac{x}{r} \tag{5.27}$$

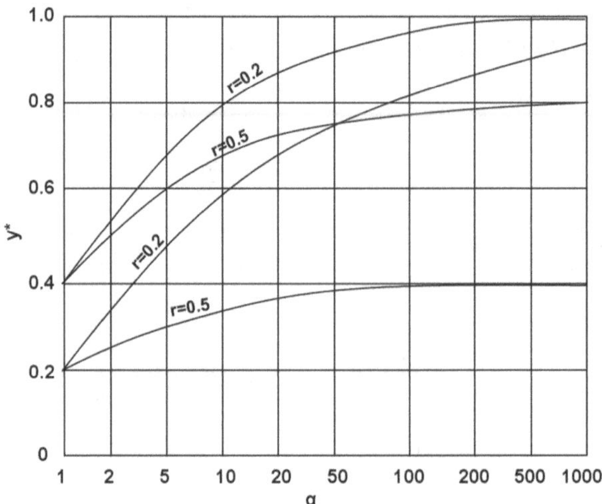

Fig. 5.18 Selectivity and pressure ratio effects on the local permeate concentration. (With permission from McGraw-Hill Companies, Inc. [12])

As an example, for a feed with 30% A, and a pressure ratio of 0.5, the highest achievable permeate concentration is 0.60, even for a highly selective membrane. Notice that a reduction in the pressure ratio leads to an increase in the purity of gas A in the permeate. In Fig. 5.18, the variation of y^* with α is shown for $r = 0.2$ and 0.5.

The permeate composition can be approximated through averaging the composition values at each end of the separator, assuming that the change in x is not too large. The membrane surface area required can be determined from the flux of the more permeable gas as:

$$A \cong \frac{V_{out} y_{out}}{\bar{P}_A (p_1 x - p_2 y)_{ave}} \tag{5.28}$$

such that V_{out} is the molar flow rate of the permeate and y_{out} is the mole fraction of the more permeable species in the permeate stream. From examination of Fig. 5.18, it is clear that the selectivity plateaus, and in general it is important to recognize that the selectivity is optimized to the process and that the pressure ratio is a crucial aspect in the practical design of membrane separation processes. The accuracy of this type of approximation can be improved by using a log-mean partial pressure difference as demonstrated by Hogsett and Mazur [53]. In this approximation, the membrane area, A may be rewritten in terms of the log-mean average of the feed composition as:

$$A \cong \frac{V_{out} y_{out}}{\bar{P}_A (p_1 x_{log-mean} - p_2 y)} \tag{5.29}$$

such that

$$x_{log-mean} = \frac{x_F - x_R}{ln\left(\frac{x_F}{x_R}\right)} \tag{5.30}$$

where x_F and x_R are the mole fractions of the more permeable species in the feed and residue streams, respectively.

5.8 Pressure Drop

There will exist frictional pressure drop and diffusion resistance on both the feed and permeate sides of the membrane. The frictional pressure drop inside hollow fiber membranes can affect the choice of optimal fiber size, since large bore diameters are often required for the case of high gas permeabilities. The fiber length has a limit, with longer fibers being impractical since the pressure drop changes as the square of the fiber length due to the increase in permeate velocity. The mass-transfer resistances in the support and boundary layers are usually negligible except in the case of ultrathin or very high relative component permeabilities. In particular, mass-transfer effects associated with pressure drop have been reported in gas separation studies of high-flux silicone membranes [54].

5.9 Membrane Modules and Flows

Membranes may be fabricated as a flat sheet or hollow fiber and require additional "packaging" for commercial application. The packaged system needs to contain as much surface area per unit volume as possible, while allowing reasonable flow distribution and gas-surface contact efficiency. The three major types of permeators include spiral-wound, hollow-fiber, and flat sheets in plate-and-frame configurations. A comparison of area-to-volume ratios for various membrane configurations is compared in Table 5.10.

Spiral-Wound Permeators An example of a spiral-wound permeator fabricated from a flat membrane film is shown in Fig. 5.19.

The module is formed by sealing two membrane sheets together on three edges separated by a spacer, with the fourth edge attached to a perforated tube and the outer surface termed the membrane leaf. Pressurized feed gas is brought into contact with the outer surface of the leaf with the gas flowing along the surface of the leaf until it reaches the perforated tube, which is held at a lower pressure than the feed. Spiral-wound membranes typically contain on the order of 3000 m^2 of membrane area per cubic meter of pressure vessel volume. Hollow-fiber permeators have approximately three or more times as much area per unit volume compared to the spiral-wound designs.

Hollow-Fiber Permeators There are three primary configurations as shown in Fig. 5.20 for commercial hollow-fiber modules, *i.e.*, 1) shell-side feed with counter-

Table 5.10 Area-to-volume ratios for typical membrane configurations

Module configuration	Approximate area-to-volume ratios (m^2/m^3)
Plate-and-frame	200
Spiral-wound	500–3,000
Hollow-fiber	1,500–10,000

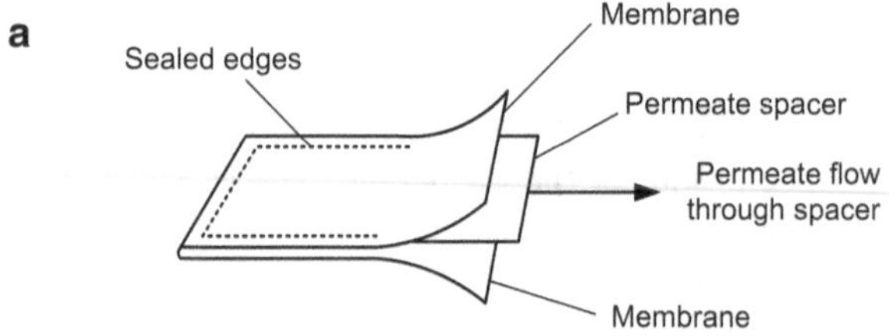

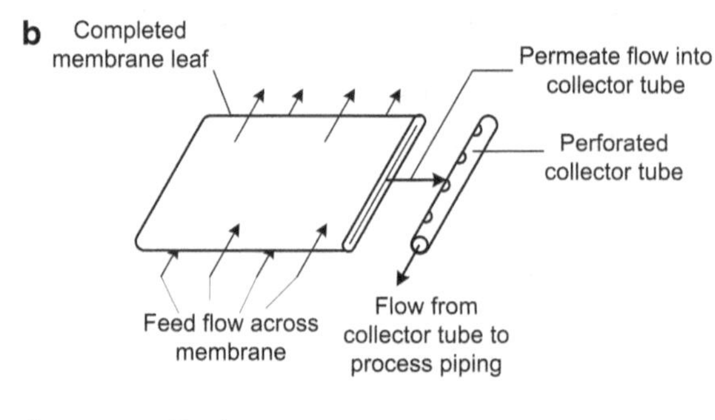

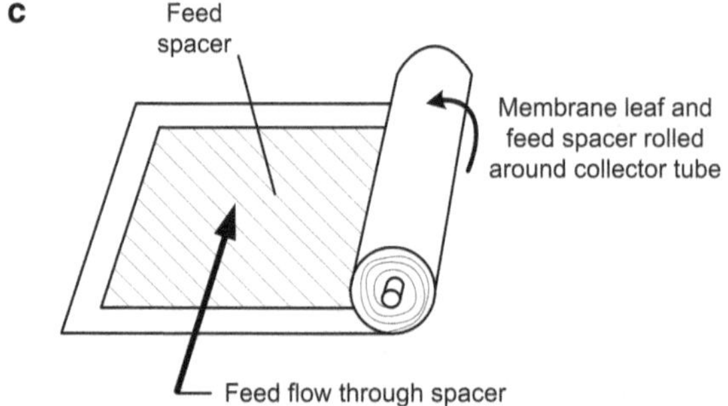

Fig. 5.19 Schematic of a spiral-wound permeator assembly [3]

current flow between the feed and permeate, 2) shell-side feed with countercurrent flow of the feed gas, and 3) tube-side feed (often referred to as bore-side) with cross flow of feed and permeate. Henry I. Mahon and researchers at Dow Chemical first developed hollow-fiber membranes in 1966. The permeators for all of these scenarios are fabricated in a similar fashion to shell-and-tube heat exchangers as demonstrated in Fig. 5.21. The hollow fibers are arranged parallel and are between 100–500 μm in outer diameter. There are many reviews that provide excellent overviews of both early and recent developments of hollow-fiber membranes [55]. Example fibers are shown in Fig. 5.22 with their traditional applications. The smaller fibers (*i.e.*, 50–200 μm) can withstand quite high transmembrane pressure differences of approximately 1000 psig or more across the membranes.

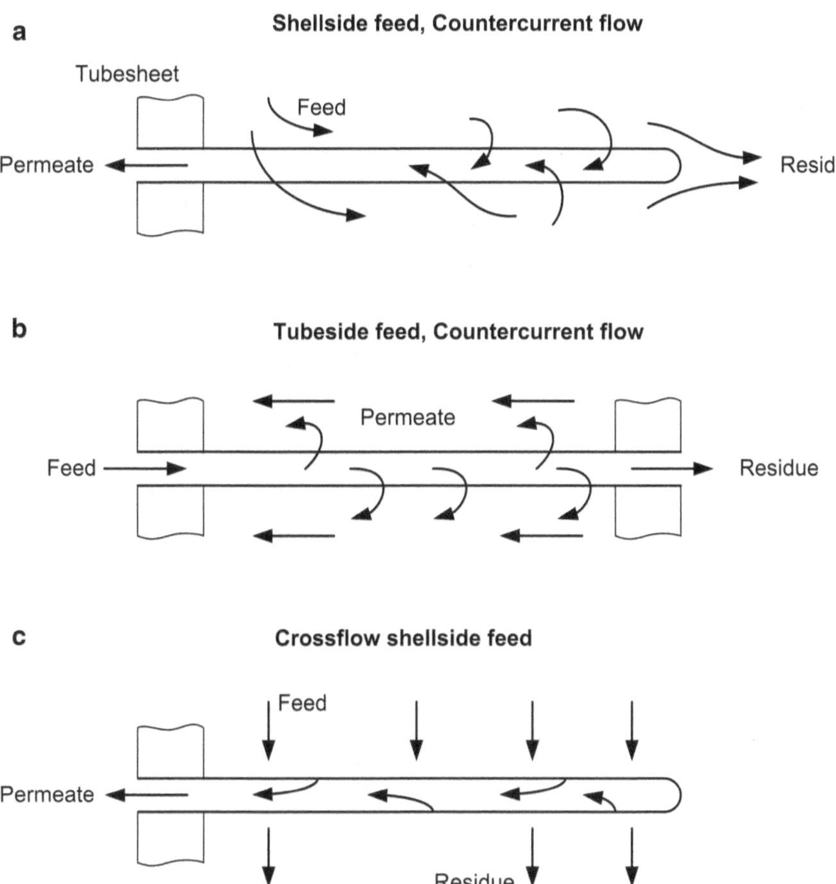

Fig. 5.20 Hollow-fiber permeator flow configurations [3]

5.10 Multiple-Stage Separator Arrangements

The majority of membrane applications for gas separations require multiple units with the largest single units approximately 0.3 m in diameter and 3–5 m in length. In particular, a hollow-fiber module of this size may contain up to a thousand square feet of membrane area with the ability to process several hundred cubic feet of gas per minute. It is important to note that the membrane areas required for CO_2 separation from a flue gas may require on the order of a million square meters. Separators may be arranged in series as shown in Fig. 5.23a. The frictional pressure drop at the feed is typically small so that several units may be placed in series without the need to recompress the feed. To obtain a higher purity exit stream, the permeate from the first module is compressed and sent to a second stage as shown in Fig. 5.23b. Several of these modules may be used in series to obtain a desired purity, but it is important

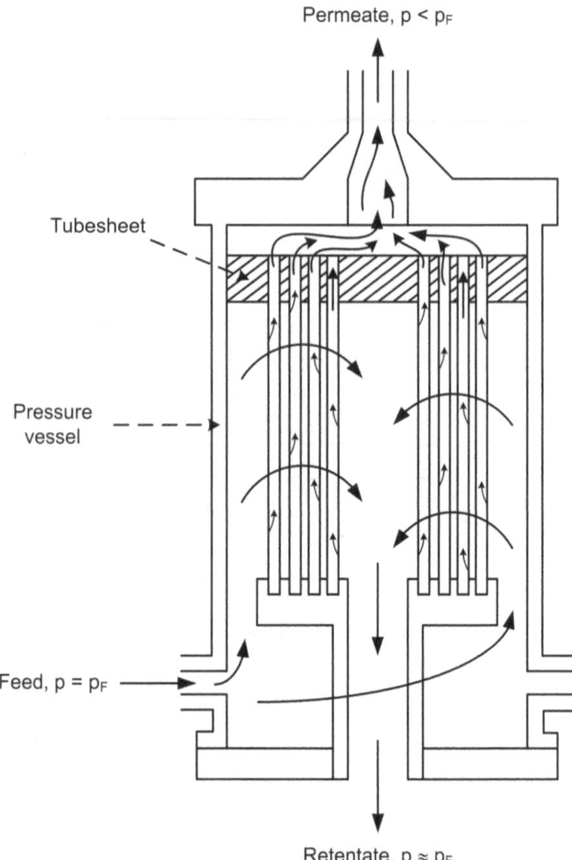

Fig. 5.21 Cross-flow feed permeator schematic [3]

to consider the costs associated with the recompression energy. Another approach, as shown in Fig. 5.23c, combines two separation modules and one recompression step in a continuous membrane column [56]. In this set-up, a portion of the permeate product from the second separator is compressed and fed back into the opposite side of the membrane, where it flows in a countercurrent fashion to the permeate. This reflux step allows for a high purity permeate to be obtained, while the reflux stream loses the more permeable gas as it flows through the separator and is subsequently combined with the feed stream in the first separator.

Researchers at Membrane Technology & Research Incorporated (MTR) have designed a contactor that increases the partial pressure of CO_2 in the flue gas. A schematic of the MTR contactor is shown in Fig. 5.24.

The incoming flue gas stream has a CO_2 partial pressure of 0.12 atm (12 mol% CO_2) and after passing through a selective membrane with an air sweep gas, the CO_2-enriched air stream is then fed into the coal-fired boiler. Without modifying the combustion process significantly, the molar percent of CO_2 in the flue gas can be enriched up to 20 mol%. Figure 5.25 shows an example of the process with

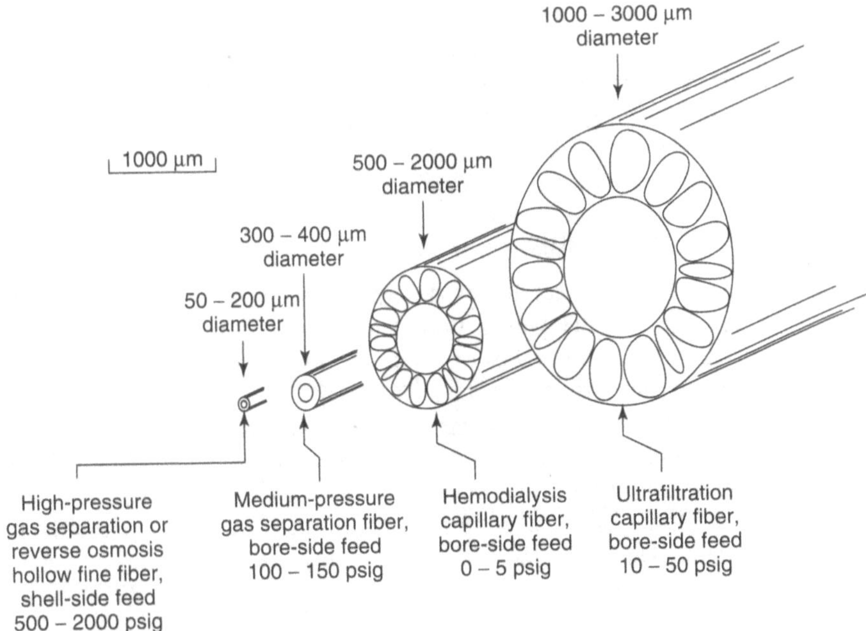

Fig. 5.22 Schematic of traditional hollow-fiber membranes. (Reprinted with permission of John Wiley & Sons, Inc. [4])

a selective purge that requires compression energy, but results in enhanced CO_2 separation compared to the case without compression energy in Fig. 5.24. Gas-fired plants typically run off of 2.5 times excess air compared to an average of 1.2 for a typical coal-fired power plant, making gas-fired plants possibly more applicable for this technology.

5.11 Work Required for Separation

Gas Compression Work In a membrane separation process, target separation performance may be determined by the following parameters: 1) the component recovery (*i.e.*, % capture), which is the ratio of the concentration of CO_2 in the permeate to the concentration of CO_2 in the feed, and 2) the permeate mole fraction (*i.e.*, % purity), assuming that the membrane is CO_2-selective and CO_2 is the desired product. For CO_2 capture, the desired recovery[4] is on the order of 0.9 and with a permeate mole fraction of CO_2 between 0.8 and 0.9 depending upon purity specifications. The analysis of a single-stage membrane module reveals that for a 10 mol% concentration of CO_2 in the feed mixture, a stage cut of 0.9 and permeate mole fraction of 0.9 cannot be achieved simultaneously, even using a membrane with an extremely high

[4] Based upon DOE target goals of 90% capture

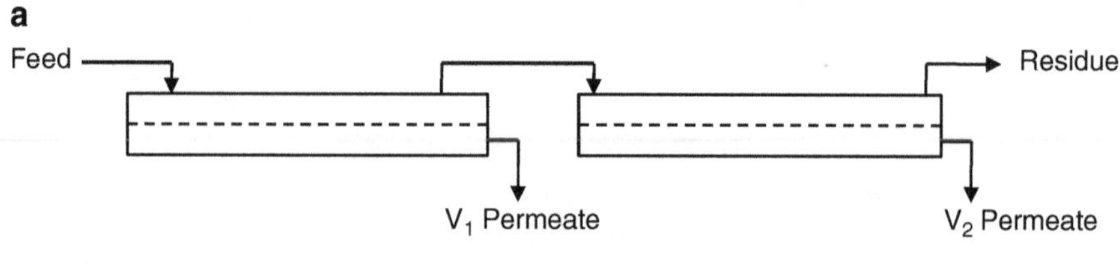

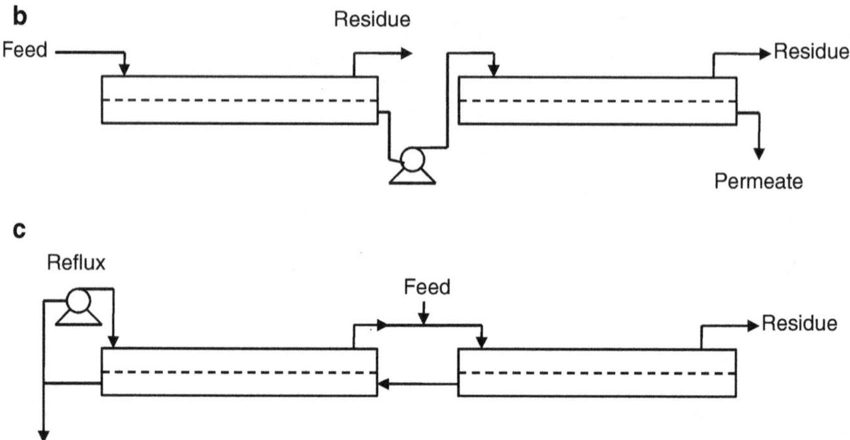

Fig. 5.23 Membrane separator arrangements **a** series flow, **b** two-stage flow, and **c** continuous membrane column. (With permission from McGraw-Hill Companies, Inc. [12])

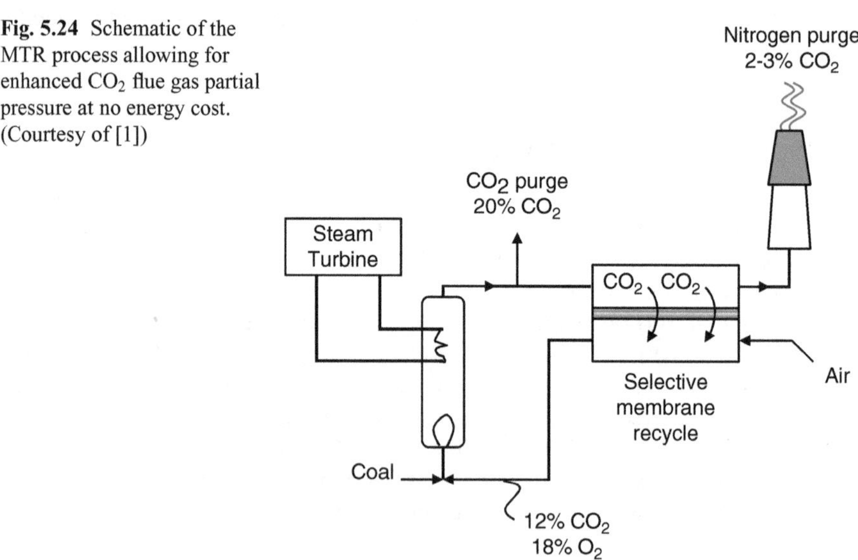

Fig. 5.24 Schematic of the MTR process allowing for enhanced CO_2 flue gas partial pressure at no energy cost. (Courtesy of [1])

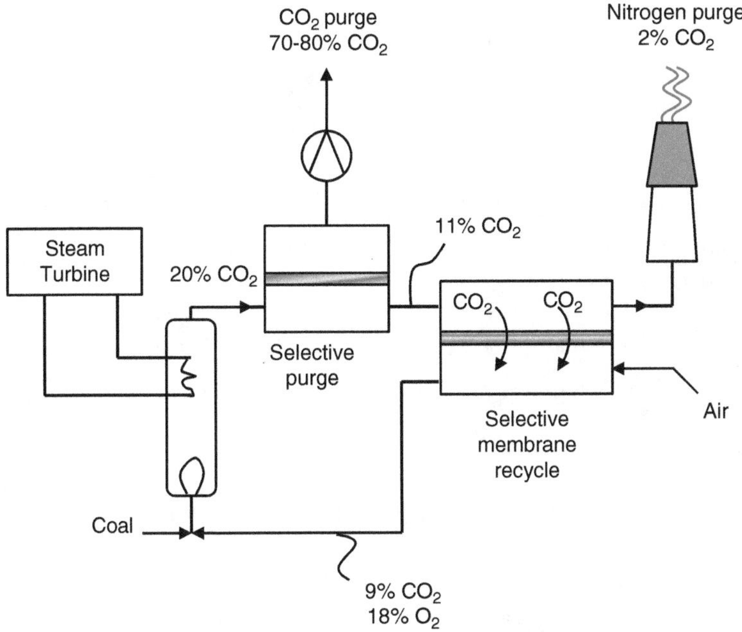

Fig. 5.25 Schematic of the MTR process with selective purge allowing for 70–80% capture through a CO_2-enhanced flue gas composition. (Courtesy of [1])

selectivity, *e.g.*, 200. A mole fraction of CO_2 in the feed would need to be at least 20 mol% to achieve the target goals. A study by Favre et al. [57] revealed that for a 10 mol% feed concentration of CO_2, to achieve a stage cut of 0.8 and a CO_2 permeate concentration of 0.8 requires a membrane with a selectivity of 120. In addition, they estimated that the work required to achieve these targets is approximately 8 MJ/kg CO_2 captured (352 kJ/mol CO_2).

To calculate the work required for a single-stage membrane separation process, one can envision the use of compression at the feed-side or vacuum conditions at the permeate-side of the membrane module.

As previously discussion in Chaps. 2 through 4, the power required for adiabatic and reversible compression of an ideal gas is:

$$P = \frac{dw_c}{dt} = \frac{\dot{m}RTk}{M(k-1)\varepsilon}\left[\left(\frac{p_2}{p_1}\right)^{(k-1)/k} - 1\right] \tag{5.31}$$

such that T is the gas temperature, M is the molecular weight, p_1 is the initial gas pressure, p_2 the final gas pressure, and k is the ratio of specific heats, c_p/c_v, which are available for selected gases in Appendix B. Typical compressor efficiencies can range between 65–85%. To determine the work required to compress a gas from a pressure of p_1 to p_2 using (Eq. 5.31), the power expression can be divided by the molar flow rate of the gas, which will result in the work required per mole of gas compressed.

Example 5.3 (a) Calculate the thermodynamic minimum work required for the following N_2/CO_2 separation process at 300 K, with feed and permeate compositions and molar flow rates of $x_{F,CO_2} = 0.10$ and $n_F = 100$ mol/s; $y_{P,CO_2} = 0.80$ and $n_P = 10$ mol/s, respectively.

(b) What is the 2nd-Law efficiency of the CO_2-selective membrane process if the actual power required for the process is 8 MJ per kg CO_2 separated?

Solution: (a) The minimum power, P (dW/dt) required for the separation process is:

$$P = RT[\dot{n}_{R,CO_2}\ln(x_{R,CO_2}) + \dot{n}_{R,CH_4}\ln(x_{R,CH_4}) + \dot{n}_{P,CO_2}\ln(y_{P,CO_2}) + \dot{n}_{P,CH_4}\ln(y_{P,CH_4}) - \dot{n}_{F,CO_2}\ln(x_{F,CO_2}) - \dot{n}_{F,CH_4}\ln(x_{F,CH_4})]$$

$$= 8.314 \cdot 300\ [2\ \ln(0.02) + 88\ \ln(0.98) + 8\ \ln(0.80) + 2\ \ln(0.20) - 10\ \ln(0.10) - 90\ \ln(0.90)]$$

$$P = 4.47 \times 10^4\ \frac{\mathrm{J}}{\mathrm{s}} = 0.044\ \mathrm{MW}$$

$$W_{min} = \left(\frac{44{,}705\ \mathrm{J}}{\mathrm{s}}\right)\left(\frac{\mathrm{s}}{8\ \mathrm{mol}\ CO_2}\right) = 5.6\ \mathrm{kJ/mol}\ CO_2\ \text{captured}$$

(b) The 2nd-law efficiency, η_{2nd} is found by:

$$\eta_{2nd} = \frac{W_{min}}{W_{actual}}$$

$$W_{actual} = \left(8\frac{\mathrm{MJ}}{\mathrm{kg}\ CO_2}\right)\left(\frac{0.044\ \mathrm{kg}\ CO_2}{\mathrm{mol}\ CO_2}\right) = 352\ \mathrm{kJ/mol}$$

$$\eta_{2nd} = \frac{5.6}{352} \sim 1.6\%$$

Example 5.4 A Kapton membrane ($\alpha_{H_2,CO_2} = 5.55$) is used in a fuel gas (60 mol% CO_2/40 mol% H_2) separation process to produce a CO_2-rich retentate stream and a H_2-rich permeate stream. (a) Determine the permeate composition of H_2 as a function of the compression ratio, r. (b) With a stage cut of 0.6, how much work is required per mol of H_2 permeated as a function of compression ratio. Assume that the compressor efficiency is 90% and the temperature of the process is 339 K.

Solution: (a) The permeate concentration, y_{P,H_2}, may be obtained from Eq. (5.24)

$$(\alpha_{H_2,CO_2} - 1)(y_{P,H_2})^2 + \left[1 - \alpha - \frac{1}{r} - \frac{x_{F,H_2}(\alpha - 1)}{r}\right] y_{P,H_2} + \frac{\alpha x_{F,H_2}}{r} = 0$$

given a selectivity, α_{H_2,CO_2} of 5.55 and H_2 feed composition, x_{F,H_2} of 0.4, yields:

r	y_{P,H_2}
0.5	0.59
0.2	0.72
0	0.78

(b) To determine the required work, the molar flow rate of the permeate stream must be known. Assuming a basis of the feed molar flow rate of 100 mol/s, a given stage cut of 0.6, a H_2 feed molar composition of 0.4 and the previously calculated H_2 permeate compositions yields the following H_2 permeate molar flow rates as a function of pressure ratio, r:

r	y_{P,H_2}	$\dot{n}_{P,H_2}$(mol/s)
0.5	0.59	35.4
0.2	0.72	43.2
0	0.78	46.8

The needed power for compression can be determined using Eq. (5.31) as:

$$P = \frac{dw_c}{dt} = \frac{\dot{m}RTk}{M\,(k-1)\,\varepsilon}\left[\left(\frac{p_2}{p_1}\right)^{(k-1)/k} - 1\right]$$

Assume the ratio of specific heats, k is 1.4, and since $\dot{m}/M = \dot{n}$, the power required for $r = 0.5$ is:

$$P = \frac{(35.4)(8.314)(339)(1.4)}{(1.4-1)(0.9)}[(0.5)^{(1.4-1)/1.4} - 1] = -69.5 \text{ kW}$$

This separation process requires 69.5 kW of vacuum power (note the negative value associated with work being done on the surrounding) applied to the permeate stream.

The work required per mol of H_2 separated is:

$$w = \left(\frac{69.5\ \text{kJ}}{\text{s}}\right)\left(\frac{\text{s}}{35.4\ \text{mol H}_2}\right) = 1.96\ \text{kJ/mol H}_2$$

This process can be repeated for different compression ratios:

r	y_{P,H_2}	P, H_2 (mol/s)	w (kJ/mol H_2)
0.5	0.59	35.4	1.96
0.2	0.72	43.2	4.11
0	0.78	46.8	12.0

5.12 Problems

Problem 5.1 A zeolite imidazolate framework-8 (ZIF-8) membrane [60] has been developed for separating CO_2 from CH_4. Experimental data indicates a gas permeance of CO_2 is 2.4×10^{-5} mol/m^{2}sPa with a CO_2/CH_4 selectivity of 5.5. Determine the transmembrane flux (in kmol/m^{2}s) for CO_2 and CH_4 if the feed gas is 25 mol% CO_2 at 295 K and 140 kPa. Assume the permeate pressure is 101.3 kPa and that a CH_4 purity of 98 mol% is required.

Problem 5.2 Oxycombustion plants are characterized by the need for large-scale air separation units (ASU) for O_2 production. A potential membrane for the ASU has an O_2 permeability, P_{O_2} of 730 Barrer and a N_2 permeability, P_{N_2} of 330 Barrer. Assume a 400-MW oxycombustion plant requires 8000 tons of O_2 per day and that air (79 mol% N_2, 21 mol% O_2) is sent to the unit at 8 bar. Additionally, assume perfect mixing on both sides of the membrane, such that compositions on both sides are uniform, and neglect pressure drop and mass-transfer resistances external to the membrane. Determine the:

a. transmembrane fluxes of O_2 and N_2 for a 0.2-μm membrane that produces 8000 tons of O_2 per day at a purity of 90 mol% at 25°C and 1.5 bar, and
b. required membrane surface area (m^2).

Problem 5.3 A spiral-wound permeator assembly with a novel polymeric 100-micron thick membrane is tested for CO_2 removal from coal-bed methane (15 mol% CO_2) in an enhanced natural gas recovery project. At a slipstream feed rate of 70 L/min (at 25 bar and 25°C), 45.4 L/min of 80 mol% CO_2 is produced as a permeate stream and a residue stream with 95 mol% CH_4. The feed and residue pressures are equal at 25 bar, assuming negligible pressure drop throughout the permeator. In addition, the membrane surface area is 100 m^2. Determine the:

a. permeabilities of CO_2 and CH_4 in addition to the membrane selectivity, and
b. required feed pressure to obtain a residue stream containing 99 mol% CH_4.

Problem 5.4 Membranes made of poly(methyl methacrylate) are highly selective toward CO_2 in natural gas purification processes (*e.g.*, $\alpha_{CO_2,CH_4} = 116$). Since the permeate stream will only be a small fraction of the total molar feed, the compressor is downstream of the membrane. The compressor efficiency is 85% and the temperature of the process is 300 K. If the feed natural gas mixture is 15% CO_2 and 300 K, determine the:

a. compression ratio, r, for y_{P,CO_2} from 0.3–0.9, (in intervals of 0.1),
b. maximum obtainable CO_2 permeate mole fraction for each compression ratio, and
c. work required per kg of CO_2 permeated for each of the calculated compression ratios.

Problem 5.5 An entrepreneur has approached you with the claim that her colleague, a polymer scientist, has developed a polymer that falls slightly above the Robeson curve for O_2/N_2 separation, with $\alpha_{O_2,N_2} = 10$ and $P_{O_2} = 10$ Barrer. The membrane thickness is 100 nm. She has an idea to produce at-home medical oxygen concentrators, which produce at least 90 mol% O_2, using this particular membrane. The goal is to have a 2-stage membrane unit using air at 298 K and 1 atm as the feed stream for the first stage. Assuming a constant compression ratio across each stage, determine the:

a. required compression ratio, r, to produce a permeate of 90% O_2 after the second stage,
b. power required for each stage per kg of O_2 produced, and
c. membrane surface area at the second stage to produce 1.5 kg of O_2 per day.

References

1. Richard W Baker, Personal Communication, Membrane Research Technology, Palo Alto, CA, 2011
2. J. Membrane Sci., 320(1–2), Robeson LM The upper bound revisited, 390–400, Copyright (2008)
3. Ho WS, Sirkar KK Membrane handbook. Van Hostrand Reinhold: New York, 1992; p 3–101
4. Baker, R., Membrane Technology and Applications, 2nd ed. (2004)
5. Rackley, S. A., Membrane Separation Systems, 171–172, Copyright Elsevier (2010)
6. Springer Science + Business Media B.V. Springer and Van Hostrand Reinhold, Membrane Handbook, 1992, Ho WS Sirkar, K.K., 64–66
7. Seader JD, Henley EJ Separation Process Principles (2006) John Wiley & Sons
8. Industrial Gas Separations, Chapter 3, ACS Symposium Series Chern RT, Koros WJ, Sanders ES, Chen SH, Hopfenberg HB V223. American Chemical Society, pp. 47–73 (Copyright 1983)
9. J Membrane Sci, Robeson LM Correlation of separation factor versus permeability for polymeric membranes, 165–185
10. Wilcox J (2011) Nitrogen-Permeable membranes and uses thereof. U.S. Patent 2011/0182797, Stanford University, Stanford, CA

11. Baker RW (2004) Membrane technology and applications, 2nd edn. John Wiley & Sons, Inc., Chichester
12. McCabe WL, Smith JC, Harriott P (2005) Unit Operations of Chemical Engineering, 7th edn. McGraw-Hill, New York
13. Stern AS (1994) Polymers for gas separations: the next decade J Membrane Sci 94 (1):1–65
14. Robeson LM (2001) Polymeric membranes for gas separation in Encyclopedia of Materials: Science and Technology. Pergamon
15. Favre E (2010) Polymeric Membranes for Gas Separation. Drioli, E and Giorno, L. (eds), in Comprehensive Membrane Science and Engineering, Elsevier: Vol 2
16. Robeson LM (1991) Correlation of separation factor versus permeability for polymeric membranes J Membrane Sci 62(2):165–185
17. Robeson LM (2008) The upper bound revisited J Membrane Sci 320(1–2):390–400
18. Lin H, Freeman BD (2004) Gas solubility, diffusivity and permeability in poly (ethylene oxide) J Membrane Sci 239(1):105–117
19. Okamoto K, Umeo N, Okamyo S, Tanaka K, Kita H (1993) Selective permeation of carbon dioxide over nitrogen through polyethyleneoxide-containing polyimide membranes Chemestry Letters 22(2):225–228
20. Robeson LM, Burgoyne WF, Langsam M, Savoca AC, Tien CF (1994) High performance polymers for membrane separation Polymer 35(23):4970–4978
21. Budd PM, Msayib KJ, Tattershall CE, Ghanem BS, Reynolds KJ, McKeown NB, Fritsch D (2005) Gas separation membranes from polymers of intrinsic microporosity J Membrane Sci 251(1–2):263–269
22. Park HB, Jung CH, Lee YM, Hill AJ, Pas SJ, Mudie ST, Van Wagner E, Freeman BD Cookson DJ (2007) Polymers with cavities tuned for fast selective transport of small molecules and ions. Science 318(5848):254
23. Barrer RM, Barrie JA, Slater J (1958) Sorption and diffusion in ethyl cellulose. Part III. Comparison between ethyl cellulose and rubber J Polym Sci 27(115):177–197
24. Michaels AS, Vieth WR, Barrie JA (1963) Diffusion of gases in polyethylene terephthalate J Appl Phys 34(1):13–20
25. Koros WJ, Chan AH, Paul DR (1977) Sorption and transport of various gases in polycarbonate J Membrane Sci 2:165–190
26. Billmeyer FW (1971) Polymer chains and their characterization. In: Textbook of polymer science. Wiley Interscience Publishers, New York, pp. 84–85
27. Petropoulos JH (1994) Mechanisms and theories for sorption and diffusion of gases in polymers. In: Polymeric gas separation membranes. Paul DR, Yampol'skii YP (eds) CRC Press: Boca Raton, pp. 17–81
28. (a) Dhingra SS, Marand E (1998) Mixed gas transport study through polymeric membranes J Membrane Sci 141(1):45–63; (b) Koros W (1980) Model for sorption of mixed gases in glassy polymers J Polym Sci Polym Phys 18(5):981–992; (c) Frisch HL (1980) Sorption and transport in glassy polymers Polym Eng Sci 20(1):2–13; (d) Barbari TA, Conforti RM(1980) Recent Theories in Gas Sorption in Polymers Polym Adv Technol 5(11):698–707
29. Sonwane CG, Wilcox J, Ma YH (2006) Achieving optimum hydrogen permeability in PdAg and PdAu alloys J Chem Phys 125:184714
30. Sieverts A, Danz W (1936) Solubility of D2 and H2 in Palladium Z Phys Chem 34:158
31. (a) Aboud S, Wilcox J (2010) A density functional theory study of the charge state of hydrogen in metal hydrides J Phys Chem C 114(24):10978–10985; (b) Kamakoti P, Morreale BD, Ciocco MV, Howard BH, Killmeyer RP, Cugini AV, Sholl DS (2005) Prediction of hydrogen flux through sulfur-tolerant binary alloy membranes. Science 307(5709):569
32. Sammells AF, Mundschau MV (2006) Nonporous inorganic membranes: for chemical processing. Wiley-VCH, Weinheim
33. Stannett V, Koros W, Paul D, Lonsdale H, Baker R (1979) Recent advances in membrane science and technology Adv Polymer Sci 32:69–121
34. Keller GEI, Anderson RA, Yon CM (1987) Adsorption. In: Handbook of separation process Technology. Rousseau RW (ed). John Wiley & Sons, New York, p. 644–696

35. Stern SA, Shah VM, Hardy BJ (1987) Structure permeability relationships in silicone polymers J Polym Sci Polym Phys 25(6):1263–1298
36. Chern RT, Koros WJ, Sanders ES, Chen SH, Hopfenberg HB (1983) In: Implications of the dual-mode sorption and transport models for mixed gas permeation, ACS Symposium Series 233. ACS Publications, Washington, p. 47
37. (a) Pick MA, Davenport JW, Strongin M, Dienes GJ (1979) Enhancement of hydrogen uptake rates for Nb and Ta by thin surface overlayers Phys Rev Lett 43(4):286–289; (b) Roa F, Way JD (2003) Influence of alloy composition and membrane fabrication on the pressure dependence of the hydrogen flux of palladium-copper membranes Ind Eng Chem Res 42(23):5827–5835
38. (a) Sonwane CG, Wilcox J, Ma YH (2006) Solubility of hydrogen in PdAg and PdAu binary alloys using density functional theory J Phys Chem B 110(48):24549–24558; (b) Kamakoti P, Sholl DS (2003) A comparison of hydrogen diffusivities in Pd and CuPd alloys using density functional theory J Membrane Sci 225(1–2):145–154
39. Roa F, Block MJ, Way JD (2002) The influence of alloy composition on the H_2 flux of composite Pd–Cu membranes Desalination 147(1–3):411–416
40. Mckinley DL (1969) Method for hydrogen separation and purification
41. Gade SK, Keeling MK, Davidson AP, Hatlevik O, Way JD (2009) Palladium-ruthenium membranes for hydrogen separation fabricated by electroless co-deposition Int J Hydrog Energy 34(15):6484–6491
42. Morreale BD, Ciocco MV, Enick RM, Morsi BI, Howard BH, Cugini AV, Rothenberger KS (2003) The permeability of hydrogen in bulk palladium at elevated temperatures and pressures J Membrane Sci 212 (1–2):87–97
43. Zhang GX, Yukawa H, Watanabe N, Saito Y, Fukaya H, Morinaga M, Nambu T, Matsumoto Y (2008) Analysis of hydrogen diffusion coefficient during hydrogen permeation through pure niobium Int J Hydrog Energy 33 (16):4419–4423
44. Yukawa H, Nambu T, Matsumoto K V-W alloy membranes for hydrogens for hydrogen purification J Alloy Compd 509, S881–S884, 2011
45. Steward SA (1983) Review of hydrogen isotope permeability through materials. Lawrence Livermore National Lab, CA
46. Watanabe N, Yukawa H, Nambu T, Matsumoto Y, Zhang GX, Morinaga M (2009) Alloying effects of Ru and W on the resistance to hydrogen embrittlement and hydrogen permeability of niobium J Alloy Compd 477(1–2):851–854
47. Way JD, Noble RD, Reed DL, Ginley GM, Jarr LA (1987) Facilitated transport of CO_2 in ion exchange membranes AIChE J 33(3):480–487
48. Rackley SA (2010) Carbon capture and storage. Butterworth-Heinemann, Burlington MA, p 167
49. (a) Hsieh HP, Bhave RR, Fleming HL (1988) Microporous alumina membranes J Membrane Sci 39(3):221–241; (b) Niwa M, Ohya H, Tanaka Y, Yoshikawa N, Matsumoto K, Negishi Y (1988) Separation of gaseous mixtures of CO_2 and CH_4 using a composite microporous glass membrane on ceramic tubing J Membrane Sci 39(3)301–314
50. Koresh JE, Soffer A (1987) The carbon molecular sieve membranes. General properties and the permeability of CH_4/H_2 mixture Separ Sci Technol 22(2):973–982
51. McCaffrey RR, Cummings DG (1988) Gas separation properties of phosphazene polymer membranes Separ Sci Technol 23(12):1627–1643
52. McCaffrey RR, McAtee RE, Grey AE, Allen CA, Cummings DG, Appelhans AD, Wright RB, Jolley JG (1987) Inorganic membrane technology Separ Sci Technol22(2):873–887
53. Hogsett JE, Mazur WH (1983) Estimate membrane system area Hydrocarbon Process62(8): 52–54
54. Lokhandwala KA, Segelke S, Nguyen P, Baker RW, Su TT, Pinnau I (1999) A Membrane process to recover Chlorine from Chloroalkali Plant Tail Gas Ind Eng Chem Res 38(10):3606–3613
55. (a) McKelvey SA, Klausi DT, Koros JW (1997) A guide to establishing hollow fiber macroscopic properties for membrane applications J Membrane Sci124(2):223–232; (b) Gabelman A, Hwang ST (1999) Hollow fiber membrane contactors J Membrane Sci159(1–2):61–106;

(c) Masourizadeh A, Ismail AF (2009) Hollow fiber liquid-gas contactors for acid gas capture: a review J Hazard Mater,171(1–3):38–53; (d) Porcheron F, Ferre D, Favre E, Nuguyen PT, Lorain O, Mercier R, Rougeau L (2011) Hollow fiber membrane contactors for CO_2 capture: From lab-scale screening to pilot-plant module conception Energy Procedia 4(763–770)

56. Hwang ST, Thorman JM (1980) The continuous membrane column Am Inst Chem Eng 26(4):558–566
57. Favre E (2007) Carbon dioxide recovery from post-combustion processes: can gas permeation membranes compete with absorption? J Membrane Sci 294(1–2):50–59
58. Paul DR, Koros WJ (1976) Effect of Partially Immobilizing Sorption Permeability and the Diffusion Time Lag J Polm Sci Pol Phys 14(4):675–685
59. Gestel TV, Sebold D, Hauler F, Meulenberg WA, Buchkremer H-P (2010) Potentialities of microporous membranes for H_2/CO_2 separation in future fossil fuel power plants: evaluation of SiO_2, ZeO_2, Y_2O_3-ZrO_2 and TuO_2-ZrO_2 sol-gel membranes. J Mem Sci 359(1–2):64–79
60. Venna SR, Jasinski JB, Carreon MA (2010) Structural Evolution of Zeolitic Imidazolate Framework-8. J Am Chem Soc 132(51):18030–18033

Chapter 6
Cryogenic Distillation and Air Separation

Various advanced coal conversion-to-electricity processes are discussed in Chap. 1 that depend on the use of a gas stream comprised primarily of oxygen; therefore, air separation into its primary components, *i.e.*, nitrogen (N_2), oxygen (O_2), and argon (Ar) are discussed within the context to CO_2 capture. One of the dominant processes used for air separation is cryogenic distillation. Cryogenic distillation may also be used as a polishing step to enhance the purity of a gas stream predominantly comprised of CO_2.

6.1 Cryogenic Distillation

The process of cooling a gas mixture to induce a phase change for effective separation is termed *cryogenic distillation*. Gas separation by cryogenic distillation involves the separation of its components based upon their differing boiling points and volatilities. The *volatility* is the tendency of the molecules at the vapor-liquid interface to escape into the vapor phase. The term "vapor" is often used when discussing equilibrium with a liquid phase, with "gas" being a more general term. The vapor pressure of a given component is defined as the pressure exerted by a vapor in equilibrium with the corresponding liquid phase. The *partition function*[1], K (or K-value) is defined as the concentration ratio of vapor to liquid at a given temperature and pressure. If a component i in a liquid mixture of components i and j has a higher K-value, this implies that its vapor pressure will be higher and that it will be present in a higher concentration than j in the gas phase. In a binary mixture, at a given temperature and pressure, the relative volatility (α_{ij}) of two components i and j is given by:

$$\alpha_{ij} = \frac{K_i}{K_j} \tag{6.1}$$

[1] Note that the definition of partition function in the field of statistical mechanics is different than this and refers to the sum over states corresponding to the energies associated with the electronic, vibrational, rotational, and translational degrees of freedom within a molecule.

J. Wilcox, *Carbon Capture*,
DOI 10.1007/978-1-4614-2215-0_6, © Springer Science+Business Media, LLC 2012

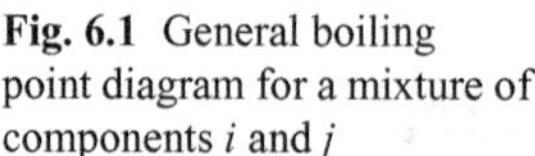

Fig. 6.1 General boiling point diagram for a mixture of components *i* and *j*

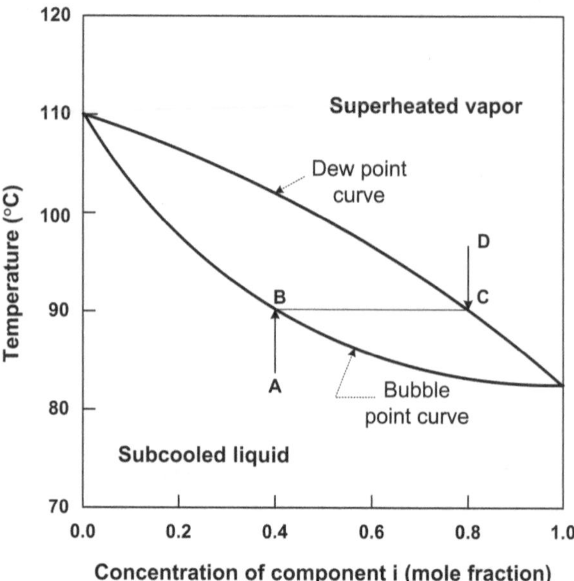

such that K is:

$$K = \frac{y}{x} \tag{6.2}$$

in which y and x are the mole fractions of a given component in the vapor and liquid phases, respectively. The boiling point of a liquid at a given temperature is determined by the vapor pressure of the liquid. Antoine's equation, as shown in Eq. (6.3), may be used to determine the boiling points of pure substances as a function of pressure. In this equation, p_0 is the vapor pressure in mm Hg and T is the temperature in °C. The component-specific coefficients (A, B and C) can be found in Appendix K. A general Txy diagram for an ideal mixture of components i and j is displayed in Fig. 6.1. The boiling point of the mixture lies somewhere in between the pure-component boiling points, T_i and T_j, respectively.

$$\log p_0 = A - \frac{B}{T + C} \tag{6.3}$$

From Fig. 6.1, when a mixture with 40% mole fraction of component i (A on Fig. 6.1) is heated to its boiling point (B on the *bubble-point curve*), the produced vapor has an equilibrium composition of 80% (C on the *dew-point curve*) at the same temperature. At this point the vapor has a higher concentration of component i compared to j and if the remaining liquid has a reduced concentration (mole fraction) of component i the boiling point of the remaining mixture will rise, with boiling eventually stopping. It is important to note that Fig. 6.1 corresponds only to ideal gas mixtures and that real, non-deal mixtures may deviate from this behavior and form azeotropes, in which the dew point and bubble point curves touch. This simplified example illustrates

Table 6.1 Approximate relative boiling points [4] of relevant gases at various pressures

Gas name	Boiling point (K)				
	10.1 kPa (0.1 atm)	50.6 kPa (0.5 atm)	101.3 kPa (1 atm)	303.9 kPa (3 atm)	607.9 kPa (6 atm)
Water (H_2O)	319	355	373	407	432
Carbon dioxide (CO_2)	solid	solid	solid	solid	221
Oxygen (O_2)	73	84	90	102	113
Argon (Ar)	solid	solid	87	99	108
Nitrogen (N_2)	solid	72	77	87	98
Hydrogen (H_2)	14.5	18	20	24	28

the distillation process that allows for the separation of a component with a higher volatility than the remaining gas mixture.

Distillation equipment consists of a packed or trayed tower nearly identical to that used in absorption-based separation processes as discussed in Chap. 3. The separation of CO_2 from N_2 is not easily achievable since both would prefer to exist as a gas at standard temperature and pressure. As previously mentioned, cryogenic distillation may be of interest in the CO_2 capture field to separate air (O_2 and N_2) for oxyfuel and chemical looping combustion, and gasification technologies. Table 6.1 provides a list of approximate boiling points of relevant gases at various pressures. It is important to keep in mind that condensation temperature of the vapor depends on its partial pressure, and the more that it is condensed throughout the process, the lower its concentration is in the vapor phase, thus lowering its condensation temperature further, with the condensation process taking place over a range of temperatures. Within a cryogenic distillation process one has to make sure that condensation always takes place at temperatures above the triple point temperature, which is 216.6 K for CO_2. For instance, below this temperature, CO_2 would sublime and form a solid, rather than condense into liquid form. The need to avoid the formation of a solid phase sets an upper bound to the amount of gas that can be separated in a single-step process.

In a cryogenic-based separation process for CO_2 capture, the gas mixture containing CO_2 is first compressed and pretreated to remove any water. The compressed fluid then flows through a series of heat exchangers to decrease the temperature prior to cryogenic distillation for the separation of liquefied CO_2 from gas species such as N_2 or H_2. After separation, liquid CO_2 may be stored for product delivery. Through this process, slightly pressurized N_2 and H_2 (or argon) gas may also be additionally treated for product delivery, with treatment depending upon the purity and pressure requirements of the application.

Equipment used in Cryogenic Separation Compressors are the major machinery employed in a cryogenic separation process. The compressor selection is dependent upon the fluid and volumetric flow rate. For large volumetric flows, axial compressors are primarily used. A quick method for the selection of a multistage compressor is available in Appendix D. For high pressures with small volumetric flows, volumetric compressors (often of the reciprocating type) are used and centrifugal compressors are used for lower volumetric flows and pressures. Combining these types of compressors to achieve desired purity may be necessary and is an approach used for air separation. This optimization depends on the volumetric flow rate of gas to be compressed. For instance, if the flow rate is very large, an axial-flow section followed by a centrifugal-flow section (each comprising several stages) may be the most convenient. Heat exchangers are also a major component of a cryogenic-based separation process with the coil-wound, shell-and-tube, and plate-and-fin being the most commonly used. Additional equipment includes distillation columns and insulation for storage and transport of the liquefied products.

6.2 Air Separation

Air separation involves the separation of air into its primary components, N_2, O_2 and often Ar. For the high quantities of O_2 required of large-scale industrial applications, *e.g.*, steel manufacture, petrochemical, and the gasification of solid feedstocks, cryogenic distillation of air is the primary technology of choice. Similarly, this is the only method currently available for the large production rates that are required for gasification and oxyfuel combustion applications with CO_2 capture.The O_2 requirements for these applications range between 3.0–20 tons/day/MW [5]. In particular, for oxyfuel applications, a traditional 1000-MW power plant requires approximately 20,000 tons of O_2 per day [6]. For an integrated gasification combined cycle (IGCC) plant, the O_2 consumption of a 1000-MW is approximately 7000 tons per day [6]. Recall from Chap. 1 that IGCC requires much less oxygen since it is only used for the partial oxidation taking place in the gasifier. The installed capacity in the U.S. is 315 GW, which represents a significant potential market for O_2 within a CO_2 capture context. In fact, air separation via cryogenic distillation has been taking place since the start of the twentieth century and has the ability to produce O_2 at 98.5% purity, beyond which would require the use of three distillation columns. The world's largest plant produces approximately 30,000 tons of O_2 per day (868 scfd). The plant was constructed by Linde AG for the Pearl Gas-to-Liquids project in Qatar and uses eight 3,750 tons of O_2-per-day ($\sim$108 scfd) trains in parallel. Other technologies for air separation include pressure swing adsorption (PSA), and membrane permeation.

Figure 6.3 shows commonly used separation processes for air separation for the production of N_2 as a function of N_2 purity and flow rate. Nitrogen production is a related topic to O_2 production since it is possible to generate O_2 as a byproduct of a N_2 production process depending on the desired O_2 purity. Figure 6.2 illustrates that for N_2 production, other technologies such as PSA and membrane technology are becoming competitive with cryogenic distillation for smaller-scale applications. In general, for the treatment of large volumes of air, cryogenic distillation is the most cost-effective technology, and for smaller volumes, PSA is the most cost-effective [7]. The N_2 (or O_2) purity generally dictates the cost associated with air separation.

The components of air are listed in Table 6.2 with their corresponding boiling points. Since the boiling points of O_2 and Ar are so similar (*i.e.*, 90.18 versus 87.28 K), Ar is the primary impurity in a high-purity O_2 gas stream.

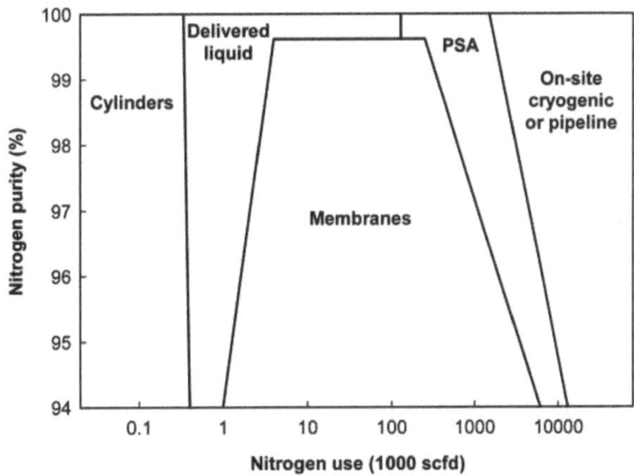

Fig. 6.2 Approximate competitive range of current N_2 production systems, in which site-specific factors may affect the system selection. (Reprinted with permission of John Wiley & Sons, Inc. [2])

Table 6.2 Composition of air

Compound	Vol%	Boiling point (K)
	Fixed components	
Nitrogen (N_2)	78.084 ± 0.004	77.36
Oxygen (O_2)	20.946 ± 0.002	90.18
Argon (Ar)	0.934 ± 0.001	87.28
Carbon dioxide (CO_2)	0.033 ± 0.003	194.68[a]
Neon (Ne)	$(1.821 \pm 0.004) \times 10^{-3}$	27.09
Helium (He)	$(5.239 \pm 0.05) \times 10^{-4}$	4.215
Krypton (Kr)	$(1.14 \pm 0.01) \times 10^{-4}$	119.81
Xenon (Xe)	$(8.7 \pm 0.1) \times 10^{-6}$	165.04
Hydrogen (H_2)	$\sim 5 \times 10^{-5}$	20.27
	Impurities	
Water (H_2O)	0.1–2.8	
Methane (CH_4)	1.5×10^{-4}	
Carbon monoxide (CO)	$(6–100) \times 10^{-6}$	
Sulfur dioxide (SO_2)	0.1–1.0	
Nitrous oxide (N_2O)	5×10^{-5}	
Ozone (O_3)	$(1–10) \times 10^{-6}$	
Nitrogen dioxide (NO_2)	$(5–200) \times 10^{-8}$	
Radon (Ra)	6×10^{-18}	
Nitric oxide (NO)	Very trace amounts	

[a] Sublimation temperature since liquid CO_2 does not exist at 1 atm

Although cryogenic distillation is the primary method for producing O_2 at scales required of CO_2 capture applications, there are other technologies that are advancing and may one day be competitive as the O_2 market increases. Next, cryogenic distillation, adsorption, and membrane technology are discussed briefly in the context of air separation for O_2 production.

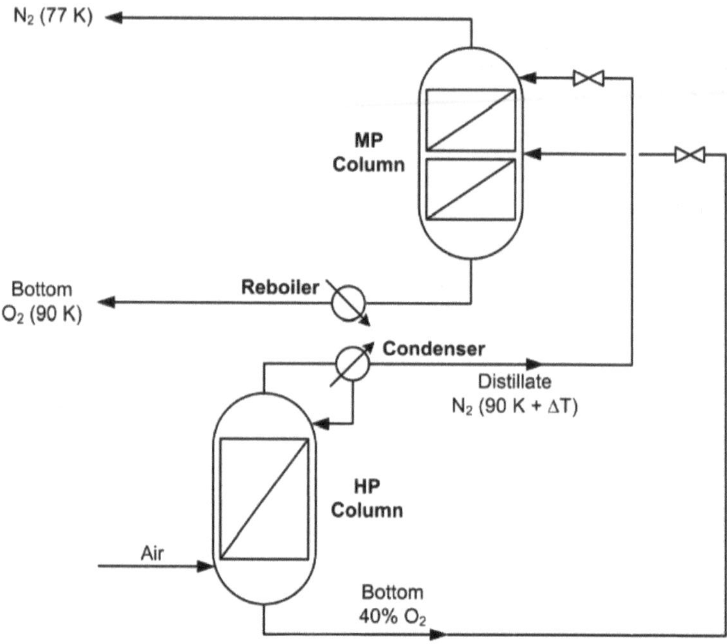

Fig. 6.3 Schematic of a double-column cryogenic air separation process. (Figure courtesy of [3])

6.2.1 Cryogenic Air Separation

An example of a traditional double-column cryogenic air separation process is illustrated in Fig. 6.3. Within this process, ambient air is pretreated through filtration and compression to remove moisture and CO_2, after which it is cooled to cryogenic temperatures using a series of heat exchangers and fed into a distillation column resulting in the gas separation into components oxygen, nitrogen, and in some cases, argon. Since the boiling points of argon and oxygen are quite similar, achieving optimum purity can be difficult and costly. The operating temperature of the distillation column is maintained between the boiling points of N_2 and O_2. It is also important to recognize the partitioning of impurities into each N_2 and O_2 streams for air separation. For instance, since the boiling points of neon and helium are lower than O_2, they will remain in the gas phase with N_2, while krypton and xenon will partition into the liquid phase with O_2 since their boiling points are higher.

In addition, cryogenic air separation systems require little refrigeration power, which includes start-up of the process and also what is required to compensate for the temperature difference at the hot end of the cold box heat exchanger. In the double-column system of Fig. 6.4, the expansion of the two flows extracted from the medium-pressure (MP) column takes place within the two-phase region, thereby making the low-pressure (LP) column operate at a lower temperature, which enables cooling of the MP (*i.e.*, 6–8 bar) column condenser.

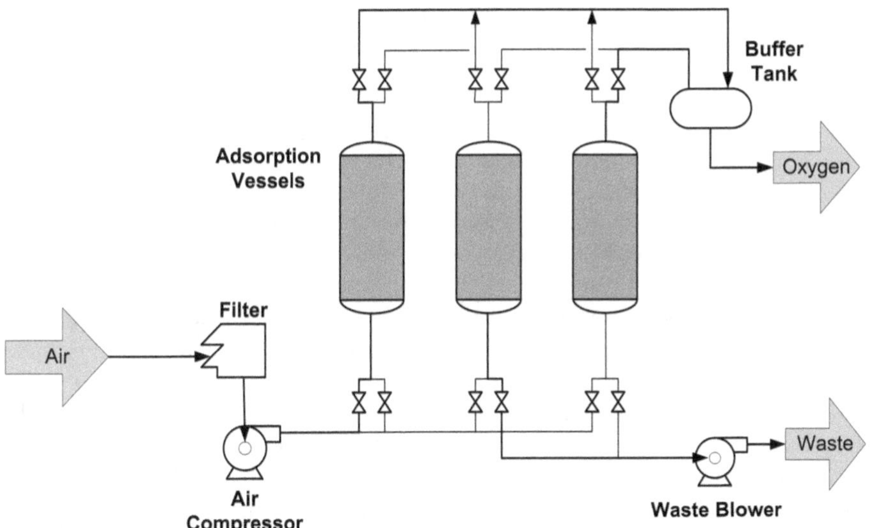

Fig. 6.4 Schematic of an adsorption-based air separation process. (With permission from Elsevier [1])

6.2.2 *Air Separation using Adsorption*

Typical sorbents used for air separation include carbon molecular sieves (CMS) and zeolites. Both sorbents are selective toward the capture of N_2, allowing O_2 to pass freely through the sorbent bed. Although recent developments have led to the realization of O_2-selective sorbents for air separation applications using metal-organic frameworks (MOFs), which are described in greater detail in Sect. 4.2 of Chap. 4 [15]. A CMS sorbent works as a molecular sieve to kinetically separate O_2 and N_2, since the kinetic diameter of O_2 is slightly smaller than N_2, thereby allowing it to diffuse through the sorbent bed faster. Recalling from the discussion on adsorption-based separation processes in Chap. 4, N_2 has a stronger quadrupole moment than O_2. In a zeolite sorbent, N_2 will selectively interact with the positive-charged cations stabilizing the partially-negatively charged aluminum atoms in the zeolite framework, thereby allowing oxygen to pass through the sorbent bed. A schematic of an N_2-selective adsorption-based separation process is shown in Fig. 6.4. This process has been described previously in Chap. 4, and will be revisited just briefly. Pressurized air enters one of the adsorber units and the sorbent bed is saturated with N_2 while producing an outlet stream rich in O_2. Once the bed is saturated with N_2, the feed air is switched to a fresh sorbent bed and regeneration of the saturated bed begins. Regeneration may take place through the addition of heat (*i.e.*, temperature-swing adsorption) or via a pressure reduction (PSA or vacuum swing adsorption). As described previously in Chap. 4, due to the faster cycling times associated with PSA processes, this is the preferred method of bed regeneration. The O_2 purity from this process is typically between 90–93%.

6.2.3 Air Separation using Membrane Technology

Unlike cryogenic-based separation processes, membrane systems allow for operation over a range of temperatures, including near-ambient conditions. This is a primary benefit of membrane technology for air separation. The two primary types of membranes used for air separation are polymeric and ion transport. The overall O_2-N_2 separation mechanism for dense polymeric membranes used in air separation is called the solution/diffusion model as previously described in Chap. 5. The solution/diffusion model is comprised of three main steps, *i.e.*, (1) gas absorbs into the polymer matrix from the upstream high-pressure side of the membrane, (2) gas diffuses across a concentration (or pressure) gradient through the polymeric separating layer, and (3) gas desorbs from the downstream low-pressure side of the polymer membrane. Separation of the various molecular species takes place due to the difference in the permeation rates of each of the gases through the membrane. The difference in the permeabilities of O_2 and N_2 is dominated by the difference in their diffusivities, since solubilities of O_2 and N_2 are essentially equivalent in common polymeric membrane materials.

Polymeric membranes may also transport N_2, CO_2, and water vapor to some extent depending on the relative selectivities of each of the components to O_2. For this reason, these membranes are mostly employed for the production of O_2-enriched air (25–50% O_2) or high-purity N_2 gas [1]. Polymeric membranes currently produce 30% of all gaseous nitrogen [8]. In such processes, polymeric membranes are generally housed within modules, with membranes in a hollow-fiber configuration, not unlike a shell-and-tube heat exchanger. Effective thickness of the active separating layers of state-of-the-air membranes may be as thin as 35 nm. The hollow fibers typically have diameters of 100–500 μm [9]. Depending on method of membrane fabrication, the thin active separating layer may be coated onto a porous hollow fiber support, or the thin separating layer and porous support may be formed in a single step as a unitary membrane structure. Examples of polymers used in such applications include polysulfone, polycarbonate, polyimides and perfluoropolymers. Additionally, polymeric membrane systems require either a feed-side compressor or a permeate-side vacuum pump to maintain the pressure difference across the membrane. The vast majority of polymeric membrane systems for air separations use feed air compression (most at pressures between 3–12 barg)[2] with the permeate side operating typically near atmospheric pressure (1 bara)[3].

Ceramic membranes are another important air separation technology that can provide high-purity O_2 (~99 vol%) from air at elevated temperatures between 800–900°C [10]. The separation mechanism of these membranes involves the transport of O_2 in its ionic form (O^{2-}) and the simultaneous flux of electrons to accomplish electronic neutrality [11]. These membranes are often classified as either pure-O_2 conducting membranes, mixed ionic-electronic conducting membranes (MIEC), or

[2] barg represents gauge pressure (p_g).

[3] bara represents absolute pressure (p_a), $p_g = p_a - p_{atm}$.

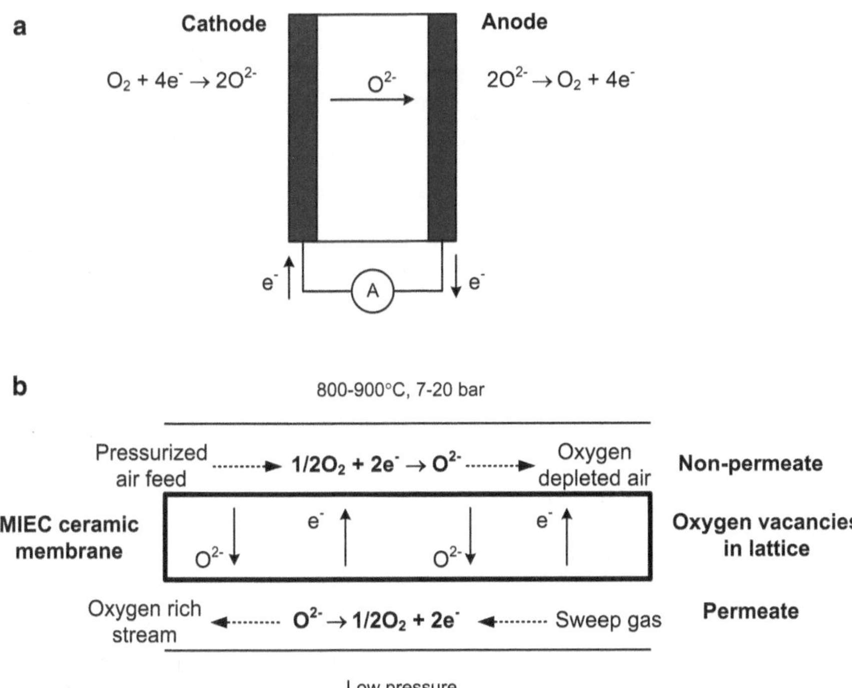

Fig. 6.5 (**a**) Schematic of oxygen anion membrane transport and (**b**) operating conditions and illustration of charge neutrality through the membrane. (With permission from Elsevier [1])

ion transport membranes. The driving force required for the separation in each case is either an electronic or a chemical potential gradient.

Solid electrolytes are an example of pure-O_2 conducting membranes with electrodes interfaced on either side of the membrane to act as a pathway for the electron transport. In these systems, the amount of O_2 generated is controlled through the application of an external current as illustrated in Fig. 6.5a. Ion transport membranes, on the other hand, do not require any electrodes or external circuit. These membranes facilitate the selective separation of O_2 by taking advantage of the electronic conductivity of O^{2-} within the membrane and the difference in the chemical potential of O_2 on either side of the membrane. The O^{2-} passes from the high O_2 chemical potential side to the low chemical potential side, while charge neutrality is achieved by the counter-current flow of electrons as shown in Fig. 6.5b [12]. Example membrane materials include perovskites ($CaTiO_3$), brownmillerites ($Ca_2(Al, Fe)_2O_5$), and fluorites (CaF_2). Oxygen-ion membranes operate in the range of 750–1000°C, with an approximate 20 bar partial pressure differential from the feed to permeate sides of the membrane.

Figure 6.6 shows a schematic of the O^{2-} air separation process, in which the air is passed through a filter for pretreatment, compressed and heated to the operating temperature and then fed into the membrane units and effectively separated from

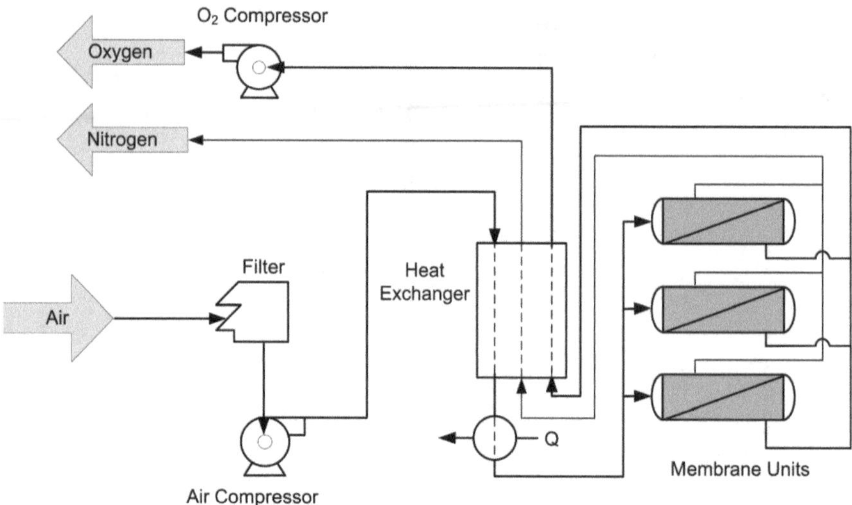

Fig. 6.6 Schematic of oxygen anion transport air separation process. (With permission from Elsevier [1])

N_2. Because of their success in producing high-purity oxygen, ion transport membranes have been the focus of many gas companies. Air Products and Chemicals has patented an ion transport membrane system working at elevated temperatures (800–900°C), which uses a flat membrane with a wafer configuration module [13]. Norsk Hydro has been improving a composite multi-channel heat exchanger including membrane modules in extruded ceramics, and Praxair has a system based on a tubular configuration [5, 14].

References

1. Smith AR, Klosek J (2001) A review of air separation technologies and their integration with energy conversion processes. Fuel Process Technol 70(2):115–134
2. Baker R (2004) Membrane technology and applications, 2nd Edn
3. Giampaolo Pelliccia, Decarbonized Electricity Production from the OxyCombustion of Coal and Heavy Oils, Dipartimento di Energetica, Politecnico Di Milano, (2006) Advisor: Stefano Consonni
4. Lide DR (2008) CRC handbook of Chemistry and Physics. CRC Press, Boca Raton, p 2736
5. Allam RJ (2009) Improved oxygen production technologies. Energy Procedia 1(1):461–470
6. Pelliccia G (2006) Decarbonized electricity production from the Oxycombustion of coal and heavy oils. Thesis (PhD) Politecnico Di Milano
7. Ruthven DM (1997) Encyclopedia of separation technology. John Wiley & Sons, Inc. New York
8. Koros WJ, Mahajan R (2000) Pushing the limits on possibilities for large scale gas separation: which strategies? J Membrane Sci 175(2):181–196
9. Coombe HS, Nieh S (2007) Polymer membrane air separation performance for portable oxygen enriched combustion applications. Energy Convers Manag 48(5):1499–1505

10. Hashim SM, Mohamed AR, Bhatia S (2010) Current status of ceramic-based membranes for oxygen separation from air. Adv Colloid Interface Sci 160:88–100
11. Sunarso J, Baumann S, Serra JM, Meulenberg WA, Liu S, Lin YS, Diniz da Costa JC (2008) Mixed ionic-electronic conducting (MIEC) ceramic-based membranes for oxygen separation. J Membrane Sci 320(1–2):13–41
12. Mancini ND, Mitsos A (2011) Ion transport membrane reactors for oxy-combustion-Part I: intermediate-fidelity modeling. Energy, 36(8):4701–4720
13. Stiegel GJ, Bose A, Armstrong P (2006) Development of ion transport membrane (ITM) oxygen technology for integration in IGCC and other advanced power generation systems. National Energy Technology Laboratory (NETL), U.S. Department of Energy
14. Hashim S S, Mohamed AR, Bhatia S (2011) Oxygen separation from air using ceramic-based membrane technology for sustainable fuel production and power generation. Renew Sustain Energy Rev 15(2):1284–1293
15. Bloch ED, Murray LJ, Queen WL, Chavan S, Maximoff SN, Bigi JP, Krishna R, Peterson VK, Grandjean F, Long GJ, Smit B, Bordiga S, Brown CM, Long JR (2011) Selective binding of O_2 over N_2 in a redox-active metal-organic framework with open iron(II) coordination sites. J Am Chem Soc 133(37):14814–14822

Chapter 7
The Role of Algae in Carbon Capture

An even stronger policy driver than climate change today is the expansion of alternatives to crude oil for transportation fuels. The U.S.[1] currently imports more than 60% of its petroleum, of which two-thirds are used for the production of transportation fuels [1]. These two drivers are likely to remain present for a long time, with indeterminate relative weights. The reduction of CO_2 via photosynthesis is a route to alternative fuels. The purpose of including a chapter on the algae route to biofuels, but not other routes, is that this process specifically involves the use of a CO_2 stream to enhance the value of the primary biological feedstock.

The conversion of algae to biofuel is a process that is receiving increased attention and may one day be an aspect of the portfolio of CO_2 mitigation solutions. Ideally, one may consider the algae-to-biofuel process to be carbon-neutral; however, this is not the case, as illustrated in Fig. 7.1, which includes a schematic of a generic algae conversion process. The CO_2 may be bubbled into an algae reactor or pond with some unknown optimal purity to achieve a desired biofuel, and may require work for separation. Mechanical energy such as pumping, blowing, and drying will also be required since CO_2, water, sunlight, and nutrients are used to cultivate the algae. Energy will then be required to extract the oil from the biomass and to refine it to a useable fuel. Biogas is a byproduct of the process and may be used in offsetting the cost by supplying heat or electricity back into the system. Additionally, the remaining biomass may be used for animal feed, assisting further in reducing cost. It is important to also recognize that within this system the biofuel will be oxidized to harness its energy potential, with the generation of CO_2 and H_2O. The process is not one in which CO_2 is permanently sequestered. Therefore, if the energy required for this system is not sourced from renewables (non-carbonized fuel), the process will have some CO_2 generation associated with it.

This chapter discusses various aspects of the CO_2-to-biofuel conversion process of algae and how algae-based biofuel compares to other biofuels in terms of land

[1] In addition to the U.S., other countries highly dependent on petroleum include Japan at 97%, India at 70%, and China at 53% as of 2009. (U.S. EIA)

J. Wilcox, *Carbon Capture*,
DOI 10.1007/978-1-4614-2215-0_7, © Springer Science+Business Media LLC 2012

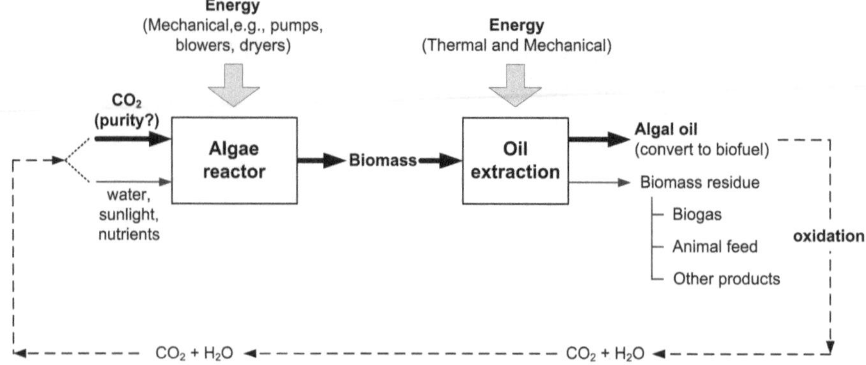

Fig. 7.1 An algae-to-biofuel conversion system

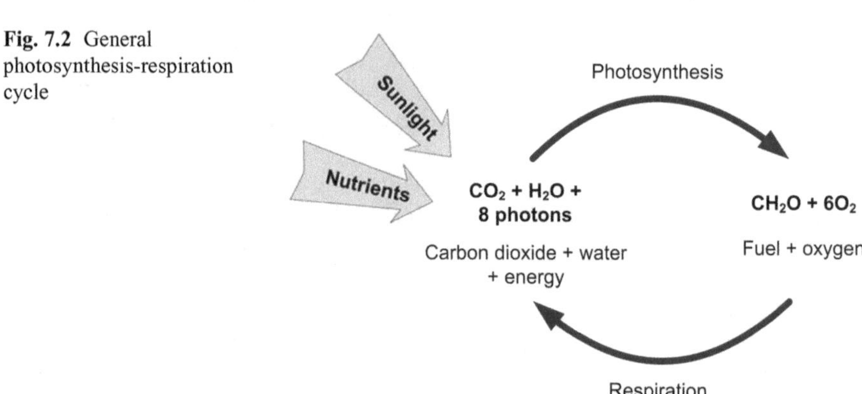

Fig. 7.2 General photosynthesis-respiration cycle

requirements and energy density. The chapter also addresses the mitigation potential of CO_2 and the current challenges associated with the process.

7.1 Microalgae to Biofuel

Microalgae[2] are water-based organisms that grow in fresh, saline, or brackish seawater or wastewater. Using sunlight and water, algae are capable of "capturing" CO_2 through its conversion to a lipid that may be extracted and processed into jet fuel, diesel, or gasoline. While these fuel types represent the ideal products of algae conversion, other near-term products include animal feed and food supplements. Figure 7.2 illustrates the CO_2 conversion process, which involves sunlight, CO_2, water, and nutrients as inputs and sugar and O_2 as outputs.

[2] Microalgae are on the order of microns and are the type considered for large-scale cultivation.

7.1.1 Theoretical Maximum of Algae-based Biofuel Production

As Fig. 7.2 shows, CO_2 in addition to water and photons results in the formation of sugar (represented simply as CH_2O) and O_2 as:

$$CO_2 + H_2O + 8 \text{ photons} \rightarrow CH_2O + O_2 \tag{7.1}$$

The chemical formula, CH_2O is simply formaldehyde, but in this context the formula is used to simply represent sugars in general. The chemical reaction of Eq. (7.1) illustrates the link between the conversion of photon energy to chemical energy through the reduction of CO_2 to "fuel" (CH_2O). For every 8 moles of photons "reacted", 1 mol of CO_2 is reduced to 1 mol of CH_2O. The details associated with this ratio are omitted here, but are discussed in more detail in the study carried out by Weyer et al. [2]. There are two primary components required to determine the maximum annual growth of biomass (*i.e.*, CH_2O in the simplest case) derived from microalgae. These include: 1) the *annual photon flux molar density* (*PFD*) and 2) the energy content of the biomass produced. Each of these terms are discussed in detail next.

Annual Photon Flux Molar Density Determining *PFD* requires knowledge of several components that relate to the percent of solar energy the algae is able to convert to chemical energy toward the formation of CH_2O and O_2 as shown in Eq. (7.1). The *PFD* can be calculated by:

$$PFD\left(\frac{\text{mol}}{\text{m}^2\text{yr}}\right) = \frac{E_{full} \cdot PAR}{E_{ph}} \tag{7.2}$$

such that E_{full} is the *full-spectrum solar energy*, *PAR* is the percent of photosynthetically active radiation, and E_{ph} represents the photon energy, which may be determined by:

$$E_{ph} = h\upsilon = h\left(\frac{c}{\lambda}\right) \tag{7.3}$$

such that h is Planck's constant (6.63×10^{-34} J s), c is the speed of light (2.998×10^8 m/s), and λ is the wavelength associated with the photon.[3]

The full-spectrum solar energy represents the total solar irradiance incident on the algae cultivation system, and is affected by cloud coverage, aerosols, ozone, and other gases in the atmosphere. The atmospheric conditions may affect the magnitude and spectral distribution of solar irradiance that reaches the surface of Earth. The Bird Clear Sky Model developed by the Solar Energy Research Institute predicts that the theoretical maximum annual solar irradiance ranges between approximately 6,000–12,000 MJ/m^{2}yr dependent on latitude [3]. The Department of Energy's EnergyPlus weather database estimates theoretical maximum annual solar irradiance to range from approximately 9,000–12,000 MJ/m^{2}yr with actual values ranging from

[3] Photon energies range from the least energetic at 160 kJ/mol and 750 nm (red), to the most energetic at 315 kJ/mol and 380 nm (violet).

approximately 5,800 (Kuala Lumpar, 3°) to 7,200 MJ/m^{2}yr (Phoenix, 33°) [4]. Only a portion of the solar spectrum may be used for photosynthesis (*PAR*) and is often defined as 400–700 nm. A common value for *PAR* used in the literature is 0.458 (45.8%).

The photon flux molar density calculated from Eq. (7.2) is reduced under the configuration of the growth set up and subsequent losses in incident solar energy, termed the *photon transmission efficiency*, ε_{pt}. Example configurations for microalgae growth include open pond and photobioreactor (PBR) systems, which will be described in further detail later in the chapter. A reasonable photon transmission efficiency is 95% in the summer months and between 85–93% in the winter months depending upon latitude [2]. An additional efficiency loss to *PFD* is the *photon utilization efficiency*, ε_{pu}, which accounts for the reduction in full photon absorption due to suboptimal growth conditions of the microalgal culture. Examples of suboptimal growth conditions include high-light levels and suboptimal temperatures that may result in *photoinhibition*, in which some of the photons are reemitted, leading to cellular damage due to heat release. From previous studies, this efficiency is estimated to range between 50–90% under low-light conditions and between 10–30% under high-light conditions [5]. Therefore, the reduced *PFD* based upon these two efficiency modifications becomes:

$$PFD_{red} = PFD \cdot \varepsilon_{pt} \cdot \varepsilon_{pu} \tag{7.4}$$

Energy Content of the Biomass Produced From Reaction (7.1), to convert from the energy associated with the photons to that of CH_2O, the molar ratio of 8:1 can be used. In general, the formation energies associated with various sugars likely to be produced, ΔE_f, are reported in the literature and range between 468.9–496.0 kJ/mol. The microalgae have to use some portion of this energy for cellular functions. Therefore, the energy associated with the sugar produced is reduced due to the *biomass accumulation efficiency*, ε_a, which ranges between 34–89%, as reported in the literature [6]. From this reduced energy, the amount of biomass generated may be calculated provided that one knows the distribution of proteins, carbohydrates, and lipid (oil) in the microalgae. Each of these biomass contributions has an approximate energy content of 16.7, 15.7, and 37.6 kJ/g, respectively [7]. In general, the lipid content within microalgae is known to range between 45 and 77% [8]. The *biomass energy content*, ΔE_b, ranges from 20.0–26.9 kJ/g [2, 9]. The total annual algae growth may be estimated from combining the previously defined terms as:

$$\text{Algae Mass Flux}\left(\frac{\text{g}}{\text{m}^2\text{yr}}\right) = \frac{E_{full} \cdot PAR}{E_{ph}} \cdot \frac{1\ \text{mol CH}_2\text{O}}{8\ \text{mol h}\upsilon} \cdot \frac{\Delta E_f}{\Delta E_b} \cdot (\varepsilon_{pt} \cdot \varepsilon_{pu} \cdot \varepsilon_a) \tag{7.5}$$

These terms with their respective ranges and units are listed in Table 7.1.

Data associated with the density of algal oil is limited and most reports base annual algal oil production estimates on the density of soybean oil, with an average density of 918 kg/m^3. Based upon this density and the theoretical limits listed in Table 7.1, the study of Weyer et al. [2] estimate a theoretical annual limit of 354,000 L/ha yr

Table 7.1 Parameters required for calculating annual mass flux of algae

Term	Symbol	Literature range	Theoretical limit [2]	Units
Full-spectrum solar energy	E_{full}	5,800–12,000	11,616	MJ/m^2yr
Photosynthetically active radiation	PAR	0.458	0.458	–
Photon energy	E_{ph}	160–315	225.3	kJ/mol
Photon transmission efficiency	ε_{pt}	0.85–0.95	1.00	–
Photon utilization efficiency	ε_{pu}	0.1–0.9	1.00	–
Formation energy of CH_2O	ΔE_f	468.9–496.0	482.5	kJ/mol
Biomass accumulation efficiency	ε_a	0.34–0.89	1.00	–
Biomass energy content	ΔE_b	20.0–26.9	21.9	kJ/g

and a range of best estimates between 40,700–53,200 L/ha yr. The following are the predictions of several other studies available in the literature: [9] 27,251–163,505; [8] 32,648; and [10] 58,711–136,928 L/ha yr.

7.1.2 *Properties of Algae-based Biofuel*

Early studies[4] on the use of microalgae for food production provide comprehensive information on the growth, physiology, and biochemistry of algae, leading to the concept of using mass-cultured algae for food production [11]. Initially in 1960, William J. Oswald and Clarence G. Golueke [12] experimentally developed the concept of using microalgae for CO_2 utilization and conversion to fuels. They proposed using large ponds located nearby power plants, with the biomass converted to methane by anaerobic digestion. In their proposed process, most of the water (containing nitrogen, phosphorus, and other nutrients) was to be recycled to the ponds, along with any CO_2 generated from the conversion process. In the 1970's and early 1980's, in response to the energy crisis, more detailed development of this concept was carried out [13]. Since the1980's, work in this field has emphasized the production of liquid fuels, in particular biodiesel [14].

Increasing world oil prices in addition to the placement of renewable energy standards may eventually lead to an increase in biofuel production worldwide. Some commodities such as sugar cane, soybeans, and vegetable oil may be used either as food or a feedstock for biofuel production. A comparison of the land area requirements of various feedstocks of biodiesel is shown in Table 7.2 [15]. The required land area in millions of hectares represents the land area required for a given feedstock to replace the current annual gasoline (motor) consumption [1] in the U.S. The final column in Table 7.2 lists the percentage of crop area required for the replacement of annual gasoline consumption in the U.S. with a given biofuel feedstock. These data

[4] Growing algae as a protein source at a large scale in open ponds was first investigated by German scientists during World War II; between 1948–1950, Stanford Research Institute developed a large-scale algae plant; in 1951, Arthur D. Little advanced the technology further through the construction and operation of a Chlorella pilot plant for the Carnegie Institute.

Table 7.2 Land area comparison requirements of various biofuel feedstocks

Crop	Oil yield (L/ha yr) [15]	Required land area (Mha)[a]	Percent of existing U.S. crop area[b]
Soybean	446	1170	711
Canola	1190	440	267
Jatropha	1890	280	170
Coconut	2690	195	119
Oil Palm	5950	88.0	54
Microalgae (30 wt% oil)	58,700	9.00	5.4

[a]assuming 9,034,000 bbl/day of U.S. motor gasoline consumption in 2010 (U.S. EIA)
[b]assuming 406,424,909 acres of cropland in 2007 (USDA)

are based upon the available cropland (406, 424, 909 acres) in 2007, obtained from the U.S. Department of Agriculture census database [16].

The conversion of CO_2 to fuel via algae versus other plants does have its advantages. For instance, since algae's productivity is greater than traditional crop plants, they fix more CO_2 per unit area. The ability of algae to grow much faster than higher-order plants, with generation times on the order of hours to days, versus years to decades for trees, allows for R&D toward improving strains or developing hybrids to be completed in a shorter time frame, *i.e.*, years versus decades. Also, algae may be grown using *non-arable* land and nonagricultural waters, such as salt water, brackish water and municipal wastewater. In particular, they can grown in *hardpan soil* or soils high in clay that are not suitable for conventional agriculture or forestry, which minimizes their competition with food and fiber production. Like other biomass, microalgal biomass is potentially suitable for conversion to liquid (*e.g.*, gasoline, biodiesel, and ethanol) and gaseous (*e.g.*, methane and hydrogen) fuels.

An additional benefit is the high energy densities of the lipids obtained from algae. For example, lipids in the form of triacylglycerols have an energy density on the order of 38 MJ/kg (34 MJ/L) [17], while ethanol has an energy density of approximately 27 MJ/kg (21 MJ/L). The high energy density of lipids makes them more akin to petroleum-based fuel, such as diesel and gasoline, having energy densities of 42.8 MJ/kg (36 MJ/L) and 42.5 MJ/kg (30 MJ/L), respectively. A comparison of the densities and energy densities of algae-based biofuels to other common fuels is shown in Table 7.3.

Having high energy density is particularly important for some applications, such as aviation fuel, in which the weight of the fuel for delivering a given amount of energy is critical. Algae-based oils have fatty-acid distributions of 36% oleic, 15% palmitic, 11% stearic, 8% heptadecenoic and 7% linoleic [18], making these lipids a suitable replacement for petroleum-based diesel. These fatty acid hydrocarbons typically range in length between 16–18 carbon atoms.

The compatibility of lipid-based biofuel with the existing transportation fuel infrastructure [19] is a substantial driver in algae development. In July 2011, the American Society for Testing and Materials (ASTM) created new provisions in ASTM D7566 that allow for the addition of up to 50% of bio-derived synthetic blending components to conventional jet fuel.

Table 7.3 Comparison of density and energy densities of various fuels[a]

Fuel	Density (kg/L)	LHV (MJ/kg)	LHV (MJ/L)
Diesel	0.84	42.8	35.8
Algae-based biofuel	*0.79*	*38.0*	*34.0*
Gasoline	0.70	42.5	29.8
Methane	0.47	50.1	23.3
Propane	0.51	45.8	23.2
Ethanol	0.78	27.0	21.1
Methanol	0.79	20.1	15.9
Ammonia	0.76	18.6	14.1
Hydrogen	0.07	120.1	8.4

[a]gas densities are at standard conditions (*i.e.*, 1 atm and 25°C).

7.2 Microalgae Production

Microalgae can differ in shape, size, chemical structure, composition, and color and can be classified broadly into the following categories:

- Bacillariophyta—diatoms
- Charophyta—stoneworts
- Chlorophyta—green algae
- Chrysophyta—golden algae
- Cyanobacteria—blue-green
- Dinophyta—dinoflagellates
- Phaeophyta—brown algae
- Rhodoohyta—red algae

It has been estimated that there are possibly several million species of algae in existence compared to approximately 250,000 terrestrial-based plant species [20]. Microalgae usage has been primarily applied for the food, *nutraceutical*, and cosmetic industries as shown in Table 7.4, in which selected application and production levels are listed. Commercial large-scale production of microalgae began with a culture of Chlorella as a food additive in Japan in the 1960's, and due to increased world demand, expanded to the U.S., India, Israel, and Australia throughout the 1970's and 1980's [21]. The history of algae cultivation thus far has been limited to high-value specialty foods and nutraceutical products. A current challenge is to reduce the capital and operating and maintenance costs of these systems to produce products (*e.g.*, biofuel) with a much larger market potential and at competitive prices. The Phase I goal of DARPA for the production of fuel from algae is set at \$0.61 per kg (*i.e.*, less than \$2/gallon), which is approximately two orders of magnitude lower than the cost of the specialty products listed in Table 7.4.

Table 7.4 Annual microalgae production [22] (2004)

Microalgae	Production (tons/dry wt.)	Producer	Application	[a]Price($/g)
Spirulina	3000	China, India, U.S., Myanmar, Japan	Nutrition & Cosmetic	60
Chlorella	2000	Taiwan, Germany, Japan	Nutrition & Cosmetic	60
			Aquaculture	90
Dunaliella salina	1200	Australia, Israel, U.S., Japan	Nutrition & Cosmetic	350–3,500
Haematococcus pluvialis	300	U.S., India, Israel	Aquaculture	90
Crypthecodinium cohnii	240	U.S.	DHA oil	70
Shizochytrium	10	U.S.	DHA oil	70
DARPA Phase I Goal				0.61

[a]2011 U.S. dollars

7.2.1 Cultivation Technology

As discussed previously, microalgae are unicellular photosynthetic micro-organisms, living in saline or freshwater environments that convert sunlight, water and CO_2 to algae-based biomass [23]. The two main species of algae are filamentous and phytoplankton. These two species, in particular phytoplankton, grow rapidly to form algal blooms. Many species exhibit rapid growth and high productivity, and have the potential for additional lipid accumulation, sometimes greater than 60% of their dry biomass [24]. Industrial reactors for algal culture at present include open ponds and PBRs based on tubular, flat-plate or other designs.

Microalgae production in closed PBR systems is more expensive than ponds, and the algae density is typically lower, but the light management is better in PBR systems over open ponds. As algae grows, excess culture overflows and is harvested. In hybrid systems both open ponds as well as PBR systems are used in combination to make use of land efficiently, minimize the parasitic energy loss, and ultimately achieve favorable economics. Figure 7.3 illustrates a tubular PBR system with parallel-run horizontal tubes [25]. A tubular PBR consists of an array of straight transparent tubes that are usually made of plastic. The solar collectors have tubes usually 0.1 m or less in diameter. Tube diameter is small because light does not penetrate too deeply in the dense culture solution that is necessary for ensuring high biomass productivity. The microalgal solution is circulated from a reservoir to the solar collector and back to the reservoir.

Figure 7.4 illustrates the open-pond system design characteristic of algae farms. The pond is designed as a "raceway," in which the algae, water, and nutrients circulate around a "racetrack." Paddle wheels provide the flow properties required to maintain proper gas-liquid exchange, nutrient mixing, and algae mixing. The paddle wheel allows for the algae to circulate from the pond surface to the subsurface

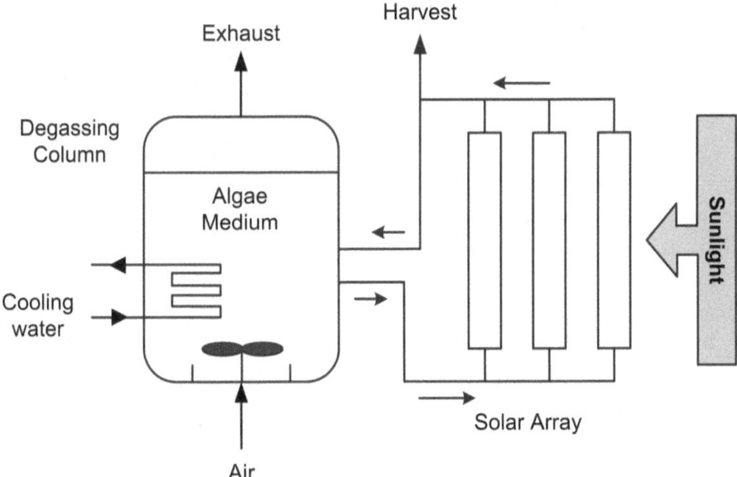

Fig. 7.3 A tubular photobioreactor with parallel-run horizontal tubes

Fig. 7.4 Schematic of open-pond system

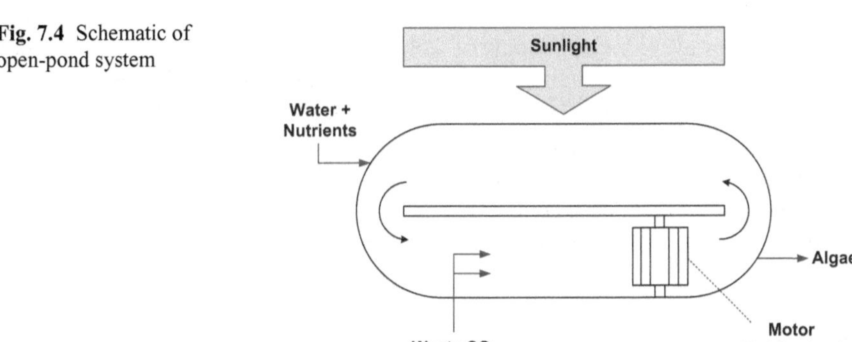

to achieve increased density. The ponds are shallow since the algae need to be exposed to sunlight and there is a limited depth in which sunlight can penetrate the algae-laden water. The ponds are operated continuously, allowing for additional water and nutrients to be constantly fed in to the pond. The pond size is measured in terms of surface area, since surface area is critical for maximizing the sunlight captured.

Algae productivity is measured in terms of biomass produced per day per unit of available surface area (*e.g.*, g dry biomass/m^2day). It has been shown that by careful control of pH and other physical conditions, more than 90% CO_2 utilization in the ponds is achievable. Raceway ponds, usually lined with plastic or cement, are approximately 15–35 cm deep to ensure adequate exposure to sunlight. In addition, they are typically mixed with paddle wheels, and are between 0.2 and 0.5 hectares in size.

7.3 CO_2 Mitigation Potential

To convert the CO_2 generated from a power plant to fuel via algae, large open ponds or PBRs nearby the power plant may be used for algae growth. In such a system, the flue gas (taken downstream of the flue gas desulfurization unit) or pure CO_2 (captured from power plants) would be bubbled into the ponds. After harvesting, the biomass would be processed to create a fossil fuel replacement, *i.e.*, a high-value liquid fuel such as biodiesel. The mitigation potential of this technology is restricted by the large requirement of land, favorable climate, conversion efficiency, and ample water requirements.

Recently, Brune et al. proposed a model for CO_2 capture from flue gas by an algae culture system [26]. Their study was based upon a 50-MW natural-gas-fired power plant with an estimated algae productivity of 20 g-dry algae per m^2 per day and a CO_2 capture rate of approximately 37 g/m^2 per day. A power plant of this capacity firing natural gas produces on average 450 tons of CO_2 per day. From their study it was estimated that a land area of 8.8 km^2 (*i.e.*, 880 hectares)[5] would be required to capture 70% of the CO_2 generated. Based upon this data, land requirement estimates may be made for a typical coal-fired power utility. For instance, an 850-MW coal-fired power plant emits on average 18,700 tons of CO_2 per day. Assuming 90% CO_2 capture, equates to a required pond surface area of approximately 500 km^2. By comparison, a typical amine-based absorption system[6] for CO_2 capture of an 850-MW power plant requires a surface area of approximately 0.04 km^2 with the power plant's footprint approximately 0.26 km^2 [27]. In addition to the challenge of size requirements, there is also parasitic energy that is required to transfer the CO_2 to the algae ponds, as well as to culture, harvest and process the algae, as demonstrated in Fig. 7.1.

7.4 Challenges of Microalgae-to-Biofuel Processes

As discussed previously, biodiesel produced from microalgae is being studied as an alternative to biodiesel production from conventional crops. Comparing microalgae and conventional crops, microalgae typically produce more oil, consume less space and may be grown on non-arable land. However, there are several primary challenges associated with advancing algae conversion processes as a CO_2 mitigation strategy. First, the energy requirements associated with the process as illustrated in Fig. 7.1 may lead to a net generation of CO_2 if carbonized fuels are used for the energy inputs. For the process to result in near neutral CO_2 emissions, renewable (non-carbonized) energy resources would have to be used to supply the mechanical and thermal energy required to drive the conversion process. In this case, the conversion of CO_2 to biofuel via microalgae would be a substitute to traditional hydrocarbon recovery processes.

---

[5] 1 hectare = 10,000 m^2

[6] Based on a previous study of a 850-MW power plant with an amine-based capture system requiring 40,400 m^2 with a power-plant footprint of 257,000 m^2.

An additional primary challenge is finding ways to increase the various efficiencies of the conversion process as listed in Table 7.1. Recall that only 48.55% of the solar spectrum may be used for photosynthesis. This can serve as the upper bound of the energy conversion from photons to chemical energy storage available in the biomass harvested from microalgae. However, as previously noted, there are several additional inefficiencies associated with the conversion process that reduces the energy further. From Table 7.1, the inefficiencies of the CO_2 conversion process are sourced from photon transmission and utilization, and biomass accumulation. Using the range of efficiencies listed in Table 7.1, the total efficiency of the CO_2 conversion process as outlined ranges between 1.3–34.5%. Assuming 48.5% is the theoretical maximum efficiency, this results in a second-law efficiency of the CO_2-to-biofuel conversion process of algae ranging between 3.8–75%. The greatest inefficiency (*i.e.*, reported in the literature to be as low as 10%) is sourced from photon utilization. Recall that this inefficiency is associated with a reduction in the complete absorption of incident photons due to the suboptimal growth conditions of the algal culture. The light conditions can significantly influence this parameter, with high light leading to efficiencies between 10–30% and low light leading to efficiencies between 50–90% [5]. Finding ways to increase the biomass accumulation efficiency, with literature estimates ranging between 34–89%, may also enhance the yield of biofuel from algae.

Aside from the inefficiency challenges, there are additional difficulties that are associated with microalgae cultivation, some of which include minimizing contamination, as many open-pond systems are exposed to the open environment and are thereby subject to disease and predation; careful control of cultivation conditions; space minimization; and efficient provision of CO_2 and light [28]. It is important to note that these limitations may vary between PBR versus open-pond systems. For additional details associated with algae-based CO_2-to-fuel conversion there are several excellent reviews available in the literature [29].

References

1. Oil Crude and Petroleum Products Explained (2011) Petroleum statistics, consumption and disposition. http://www.eia.gov/energyexplained/index.cfm?page=oil_home – tab2. Accessed 7 Aug 2011
2. (a) Weyer KM, Bush DR, Darzins A, Willson BD (2010) Theoretical maximum algal oil production Bioenerg Res 3(2):204–213; (b) Bolton JR, Hall DO (1991) The maximum efficiency of photosynthesis. Photochem Photobiol 53(4):545–548; (c) Govindjee R, Rabinowitch E (1968) Maximum quantum yield and action spectrum of photosynthesis and fluorescence in chlorella. Biochim Biophys Acta 162;539–544
3. Bird HE, Hulstrom RL (1981) Simplified clear sky model for direct and diffuse insolation on horizontal surfaces, Technical Report No. SERI/TR-642-761, Solar Energy Research Institute, Golden
4. EnergyPlus Weather Data. http://apps1.eere.energy.gov/buildings/energyplus/cfm/weather_data.cfm
5. Goldman JC (1979) Outdoor algal mass cultures–II. Photosynthetic yield limitations. Water Res 13(2):119–136

6. (a) Falkowski PG, Dubinsky Z, Wyman K (1985) Growth-irradiance relationships in phytoplankton. Limnol Oceanogr 30:311–321; (b) Sukenik A, Levy RS, Levy Y, Falkowski PG, Dubinsky Z (1991) Optimizing algal biomass production in an outdoor pond: a simulation model. J Appl Phycol 3(3):191–201
7. Rebolloso-Fuentes MM, Navarro-Perez A, Garcia-Camacho F, Ramos-Miras JJ, Guil-Guerrero JL (2001) Biomass nutrient profiles of the microalga Nannochloropsis. J Agric Food Chem 49(6):2966–2972
8. (a) Hu Q, Sommerfeld M, Jarvis E, Ghirardi M, Posewitz M, Seibert M, Darzins A (2008) Microalgal triacylglycerols as feedstocks for biofuel production: perspectives and advances. Plant J 54(4):621–639; (b) Rodolfi L, Chini Zittelli G, Bassi N, Padovani G, Biondi N, Bonini G, Tredici MR (2009) Microalgae for oil: strain selection, induction of lipid synthesis and outdoor mass cultivation in a low cost photobioreactor. Biotechnol Bioeng 102(1):100–112
9. Sheehan J, Dunahay T, Benemann J, Roessler P (1998) Look back at the U.S. Department of Energy's Aquatic Species Program: biodiesel from algae. Close-Out Report, NREL
10. Chisti Y (2007) Biodiesel from microalgae. Biotechnol Adv 25(3):294–306
11. (a) Burlew JS (1953) Algal culture from laboratory to pilot plant. Carnegie Institute, Washington, DC, pp 235–272; (b) Soeder CJ (1986) A historical outline of applied algology. CRC Press, Boca Raton
12. Oswald WJ, Golueke CG (1960) Solar energy conversion with microalgae systems. Adv Appl Microbiol 11:223
13. Benemann J, Oswald W (1996) Systems and economic analysis of microalgae ponds for conversion of CO_2 to biomass. Department of Energy, Pittsburgh
14. Benemann JR (1997) CO_2 mitigation with microalgae systems. Energy Conv Manag 38:S475–S479
15. Demirbas A (2011) Biodiesel from algae, biofixation of carbon dioxide by microalgae: a solution to pollution problems. Appl Energy 88:3541–3547
16. United States Department of Agriculture, E. R. S. (2011) The economics of food, farming, natural resources, and rural America farm characteristics, census of agriculture. http://www.ers.usda.gov/statefacts/US.HTM. Accessed 7 Aug 2011
17. Nelson LD, Cox MM (2005) Principles of biochemistry, 4th edn. WH Freeman and Company, New York, p 1119
18. Demirbas A, Demirbas MF (2010) Algae energy: algae as a new source of biodiesel. Springer, London, p 199
19. Demirbas A (2009) Progress and recent trends in biodiesel fuels. Energy Convers Manage 50(1):14–34
20. Norton TA, Melkonian M, Andersen RA (1996) Algal biodiversity. Phycologia 35(4):308–326
21. (a) Spolaore P, Joannis-Cassan C, Duran E, Isambert A (2006) Commercial applications of microalgae. J Biosci Bioeng 101(2):87–96; (b) Borowitzka MA (1999) Commercial production of microalgae: ponds, tanks, tubes, and fermenters. J Biotechnol 70(1–3):313–321
22. Brennan L, Owende P (2010) Biofuels from microalgae—a review of technologies for production, processing, and extractions of biofuels and co-products. Renew Sust Energy Rev 14:557–577
23. Ozkurt I (2009) Qualifying of safflower and algae for energy. Energy Educ Sci Technol A 23:145–151
24. Schenk PM (2008) Thomas-Hall SR, Stephens E, Marx UC, Mussgnug JH, Posten C, Kruse O, Hankamer B (2008) Second generation biofuels: high-efficiency microalgae for biodiesel production. Bioenerg Res 1(1):20–43
25. Demirbas A (2009) Energy concept and energy education. Energy Educ Sci Technol B 1:85–101
26. Brune DE, Lundquist TJ, Benemann JR (2009) Microalgal biomass for greenhouse gas reductions: potential for replacement of fossil fuels and animal feeds. J Environ Eng-ASCE 135(11):1136–1144
27. EPRI's Pulverized Coal Post-Combustion CO_2 Capture Retrofit Study Summary (2010) Report No. 1019680. Electric Power Research Institute, Palo Alto, CA

28. (a) Marxen K, Vanselow KH, Lippemeier S, Hintze R, Ruser A, Hansen UP (2005) A photobioreactor system for computer controlled cultivation of microalgae. J Appl Phyco 17(6):535–549; (b) Grima EM, Fernandez FGA, Camacho FG, Chisti Y (1999) Photobioreactors: light regime, mass transfer, and scaleup. J Biotechnol 70:231–247
29. (a) Xu L, Weathers PJ, Xiong XR, Liu CZ (2009) Microalgal bioreactors: challenges and opportunities. Eng Life Sci 9(3):178–189; (b) Pienkos PT, Darzins A (2009) The promise and challenges of microalgal derived biofuels. Biofuel Bioprod Bioresour 3(4):431–440

Chapter 8
The Role of CO_2 Reduction Catalysis in Carbon Capture

In addition to the algae-mediated process discussed in Chap. 7, to generate hydrocarbon-based fuels and useful chemicals from CO_2, it is also possible to use electrochemical and photocatalytic processes to carry out CO_2 reduction. As previously mentioned in Chap. 7, today, an even stronger policy driver than climate change is the expansion of alternatives to crude oil for transportation fuels. Another driver for advancing electrochemical and photocatalytic reduction of CO_2 is that it may allow for the storage of stranded energy from resources such as wind, solar, tidal, and geothermal in the form of chemical energy within the bonds of hydrocarbons.

Figure 8.1 illustrates an example of an overall system approach to CO_2 reduction with the hydrogen sourced from water[1] and the reaction toward hydrocarbon formation taking place in an electrochemical cell. Ideally, one may consider this process to be carbon-neutral; however, this is not the case, as shown in Fig. 8.1. The CO_2 may be introduced into the fuel cell in a semi-pure form or diluted in a flue gas stream. However, catalytic surfaces are very sensitive to poisoning via undesired gas adsorption, surface oxidation, or particulate matter accumulation. Although experience has been gained in the application of selective catalytic reduction (SCR) units for NO_x emissions, which are located just after the exit of the power plant boiler, upstream of the particulate control device, for ideal operation and application it may not be practical to envision the implementation of a catalytic CO_2 reduction system without some degree of scrubbing or air filtration.

In addition, the reduction of CO_2 is an uphill process that requires energy, usually derived from electricity or light as illustrated in Fig. 8.1. If the energy transferred is not sourced from renewables (*i.e.*, non-carbonized fuel), the process will have some CO_2 generation associated with it. In electrochemical systems CO_2 is reduced through an applied bias or current using metal electrodes, while in photocatalysis sunlight is used to reduce CO_2 using semiconducting catalytic electrodes. It is important to also recognize that within this context, the produced fuel will be oxidized to harness its energy potential, with the generation of CO_2 and H_2O. Similar

[1] To form hydrocarbons from CO_2 reduction, the source of hydrogen does not necessarily have to be water and may be of the form of H_2, H_3O^+, etc.

J. Wilcox, *Carbon Capture*,
DOI 10.1007/978-1-4614-2215-0_8,

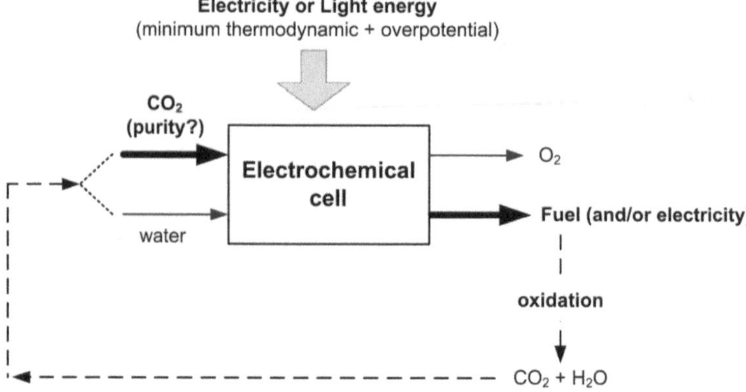

Fig. 8.1 Schematic of CO_2 catalytic reduction process

to CO_2 reduction via algae, this is not a process in which CO_2 is permanently sequestered. This chapter focuses primarily on electrochemistry with a brief mention of photochemistry, although hybrid processes of the two are also possible.

The following two approaches may be used to convert CO_2 to useful products: 1) the conversion of CO_2 and H_2O into a synthesis gas (*i.e.*, CO and H_2, also known as syngas), from which point Fischer-Tropsch catalysis (see Chap. 1) can be used to further convert the syngas to hydrocarbons or alcohols, or 2) the conversion of CO_2 and H_2O to a wide range of products such as CO, hydrocarbons, alcohols, etc. The electrocatalytic process may be carried out *heterogeneously*, in which the catalyst is in a different phase than the reactants or *homogeneously*, in which the catalyst is in the same phase as the reactants. Overall, the reduction of CO_2 and oxidation of water to form a CO_2-anion intermediate, and protons (H^+), respectively, occurs across a metal-based catalytic electrode surface separated by an *electrolyte*, which facilitates the flow of protons, as illustrated in the electrochemical cell of Fig. 8.2. Although not covered in this short discussion on CO_2 reduction, the oxygen evolution reaction from water is also a well-studied topic [2] that will likely play a role as water may serve as a source of protons for fuel synthesis with reduced CO_2.

Reactions taking place in electrochemical cells are termed *reduction-oxidation* (redox) reactions, in which the release of electrons occurs in the oxidation step and a gain of electrons occurs in the reduction step. In an electrochemical cell, the electron source is termed the *anode*, from which a chemical species is oxidized and delivers electrons to an external load, while the electron sink is termed the *cathode*, from which a chemical species is reduced, acquiring electrons exiting from the load. In electrochemical cells, the full reaction is separated into two half-cell reactions that occur in physically separate regions of the device, interconnected by an electrolyte that conducts ions, but not electrons as illustrated in Fig. 8.2.

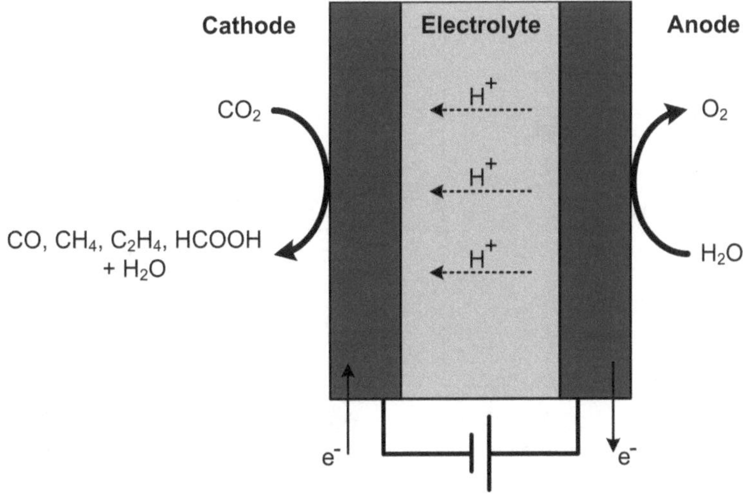

Fig. 8.2 Schematic illustrating the process of electrochemical reduction of CO_2

8.1 Electrochemical Catalytic Reduction of CO_2

To reduce CO_2 electrochemically, proton and electron sources are required. Protons may be produced at the anode from H_2O oxidation, with the simultaneous evolution of oxygen, and are transferred through the electrolyte to the cathode to react with reduced CO_2 to form hydrocarbon products (*e.g.*, carbon monoxide (CO), methane (CH_4), ethylene (C_2H_4), formic acid (HCOOH)), and water.

8.2 Electrocatalysis Tutorial

Electrochemical reactions are different from chemical reactions in that they involve the transfer of free electrons between an electrode surface and a chemical species, rather than between chemical species themselves. It is important to consider both the thermodynamics and kinetics of an electrochemical process. Thermodynamics can predict whether a reaction is energetically spontaneous and places an upper bound on the maximum electrical potential that may be generated, or that is needed to drive a given reaction. Any real electrochemical process will perform below its thermodynamic limit.

Thermodynamics of Electrochemical Reactions The applied *voltage potential,* V_{app}, is a measure of the electronic energy and is measured with respect to a reference electrode, *e.g.*, Ag/AgCl, with the measured voltage indicating the position of the *Fermi level* in the metal. Controlling the electrode potential allows for the control

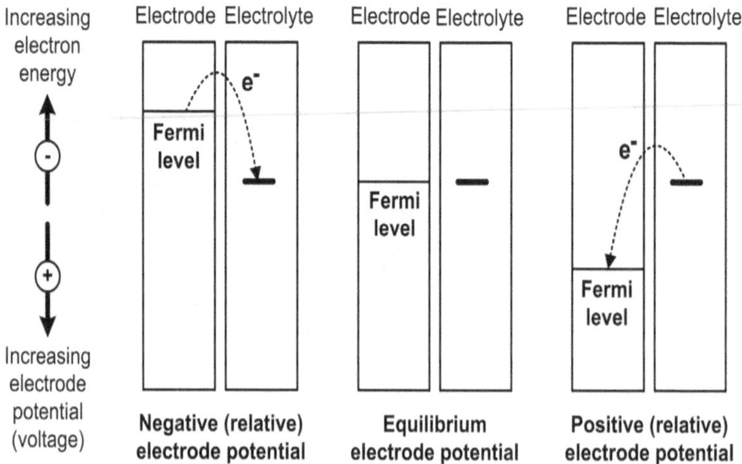

Fig. 8.3 Electrode potential can be tuned to induce reduction (*left*) and oxidation (*right*). The thermodynamic equilibrium electrode potential (*middle*) corresponds to the scenario in which oxidation and reduction processes are balanced. (Reprinted with permission of John Wiley & Sons, Inc. [1])

of the electronic energy in an electrochemical system, which thereby influences the reaction direction. This is illustrated in Fig. 8.3 in which the electrode potential is used to induce reduction (Re) or oxidation (Ox) between chemical species in the following general electrochemical reaction:

$$\mathrm{Ox} + \mathrm{e}^- \leftrightarrow \mathrm{Re} \tag{8.1}$$

If the electrode is more negative than the equilibrium potential, the reaction leans toward the formation of a reduced chemical species. In other words, a more negative electrode forces the electrons out of the electrode. In the case in which the electrode is more positive than the equilibrium potential, the reaction will lean toward the formation of oxidized chemical species since a more positive electrode "attracts" electrons to the electrode [3]. In electrochemistry, the use of voltage potential is crucial to controlling reactivity.

The *overpotential*, η, is the additional energy required to force the electrode reaction to proceed, and is measured by the difference between the applied electrode potential, E, and the standard potential, E_0. With catalysis, CO_2 conversion becomes more thermodynamically favored through a reduction in the overpotential, *i.e.*, $\eta_{\mathrm{cat}} < \eta_{\mathrm{no\ cat}}$ as illustrated in Fig. 8.4.

A critical parameter in the CO_2 reduction process is the *energetic efficiency* (ε_{en}), which is defined as the recoverable energy contained in the fuel product, and directly relates to the energy cost of the production. The energetic efficiency for formation

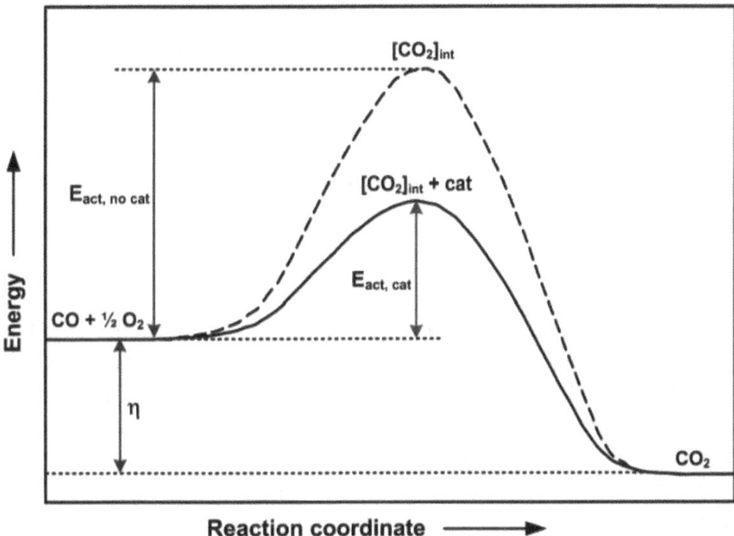

Fig. 8.4 Qualitative reaction scheme for the electrochemical reduction of CO_2 with (*solid*) and without (*dotted*) a catalyst

of a given product is defined as:

$$\varepsilon_{en} = \frac{E_0}{E_0 + \eta}(\varepsilon_{Far}) \tag{8.2}$$

$$\varepsilon_{Far} = \frac{n_e F n}{q} \tag{8.3}$$

such that E_0 is the standard potential, η is the overpotential, ε_{Far} is the Faradaic efficiency, n_e is the number of electrons, F is Faraday's constant, n is the number of moles of product, and q is total charge. The product n_eFn also represents charge, and is equivalent to the total charge going toward the formation of a specific product. The *Faradaic efficiency* is defined as the ratio of the number of electrons going to a desired product to the total number of electrons transferred in the circuit. According to Eqs. (8.2) and (8.3), a combination of high Faradaic efficiency and low overpotential will result in a high energetic efficiency. The standard potential can be calculated for a given reaction from the Gibbs free energy change of reaction, ΔG_{rxn}, provided the reaction entropy and enthalpy values are known. Gas-phase reaction data can be readily obtained from the NIST database as a function of temperature and pressure. Based on the second law of thermodynamics, not all of the chemical energy produced in an electrochemical cell will be useable, and the energy not converted to work will be transferred from the system to the surroundings in the form of heat. The heat transferred, Q, is equal to the product of the temperature, T, and reaction entropy change, ΔS_{rxn}. This term (*i.e.*, $T\Delta S_{rxn}$) is equivalent to the overpotential of a given reaction. Equation (8.2) may be written in thermodynamic terms as:

$$\varepsilon_{en} = \frac{\Delta G_{rxn}}{\Delta G_{rxn} + T\Delta S_{rxn}}(\varepsilon_{Far}) = \frac{\Delta G_{rxn}}{\Delta H_{rxn}}(\varepsilon_{Far}) \tag{8.4}$$

In addition, the maximum theoretical electrical energy produced per mole of product consumed is determined by:

$$W_e = n_e F \Delta G_{rxn} = n_e F E_0 \tag{8.5}$$

Kinetics of Electrochemical Reactions Since electrochemical reactions take place at the interface between an electrode and an electrolyte, the current produced is typically directly proportional to the interfacial surface area. In general, doubling the surface area should double the rate of reaction. However, as the process is scaled up there will be a point in which mass-transfer limitations may dominate. Another important parameter is the *current density*, i, typically expressed in units of A/m^2, and represents the electric current per unit area of electrode material. The current density can be directly linked to the reaction rate, which dictates the reactor size and ultimately the capital cost of the process, while the *exchange current density*, i_0, reflects the rates of the backward and forward reactions, with the net current equal to zero at equilibrium. Since measurements are carried out on one half-reaction at a time, the electrode that is participating in the reaction of focus is termed the working electrode, while the second electrode is termed the reference electrode, often termed as the standard hydrogen electrode (SHE) or normal hydrogen electrode (NHE). When voltages are reported it is always with respect to a reference electrode [4].

The forward and reverse reaction rates follow an Arrhenius-type law such that,

$$r_f \propto e^{-\frac{\Delta G_{f,act}}{RT}} \tag{8.6}$$

and

$$r_r \propto e^{-\frac{\Delta G_{r,act}}{RT}} \tag{8.7}$$

such that ΔG_{act} represents the Gibbs energy of activation of either the forward or reverse reaction. In 1930, John Butler and Max Volmer approximated a simplified approach to model the features of the empirical-based Tafel equation, commonly known as the BV approximation. In 1905, Julius Tafel presented the following equation to model the electrochemical transfer kinetics, which relates the overpotential to an observed current density:

$$\eta = a + b \ \log i \tag{8.8}$$

such that the constant a represents the deviation of the electrode potential from its thermodynamic equilibrium potential, while the constant b is termed the Tafel slope, which typically ranges between 30 and 120 mV, indicating the extent at which the electrode potential changes. Often times η is plotted versus i, and the y-intercept of the

Tafel line, *a* is used to compare the intrinsic activity of different electrode materials with respect to a given half-cell electrochemical reaction. A smaller intercept is indicative of a lower thermodynamic driving force required for achieving a desired reaction rate.

Within the BV approximation, ΔG_{act} is assumed to change linearly with changes in the thermodynamic driving force (*i.e.*, overpotential, $\eta = E - E_0$), such that:

$$\frac{\partial \Delta G_{f,act}}{\partial (E - E_0)} = \alpha F \tag{8.9}$$

and

$$\frac{\partial \Delta G_{r,act}}{\partial (E - E_0)} = -(1 - \alpha) F \tag{8.10}$$

such that F is Faraday's constant, and α is a dimensionless parameter termed the transfer coefficient, and represents the fraction of the driving force that leads to reduction, while $(1-\alpha)$ represents the fraction of the driving force that leads to oxidation. Integration of these two equations yields the following linear free-energy relationships:

$$\Delta G_{f,act} = \Delta G_0 + \alpha F(E - E_0) \tag{8.11}$$

$$\Delta G_{r,act} = \Delta G_0 - (1 - \alpha) F(E - E_0) \tag{8.12}$$

In Eqs. (8.11) and (8.12), ΔG_0 represents the standard Gibbs energy of activation of the process if the driving force or overpotential is zero. Substitution of Eq. (8.11) into Eq. (8.6) and Eq. (8.12) into Eq. (8.7) yields the following forward and reverse reaction rates, respectively:

$$r_f(E) \propto k_0 e^{-\frac{\alpha F}{RT}(E - E_0)} \tag{8.13}$$

and

$$r_r(E) \propto k_0 e^{-\frac{(1-\alpha)F}{RT}(E - E_0)} \tag{8.14}$$

such that k_0 is the standard rate constant.

The overall BV equation can be expressed as:

$$i = i_0 \left(-e^{-\frac{\alpha F}{RT}(E - E_0)} + e^{-\frac{(1-\alpha)F}{RT}(E - E_0)} \right) \tag{8.15}$$

where $i_0 = nFk_0c$, such that c is the concentration of reactants or products under the assumption that they are equal, n is the number of electrons transferred in the electrochemical reaction, and the overpotential, $\eta = E - E_0$ [5].

Many various metal-based catalysts have been investigated for CO_2 reduction to hydrocarbon-based fuels and chemicals. The metals that have received the most attention include copper, platinum, nickel, and gold. This chapter is not intended to be a review, but just an introduction into some of the general concepts behind electrochemical reaction processes. In particular, Cu as an electrode catalyst material is discussed, as it is an earth-abundant metal (*i.e.*, readily-available) and performs reasonably well as a CO_2 electrochemical reduction catalyst. For additional information, excellent reviews on both heterogeneous [6] and homogeneous [7] electrochemical catalysis for CO_2 reduction are available in the literature.

Copper has been widely accepted as a unique metal for CO_2 reduction due to its ability to produce hydrocarbon fuels such as methane and ethylene. Hori et al. [8] have found that the direct reduction of CO_2 to methane and ethylene takes place with a current density of 5–10 mA/cm^2, with current efficiencies up to 69% at 0°C. In addition, the distribution of products from the electrochemical reduction of CO_2 depends strongly upon the electrolytes used [9]. Ethylene and alcohols are, for example, preferentially formed in solutions of potassium chloride (KCl), potassium sulfate (K_2SO_4), potassium perchlorate ($KClO_4$), and dilute bicarbonate (HCO_3^-), while CH_4 is preferentially produced in relatively concentrated HCO_3^- and phosphate solutions. The pH at the Cu electrode is also an important factor for the selectivity of products due to the release of OH^- anions resulting from the electrode reactions [9]. The increased pH near the electrode surface may be lowered through neutralization by $KHCO_3$ [10]. Both cationic and anionic species significantly affect the selectivity of products from the electrochemical reduction of CO_2 on Cu in methanol (CH_3OH).

Copper has also been used as a promoter to enhance the CO_2 reduction process of metal-oxide-based electrodes. For instance, Arakawa et al. [11] measured the conversion of CO_2 to methanol under the presence of H_2 over Cu–ZnO catalysts. High methanol selectivity was achieved under operating conditions of high pressure and low temperature. The formation of methanol was proportional to the surface area of the Cu metal of the Cu–ZnO catalyst, highlighting the importance of Cu in the reaction process. Illustrated in Fig. 8.5 is an example of a proposed mechanism of CO_2 reduction to the formation of CH_3OH. Within this mechanism, the direct formation of CH_3OH from CO_2 in the presence of H_2 occurs via a bidentate carbonate species, formate species and methoxy species as proposed by Arakawa et al. [11].

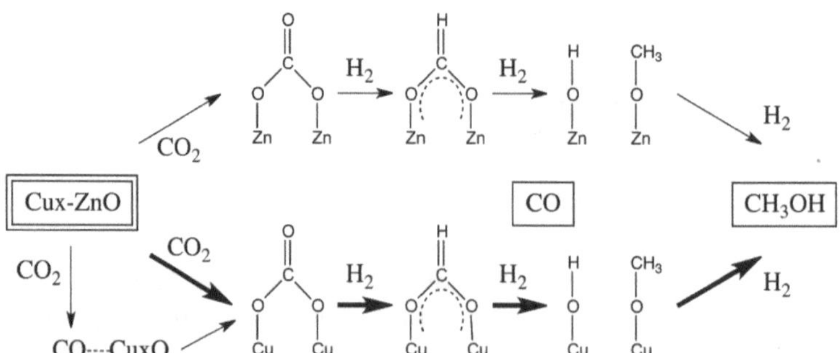

Fig. 8.5 A proposed reaction mechanism for the direct formation of CH_3OH from CO_2 in the presence of H_2 over the Cu–ZnO/SiO_2 catalyst [11]. *Thicker solid lines* represent a predominant pathway

In general, the electrochemical reduction of CO_2 is a very complex process and its mechanism is not yet fully understood. Copper is unique among other metals in terms of its ability to produce hydrocarbons at reasonable current densities. Electrode materials such as cadmium, tin, and mercury having high hydrogen overpotentials may reduce CO_2 with high current efficiencies, but since they are unable to cleave the CO bond within CO_2, they are poor catalysts and lead primarily to the formation of formate (*i.e.*, $HCOO^-$). Low hydrogen overpotential metals such as platinum, nickel, and titanium exhibit the opposite behavior, in that they reduce CO_2 to CO while overbinding CO, thereby inhibiting hydrocarbon formation. Hence, copper is a promising electrode material for the CO_2 electrochemical reduction process, but the challenge is that the required overpotential is still too high. Understanding the reaction mechanism associated with CO_2 reduction on copper will aid in the design and synthesis of more efficient and effective catalysts to carry out this reaction. Electronic structure calculations based upon density functional theory [12] are significantly aiding in the advancement of catalyst design through providing an increased understanding of the underlying mechanisms that take place at the interface between CO_2 and the electrode surface. In addition to electrochemical reduction of CO_2, photocatalytic reduction is also an area of current research and will be discussed next.

8.3 Photocatalytic Reduction

In the case of photocatalytic reduction, as illustrated in Fig. 8.6, the absorption of a photon with energy equal to or greater than the *band gap* of a semiconductor produces what is termed an electron-hole pair, which is formed from the photon excitement of an electron (e^-) from the valence band to the conduction band, leaving behind a positive hole (h^+) in the valence band [13]. The electron–hole pairs are then transferred to CO_2 and H_2O, as depicted in Fig. 8.6. The mechanism by which this reactivity occurs is complex, with a number of proposed mechanisms available in the literature.

A common material used in the photocatalytic reduction of CO_2 is titania (TiO_2). Titania (anatase) has a band gap of 3.2 eV and is active only in the ultraviolet-visible spectral (UV) region. Photocatalysis over TiO_2 is initiated by the absorption of a

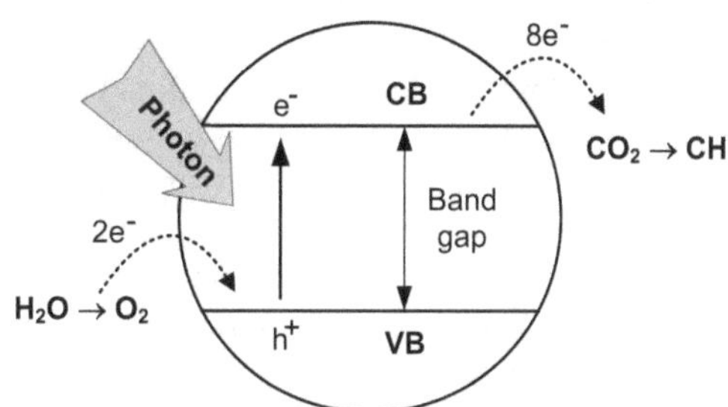

Fig. 8.6 Schematic diagram of photocatalytic reduction of CO_2 by a semiconductor; e^- and h^+ represent a conduction band electron and a valence band hole, respectively

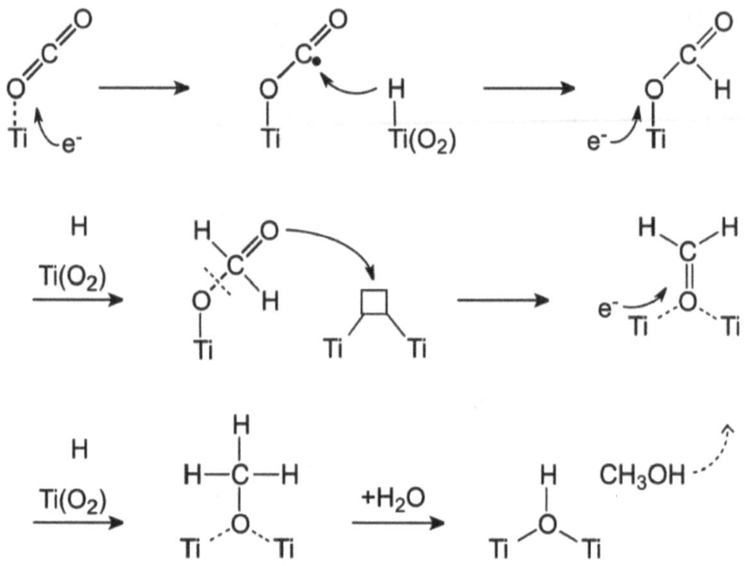

Fig. 8.7 Proposed mechanism of photocatalytic CO_2 reduction over TiO_2 [14]

photon with energy, $h\nu$ equal to or greater than the band gap of TiO_2. An example pathway shown in Fig. 8.7 has been proposed by Wu [14] and illustrates a possible mechanism of CO_2 reduction toward hydrocarbon formation under UV radiation. By adding an electron and one hydrogen atom to the adsorbed CO_2, formate (HCOO) is formed, with the addition of another hydrogen atom leading to the formation of dioxymethylene (H_2COO), which then migrates to the adjacent oxygen vacancy to form formaldehyde (H_2CO) by accepting an electron. Another hydrogen atom is added to H_2CO, resulting in the formation of an adsorbed methoxy (CH_3O) species. Finally, CH_3OH is formed due to the CH_3O interaction with surface-bound H_2O, leaving behind a surface-protonated oxygen atom.

8.4 Challenges in Catalytic Approaches to CO_2 Reduction

Similar to electrochemical CO_2 reduction processes, indeed, the photocatalytic reduction of CO_2 is challenging and still in its infancy in terms of material design for technological advancement. Within these fields, the development of inexpensive, stable catalysts that operate with minimal overpotential and that are highly selective for a specific product are the key aspects to advancing this field. To produce fuel by these methods at costs that are competitive with the existing fuel market, the current densities of today's electrode materials must be improved. Advances in the field of catalysis will be the key to reducing the overpotentials, improving product selectivities, and increasing CO_2 reduction rates.

References

1. O'hayre R, Cha SW, Colella W, Prinz FB (2009) Fuel cell fundamentals. John Wiley & Sons, Inc, New York
2. (a) Jaramillo TF, Jørgensen KP, Bonde J, Nielsen JH, Horch S, Chorkendorff I (2007) Identification of active edge sites for electrochemical H_2 evolution from MoS_2 nanocatalysts. Science 317(5834):100; (b) Surendranath Y, Kanan MW, Nocera DG (2010) Mechanistic studies of the Oxygen evolution reaction by a cobalt-phosphate catalyst at neutral pH. J Am Chem Soc 132;16501–16509; (c) Yeo BS, Klaus SL, Ross PN, Mathies RA, Bell AT (2010) Identification of hydroperoxy species as reaction intermediates in the electrochemical evolution of Oxygen on Gold. Chemphyschem 11;1854–1857; (d) Bell AT (2003) The impact of nanoscience on heterogeneous catalysis. Science 299(5613):1688; (e) Nann T, Ibrahim SK, Woi PM, Xu S, Ziegler J, Pickett CJ (2010) Water splitting by visible light: a nanophotocathode for hydrogen production Angew Chem Int Ed Engl 49(9):1574–1577; (f) Cowan AJ, Tang J, Leng W, Durrant JR, Klug DR (2010) Water Splitting by Nanocrystalline TiO_2 in a complete photoelectrochemical cell exhibits efficiencies limited by charge recombination. J Phys Chem C 114(9):4208–4214
3. O'hayre R, Cha SW, Colella W, Prinz FB (2009) Fuel cell fundamentals, 2nd edn. John Wiley & Sons, Inc., New York
4. Bard AJ, Faulkner LR (2001) Electrochemical methods, fundamentals and applications, 2nd edn. John Wiley & Sons, Inc., New York
5. Strasser P, Ogasawara H (2008) Surface Electrochemistry. In: Chemical bonding at surfaces and interfaces Nilsson, A, Pettersson, L, Nörskov, JK (eds). Elsevier, Amsterdam
6. (a) Whipple DT, Kenis PJA (2010) Prospects of CO_2 Utilization via direct heterogeneous electrochemical reduction. J Phys Chem Lett 1:3451–3458; (b) Hori Y (2008) electrochemical CO_2 reduction on metal electrodes. In: modern aspects of electrochemistry Vayenas CG, White RE, Gamboa-Aldeco ME (eds). Springer, New York, pp 89–189
7. (a) Benson EE, Kubiak CP, Sathrum AJ, Smieja JM (2008) Electrocatalytic and homogeneous approaches to conversion of CO_2 to liquid fuels. Chem Soc Rev 38(1):89–99; (b) Savéant JM (2008) Molecular catalysis of electrochemical reactions. Mechanistic aspects. Chem Rev 108(7):2348–2378
8. (a) Hori Y, Kikuchi K, Suzuki S (1985) Production of CO and CH_4 in electrochemical reduction of CO_2 at metal electrodes in aqueous hydrogencarbonate solution. Chem Lett (11):1695–1698; (b) Hori Y, Kikuchi K, Murata A, Suzuki S (1986) Production of methane and ethylene in electrochemical reduction of carbon dioxide at copper electrode in aqueous hydrogencarbonate solution. Chem Lett 15:897–898
9. Hori Y, Murata A, Takahashi R (1989) Formation of hydrocarbons in the electrochemical reduction of carbon dioxide at a copper electrode in aqueous solution. J Am Chem Soc Farad Trans 1 85:2309–2326
10. De Jesús-Cardona H, del Moral C, Cabrera CR (2001) Voltammetric study of CO_2 reduction at Cu electrodes under different $KHCO_3$ concentrations, temperatures and CO_2 pressures. J Electroanal Chem 513(1):45–51
11. Arakawa H, Dubois JL, Sayama K (1992) Selective conversion of CO_2 to Methanol by catalytic-hydrogenation over promoted copper catalyst. Energy Convers Manag 33(5–8):521–528
12. Peterson AA, Abild-Pedersen F, Studt F, Rossmeisl J, Norskov JK (2010) How copper catalyzes the electroreduction of carbon dioxide into hydrocarbon fuels. Energ Environ Sci 3 (9):1311–1315
13. Koci K, Obalova L, Lacny Z (2008) Photocatalytic reduction of CO_2 over TiO_2 based catalysts. Chem Pap 62(1):1–9
14. Wu JCS (2009) Photocatalytic reduction of greenhouse gas CO_2 to fuel. Catal Surv Asia 13(1):30–40

Chapter 9
The Role of Mineral Carbonation in Carbon Capture

Mineral carbonation takes place through the reaction of CO_2 with an alkalinity source that includes divalent cations such as calcium (Ca^{2+}) and magnesium (Mg^{2+}) as well as hydroxyl anions to form stable carbonate minerals [1]. Similar to algae-based (Chap. 7) and electrochemical CO_2 reduction (Chap. 8) processes, mineral carbonation has the potential to couple the capture with the long-term storage of CO_2. Although most studies of mineral carbonation to date have focused on the carbonation of a pure CO_2 gas (*i.e.*, assuming a previous capture step), CCS in a single-step mineral carbonation process may one day be possible. Therefore, it is important to consider this concept in the broader portfolio of CO_2 capture technologies. Figure 9.1 shows a possible mineral carbonation stream in which an alkalinity source reacts with CO_2 to form mineral carbonate. Energy requirements include thermal or mechanical treatment of the alkalinity source to improve its reactivity toward mineral carbonation. Determining the end use of the mineral carbonate may provide an upper limit on how much energy one is willing to spend on the carbonation process.

Alkalinity is defined as the ability of a given solution to neutralize acid. Chapter 3 discusses CO_2 carbonation chemistry in detail as it applies to CO_2 absorption separation processes. Figure 3.6 exhibits the carbonate speciation as a function of pH, indicating that a high pH is required for carbonate formation. Alkalinity may be sourced from naturally available minerals, such as magnesium- and calcium-rich silicates, in addition to industrial byproducts, such as fly ash, steel slag, and cement kiln dust. This chapter is divided into three sections covering the natural carbonate formation process, carbonation via natural alkaline mineral sources (*e.g.*, ultramafic minerals), and carbonation via industrial waste alkaline sources. The current challenges associated with each process are discussed in addition to the potential scale of CO_2 mitigation possible under ideal conditions. Special attention is given to the conversion of carbonate minerals produced from alkaline industrial by products and CO_2 to useful products, such as aggregate for the construction industry, since this may potentially serve as a driving force for advancing research in this area.

J. Wilcox, *Carbon Capture*,

DOI 10.1007/978-1-4614-2215-0_9,

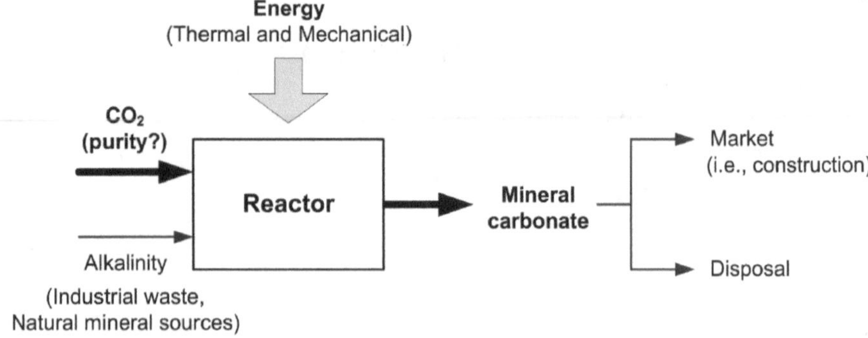

Fig. 9.1 Schematic of combined CO_2 capture and storage process

9.1 Natural and Accelerated Weathering

Carbon dioxide reacts with metal cations, such as the calcium (Ca^{2+}) and magnesium (Mg^{2+}) ions present in naturally abundant silicate minerals (*e.g.*, olivine, serpentine, wollastonite, and Ca-plagioclase), in a thermodynamically favorable process to produce carbonate minerals. *Natural weathering*, the chemical and mechanical alteration of minerals due to physical, chemical, and biological processes, increases the conversion toward carbonate formation by breaking down the silicate structure and releasing the alkaline earth cations (Ca^{2+} and Mg^{2+}). In addition to promoting mineral carbonation, chemical weathering includes *dissolution* and *hydrolysis* that lead to the fixation of CO_2 into bicarbonate or carbonate ions in solution. The following reaction proceeds spontaneously toward the production of a bicarbonate solution with dissolved CO_2 in water and calcite:

$$CO_2(g) + H_2O(l) + CaCO_3(s) \leftrightarrow Ca^{2+}(aq) + 2HCO_3^{-}(aq) \qquad (9.1)$$

and is a simplified reaction representing the dissolution of coral in seawater [3].

Given that natural weathering limits the mineral carbonation process in nature, various research groups have proposed the possibility of CO_2 sequestration through carbonation processes enhanced by accelerated weathering. For example, Rau et al. [5] have proposed a CO_2 mitigation strategy that involves accelerated weathering [4] through the mineralization of CO_2 with limestone-processing waste byproducts to form calcium bicarbonate, which may be distributed into the ocean with potential positive environmental impacts. A related approach based upon electrochemical weathering and associated with CO_2 storage in deep-sea sediments has been evaluated by House et al. [6]. In this engineered approach, the weathering kinetics are accelerated through the addition of HCl in the ocean, which is a stronger acid than carbonic acid, one of the primary acids responsible for natural silicate weathering. Both approaches examine different means of accelerating the natural weathering process to shift the thermodynamic equilibrium of reactions that lead to carbonate formation, thereby enhancing the uptake and storage of CO_2. Aside from ocean-based mineral carbonation processes, both natural and industrial-based sources of alkalinity exist for CO_2 capture and storage.

9.2 Natural Alkaline Sources

Mineral carbonation using natural alkalinity sources has limitations that include slow reaction kinetics in addition to the potential energy costs associated with increasing the mineral chemical reactivity through mechanical and/or chemical modification resulting in increased surface area. Benefits associated with mineral carbonation include: 1) the possibility of CCS as a single process, 2) permanent sequestration with minimal leakage potential, 3) stabilization of harmful pollutants associated with the mineral matter, and 4) the potential energy offset since mineral carbonation processes are thermodynamically favorable.

A simplistic overall general chemical reaction associated with mineral carbonation is:

$$MO + CO_2 \leftrightarrow MCO_3 + \text{heat} \tag{9.2}$$

with M representing a divalent cation. The basic gas-solid reactions along with their respective reaction enthalpy changes at 298 K and 1 atm are:

$$\text{CaO (s)} + \text{CO}_2\text{ (g)} \rightarrow \text{CaCO}_3\text{ (s)} \quad -179.19\ \text{kJ/mol CO}_2 \tag{9.3}$$

$$\text{MgO (s)} + \text{CO}_2\text{ (g)} \rightarrow \text{MgCO}_3\text{ (s)} \quad -116.93\ \text{kJ/mol CO}_2 \tag{9.4}$$

$$\text{FeO (s)} + \text{CO}_2\text{ (g)} \rightarrow \text{FeCO}_3\text{ (s)} \quad -75.04\ \text{kJ/mol CO}_2 \tag{9.5}$$

A comparison of these enthalpy changes with the following methane oxidation reaction, which is quite similar to the heat generated from burning natural gas to produce steam for electricity generation, illustrates that the thermodynamic drive of hydrocarbon oxidation is significantly greater than that of mineral carbonation:

$$\text{CH}_4\text{ (g)} + 2\text{O}_2\text{ (g)} \rightarrow \text{CO}_2\text{ (g)} + 2\text{H}_2\text{O (g)} \quad 802.31\ \text{kJ/mol CO}_2 \tag{9.6}$$

In particular, the thermodynamically favorable carbonation reactions involving Mg- or Ca-containing minerals have the following reported enthalpies of reaction at 298 K and 1 atm: wollastonite – 90 kJ/mol CO_2, olivine – 89 kJ/mol CO_2, and serpentine – 64 kJ/mol CO_2. The reader may refer to the literature for additional estimates of reaction enthalpies and reaction free energies associated with mineral carbonation of CO_2 [7]. As mentioned previously, carbonates are environmentally stable and benign minerals. For instance, the stability of calcium and magnesium carbonate in acidic aqueous environments such as acid rain water has been investigated by Teir et al. [8] who was found that Ca and Mg carbonates are stable and adequately resistant with minimal effects to their potential mineral carbonate storage environment.

Mineral carbonation may take place directly or indirectly and as either a gas-solid reaction or in an aqueous solution. In the indirect process, calcium or magnesium is first extracted from a mineral in aqueous solution, followed by carbonation, while in the direct process carbonation occurs in the absence of cation extraction. The extent of reaction is typically measured in terms of the *carbonation conversion*, or the percentage of calcium or magnesium components that react to produce carbonate.

9.2.1 Direct Carbonation

In early gas-solid experiments, it was found that in the case of serpentine having particle sizes between 50–100 μm, carbonation conversions of between 25–30% were only achieved after 2 hours of exposure to high temperature (823 K) and pressure (34 MPa) conditions [1a]. A recent investigation carried out by Dufaud et al. [9] reported a carbonation conversion in the case of Mg-silicates of between 3 and 57% with supercritical CO_2 at 773 K and up to 100 MPa after 4 hours. In the late 1990s, studies were carried out on aqueous-phase direct carbonation that resulted in enhanced conversion. The most successful mineral carbonation route reported in this study includes the use of 0.64 M $NaHCO_3$ and 1 M NaCl in aqueous solution for olivine, serpentine, and wollastinite at the following temperature and pressure conditions, respectively: 458 K and 15 MPa; 428 K and 15 MPa; and 473 K and 4 MPa. Mineral carbonation in the aqueous phase significantly reduces the harsh temperature and pressure conditions required to achieve any significant conversion. Also, in the case of serpentine, heat treatment between 888–903 K was determined more effective than mechanical processing, (grinding down to finer particle sizes) [10]. Evidence suggests that dehydroxylation of serpentine resulting from high temperature treatment is the key to enhancing the kinetics of the carbonation reaction. In the next section the surface chemical reactions that control the reactivity of alkaline-earth silicate minerals with CO_2 as well as the role that water may play in the carbonation mechanism are considered.

9.2.2 Indirect Carbonation

The first studies of indirect carbonation have included the extraction of MgO from Mg-silicates with particular focus on the effect of water. Hydrated MgO, *i.e.*, $Mg(OH)_2$ was found to have significantly enhanced reactivity. Powdered $Mg(OH)_2$ with average particle size of 20 μm exhibited a 90% conversion after just 30 min. at 838 K and 5.2 MPa. Investigations carried out by Butt et al. found that the conversion was limited in cases in which the partial pressure of CO_2 was too high, and that its presence was found to inhibit surface dehydroxylation from $Mg(OH)_2$ over the temperature range in which $MgCO_3$ is thermodynamically stable.

A number of aqueous-phase-based indirect mineral carbonation studies have focused on enhancing the reactivity of Mg-silicates through the addition of weak acids and additives that increase silicate dissolution, including citrates, oxalates, and ethylenediaminetetraacetic acid (EDTA) [11]. In general, faster rates of dissolution

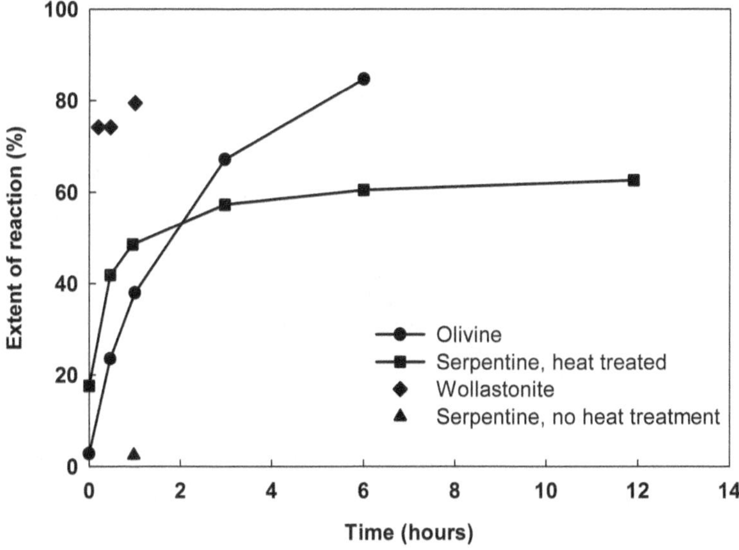

Fig. 9.2 Extraction at 185°C and 150 atm of CO_2 in NaCl $NaHCO_3$ solution [2]

lead to a higher mineral carbonation conversion [7]. For additional information on mineral carbonation for CCS applications, thorough reviews have been carried out by Kelemen et al. [12], Oelkers and Cole [13], and Zevenhoven and Fagerlund [10].

9.2.3 Ultramafic Mineral Carbonation

The U.S. DOE's National Energy Technology Laboratory (NETL) located in Albany (formerly known as the Albany Research Center), carried out an extensive study on *ex situ* aqueous mineral carbonation, focusing specifically on the reaction of Ca-, Fe-, and Mg-silicate minerals with gaseous CO_2 [2]. Nearly 700 individual experiments were conducted to investigate the effect of temperature, CO_2 partial pressure, solution chemistry, particle size, and amount of chemically bound water during heat treatment on the carbonation reaction rate. Figure 9.2 illustrates the reactivity of serpentine, olivine, and wollastonite ground to less than 75 μm. Wollastonite was found to be the most reactive of the three minerals, with just over 70% conversion reached in approximately 1 hour. Heat treatment of serpentine improved its reactivity, plateauing at a conversion of approximately 65%. In addition, it has been determined [14] that olivine reactivity is dictated primarily by surface area, so that its reaction rate may be enhanced through additional grinding down to finer particles.

Figure 9.3 shows a map of the ultramafic mineral sources available in the U.S. and their relation to point-source CO_2 emissions. Region 1 is comprised of 100% olivine, while regions 2–4 are comprised of 100% serpentine (lizardite). Regions 5, 6, and 7 contain 70% olivine, 100% serpentine (antigorite), and 50% wollastonite,

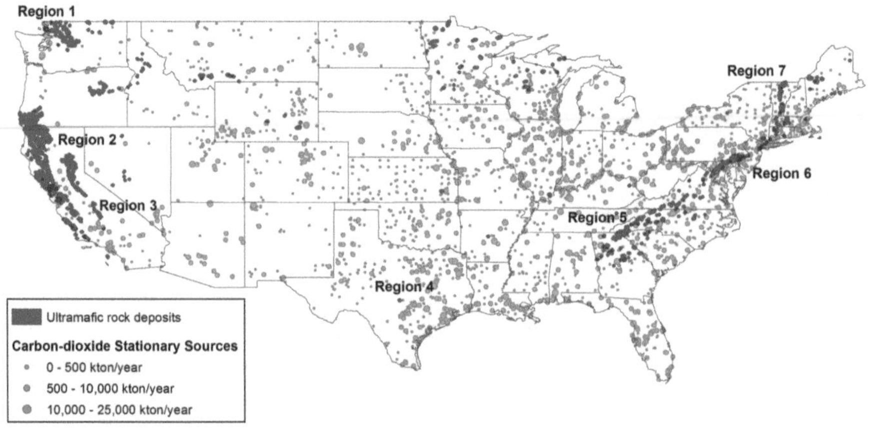

Fig. 9.3 Map indicating regions of ultramafic rock deposits in red. The regions have been slightly enhanced for visibility. CO_2 stationary sources are also shown in gray on the basis of ktons emitted per year

Table 9.1 Energy consumption of mineral-carbonation processes [2]

Region	Energy (GWh/1000 s)	CO_2 carbonated (Mt)[a]	Energy consumed (GWh/Mt CO_2 carbonated)			CO_2 avoided (Mt)		Carbonation cost (\$/t CO_2 avoided)	
			std[b]	act[c]	total	std	act	std	act
1	18	18	300	333	633	13	7	78	167
2	9	10	180	2022	2202	8	<0	537	–
3	9	10	180	2251	2431	8	<0	538	–
4	72	72	180	2022	2202	59	<0	521	–
5	184	187	320	333	653	126	63	81	173
6	220	231	180	829	1009	187	<0	309	–
7	75	76	190	239	429	62	43	112	110

[a]Total CO_2 emissions based on coal consumption and carbon content by region; CO_2 carbonated assumes 100% of estimated emissions

[b]Energy consumption for complete carbonation includes standard pretreatment, which includes mining, transportation, and grinding

[c]Energy consumption for activated pretreatment includes additional grinding beyond 75 μm in addition to heat treatment

respectively. The study carried out by Gerdemann et al. [2] of NETL also included an energy analysis that estimates the cost of mineral carbonation of olivine and wollastonite ores at \$80 and \$112 per ton of CO_2 avoided, respectively. Table 9.1 reports the energy consumption for the mineral-carbonation process with the amount and cost of CO_2 avoided, assuming optimal the mineral carbonation conditions reported in Table 9.2. The cost of CO_2 avoided as reported in Table 9.1 does not include the cost of separating the CO_2 from the flue gas of a power plant (*i.e.*, the capture process). If partial pressures up to 40 atm at a minimum (*i.e.*, wollastonite) are required to achieve optimal conversion, compression of the flue gas up to these pressures may be required for CCS to occur as a single process.

Table 9.2 Optimal carbonation conditions by mineral [2]

Mineral	Carbonation conditions		
	T (°C)	p_{CO_2} (atm)	Carrier solution
Olivine	185	150	0.64 M $NaHCO_3$, 1 M NaCl
Wollostonite	100	40	Distilled water
Serpentine[a]	155	115	0.64 M $NaHCO_3$, 1 M NaCl

[a]heat treated for 2 hours at 630°C before testing

9.3 Brine Carbonation

Approximately 20 billion barrels of brine are produced annually in the U.S. as a waste byproduct of oil and natural gas recovery processes [15]. Brines are saline-based solutions that contain divalent cations available for mineral carbonation with CO_2, but as previously discussed, due to the slow kinetics the carbonation conversion is limited. Increasing the pH of the brine through the addition of fly ash or other alkaline byproducts produced from coal combustion, such as flue gas desulfurization (FGD) byproducts, may result in enhanced mineral carbonation conversion. Specifically, Soong et al. [16] found enhanced brine carbonation through the addition of highly alkaline FGD waste within a 2-hour period using a single-stage process at 293 K and 0.136 MPa.

The pH of brine solution alone is too low (*i.e.*, 3–5) for the precipitation of mineral carbonates. A low pH is equivalent to a relatively high concentration of protons, making the formation of mineral carbonate, (MCO_3) infeasible. Through the introduction of FGD waste (or some other alkaline-rich material), which is comprised primarily of MgO, CaO, and SiO_2, the pH of a brine solution can increase to favorable carbonate precipitation levels (*e.g.*, 7.8). The rise in pH leads to an increase in hydroxyl anion (OH^-) concentration, and a subsequent increase in free cation, (Ca^{2+} and Mg^{2+}) concentration, leading to conditions suitable for mineral carbonate precipitation. This concept has been patented [17] and applied for the production of aggregate that could be used in concrete and other building materials; however, the application is limited by the availability of brine, alkalinity, and an available aggregate market for construction applications. A primary advantage of this process over mineral carbonation with ultramafic minerals is the absence of the energy-intensive dissolution step in addition to the mechanical energy required for grinding the materials fine enough for significant carbonation conversion.

9.4 Industrial Alkaline Sources

Common industrial-sourced alkaline byproducts include coal fly ash (FA) [2], and steel-making slag (SS) including electric arc furnace (EAF) dust [18], waste concrete [18a, 19] and cement kiln dust (CKD) [19, 20], municipal waste incinerator

Table 9.3 Industrial byproduct alkalinity sources for mineral carbonation

Industrial process	Alkaline byproduct
Coal combustion	Fly ash
Iron and steel making	Steel slag, blast furnace slag, electric arc furnace slag
Cement industry	Cement kiln dust, waste cement
Asbestos mining	Asbestos mine tailings
Aluminum industry	Bauxite residue
Waste incineration	Municipal waste incinerator ash

(MSWI) ash [21], asbestos mine tailings [2], and bauxite residue [22], as listed in Table 9.3, with their respective industrial sources. Some of the more common industrial-based alkalinity sources listed in Table 9.3 such as CKD, FA, and SS are discussed in greater detail next. Although the morphology and composition of the different industrial alkalinity sources are various, in general these byproducts contain significant amounts of reactive metal oxides (primarily CaO and MgO). The pretreatment of industrial-sourced alkalinity is minimal compared to that of naturally sourced ultramafic minerals since they are already micron-sized and fairly reactive, thereby eliminating the need for a dissolution step.

In addition to the potential CO_2 mitigation associated with the mineral carbonation of industrial alkalinity sources, this process adds significant environmental improvement vis-à-vis the handling of industrial byproducts that may otherwise be considered as waste materials. For instance, the mineral carbonation process has been shown to immobilize trace metals in alkaline waste byproducts [23].

Cement kiln dust is an alkali-rich dust produced during cement manufacturing at a ratio of approximately 0.15–0.20 tons CKD per ton of cement [24]. Approximately 5.2 Mt of CKD are produced annually in the U.S. The typical weight percent ranges of calcium-oxide (CaO) and magnesium-oxide (MgO) in CKD are 38–50 and 0–2, respectively [25]. Cement kiln dust is a fine-grained solid, with particle size on the order of several μm [19] and is an ideal source of alkalinity for mineral carbonation due to its composition and small particle size [2]. Huntzinger et al. investigated the carbonation of CKD at 98% relative humidity, 25°C, and 1 atm, with a partial pressure of CO_2 of 0.8 atm. They found that the degree of carbonation correlates directly with the mass fraction of calcium oxide and hydroxide content of the CKD [19]. The degree of carbonation at a given time, t, is defined as the mass of CO_2 taken up by the sample, $M_{CO_2}(t)$, divided by the maximum theoretical carbonation of the sample. The average degree of carbonation was approximately 77% over 8 days, with 90% of the carbonation occurring in less than 2 days [20b].

As a residue generated from the combustion of coal, FA is typically captured post-combustion using air pollution control devices such as fabric filters or electrostatic precipitators. Fly ash comprises approximately 60% of all coal combustion waste, and in the U.S. alone, coal-fired power plants produce approximately 42.4 Mt of FA annually [26]. Fly ash is a complex, partly amorphous, and chemically heterogeneous material, with its physicochemical properties dependent on the composition of the feed coal and the operating conditions of the coal-fired power plant. In the U.S., coal is ranked within four broad categories, listed in order of increasing rank (purity):

lignite, subbituminous, bituminous, and anthracite. Although FA is often classified based on these ranks, its composition within these categories varies greatly due to coal heterogeneity. In general, inorganic minerals comprise approximately 90–99% of fly ash, while organic compounds makeup up approximately 1–9% [27]. The inorganic minerals consist primarily of silicon dioxide (SiO_2) and CaO, along with other metal oxides such as Fe_2O_3 and MgO. The typical weight percent ranges of CaO and MgO in FA are 1–37 and 1–15, respectively [16a, 28]. In general, Ca-rich minerals are much less abundant than alumino-silicates and Fe-oxides in high-rank coals; however, in lower rank coals, Ca-rich minerals dominate the inorganic crystalline fraction of FA [2]. Montes-Hernandez et al. investigated the aqueous carbonation of FA with an average diameter of 40 μm and found that 82% of the FA CaO content is converted to $CaCO_3$ after reacting for 2 hours at 30°C [29]. The authors also report that carbonation conversion is independent of initial CO_2 pressure, but did investigate high-pressure conditions ranging from approximately 10–39 atm of CO_2. Additional investigations have considered the carbonation of FA or FA-brine mixtures, and have also found that CaO present in FA is readily converted to $CaCO_3$ [16a, 30].

Steel slag (SS) is a byproduct of iron and steel manufacturing and includes the impurities separated from iron during ore smelting. Slag is comprised of a heterogeneous mixture of crystalline components, including iron oxides, calcium and magnesium hydroxides, oxides, and silicates, and quartz [31]. Slag content varies depending on the ore and the smelting process; specifically, the type of furnace used, *e.g.*, blast furnace, basic oxygen furnace, EAF, and ladle furnace (LF). It is worth noting that data are rare on the actual average production of slag, as the amount of slag produced is not routinely measured. The amount of slag produced varies depending on the overall composition of the raw furnace feed, in particular the iron ore feed grade, and the type of furnace used. Approximately 0.2 tons of steel slag is produced for each ton of iron produced [31]; however, a significant portion of the slag is entrained metal and recovered during slag processing, and the amount of marketable slag remaining after entrained steel removal is usually equivalent to between 10–15% of the crude steel output [32]. The SS production (after metal removal) in the U.S. is estimated at 10.3 Mts.

Steel slag is an ideal feedstock for mineral carbonation due to its high alkalinity and, more specifically, high calcium content. The typical weight percent ranges of CaO and MgO in SS are 32–58 and 3.9–10.0, respectively [18e, 31]. Previous investigations suggest that SS is a viable feedstock for cost-effective CO_2 storage via mineral carbonation [18a, 18c, 33]. Bonenfant et al. [18e] investigated aqueous carbonation of EAF and LF slag suspensions and found storage capacities of 0.02 and 0.25 t-CO_2/tSS, respectively. Huijgen et al. [18c] investigated aqueous-phase mineral carbonation of SS and found that the reaction rate depends primarily on particle size and temperature. The authors report 74% extent reacted after 30 min. for SS with particle size less than 38 μm at approximately 18.7 atm CO_2 and 100°C. Stolaroff et al. [6] estimated a SS carbonation potential of 0.27 ton CO_2 sequestered per ton SS, assuming that 75% of Ca content reacts with CO_2, with an estimated cost of $8/ton CO_2 sequestered. Selected mineral carbonation reaction conditions are reported in Table 9.4.

Table 9.4 Selected mineral carbonation reaction conditions for CKD, FA, and SS

Alkaline waste product	U.S. Annual production (Mt)	Carbonation conditions		
		T (°C); p_{CO_2} (atm)	Conversion (%)	Particle size (μm)
Cement kiln dust	5.2	25; 0.8	77	~3
Fly ash	42.2	30; 10–39	82	~40
Steel slag	10.3	100; 18.7	74	~38

Other industrial-based byproducts proposed as resources for mineral carbonation include asbestos-mining tailings, bauxite residue, and municipal solid waste incineration (MSWI) ash. The carbonation of asbestos-mining tailings has the potential to not only sequester CO_2 but also to break down the potentially hazardous form of asbestos generated as a result of the tailings. Bauxite residue, also known as "red mud" waste, is a byproduct of the Bayer aluminum manufacturing process with high alkalinity and high pH [22, 34]. In general, the production of 1 ton of alumina generates on the order of 1.0–1.5 tons of bauxite residue, which results in a global production of approximately 70 Mt/yr [35] and an estimated 3 Mt/yr in the U.S. [36] Sahu et al. [37] investigated the neutralization of red mud through carbonation and found that red mud has a CO_2 storage capacity of approximately 56 wt%. In the U.S., approximately 7% of municipal solid waste (MSW) is incinerated and 8.5–9 Mt/yr of MSWI ash is produced [38]. Prigiobbe et al. [39] found that carbonation of MSWI ash occurs at flue gas conditions and suggest that the CO_2 of industrial flue gases may be captured and stored in a single step using MSWI ash and other alkaline-rich industrial residues. Due to the abundance of the alkaline byproduct sources from coal-fired power, cement manufacturing, and steel production industries, it is interesting to consider the amount of carbonate product formed, whether there exists a market for the product, and the CO_2 mitigation potential of this route.

9.4.1 Aggregate Production from Mineral Carbonation

The carbonate produced through mineral carbonation has the potential for beneficial reuse as synthetic aggregate. In addition to serving as a sink for CO_2, this synthetic aggregate has the potential co-benefit of preventing CO_2 emissions associated with mining aggregate.

Natural aggregates are traditionally sourced from either crushed stone or sand and gravel, with the various aggregates frequently interchanged with one another. The estimated annual output of crushed stone produced for consumption in the U.S. in 2010 was 1.19 billion tons (Gt), a 2% increase compared with that of 2009 [40]. The estimated annual output of construction sand and gravel produced for consumption in 2010 was 820 Mt, a 2% decrease compared with that of 2009. The estimated annual output of natural aggregates produced for consumption in the U.S. in 2010 was 2 Gt, with a market value of 17.5 billion dollars [40]. Additionally, 29 Mt of aggregate was produced from recycled asphalt and cement in the U.S. in 2009. Natural aggregates are primarily used in the construction industry, and account for approximately half of the U.S. mining industry [41].

Natural aggregates are widely used throughout the U.S.; however, they are not universally available and some areas lack quality and/or practically-accessible natural aggregate. For economic reasons, pits or quarries are located near population centers. Conversely, existing land uses, zoning, or regulations may prevent commercial exploitation of aggregate near residential areas. Transportation costs are an important factor in the natural aggregate industry. Generally, aggregate is transported by rail or ship for distances greater than 50 miles, and by truck for shorter distances. Rates for shipping aggregate by ship are approximately 1–1.2 cents per ton per kilometer (¢/t-km); by truck, typical rates are 21.7 ¢/t-km for the first mile and 4.3 ¢/t-km for additional miles [41].

9.4.2 Alkaline Availability

The carbonation capacity for a given alkalinity source depends on the total alkalinity available, the reactivity of the alkaline components, the kinetics of the reaction, and the reaction conditions. Available alkalinity may be used as a direct measure of the carbonation capacity of a given resource, and is a function of the maximum theoretical carbonation capacity of the resource and the expected extent reacted. The National Energy Technology Laboratory Carbon Sequestration Database has been used to determine CO_2 emissions by source for coal-fired power plants, cement kilns, and steel plants. The production rate of FA, CKD, and SS alkalinity has been estimated based on available data, as described next.

The alkalinity sourced from fly ash was estimated based upon the type of coal burned. The coal types considered include Appalachian Low-Sulfur bituminous, Appalachian Medium-Sulfur bituminous, Wyoming Powder River Basin subbituminous, Wyodak bituminous, North Dakota lignite, and Illinois #6 bituminous. Information regarding the U.S. power plants locations in addition to capacity and type of coal burned was determined from electricity data files available from the U.S. Energy Information Administration. The coal composition, including ash content and distribution of calcium and magnesium oxides, differs among the various coal types and was determined from the internal fuel library of the Integrated Environmental Control Module developed by Rubin and colleagues at Carnegie Mellon University [42]. For each of the coal types considered, the ratio of fly ash produced to CO_2 emitted was calculated using this software package on the basis of a 500-MW power plant. On average, power plants generate approximately 10–13 tons of ash per hour, which is small in comparison to the approximate 435 tons of CO_2 generated per hour. The rate of fly ash production by each power plant was calculated by multiplying the CO_2 emissions from the plant by the appropriate fly ash production to CO_2 emissions ratio. Based on previous work of Montes-Hernandez et al., the expected extent reacted for FA is approximated at 82% [29].

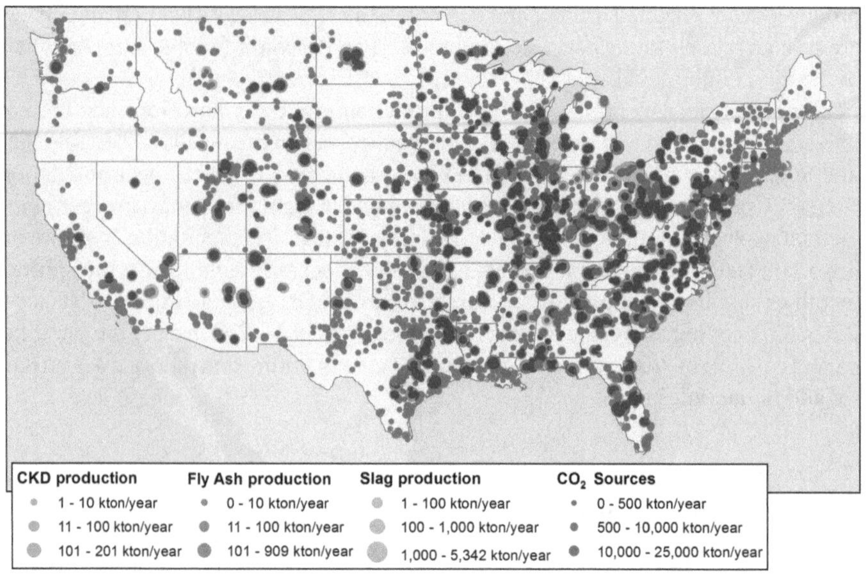

Fig. 9.4 Stationary CO_2 sources and industrial alkaline sources in the Continental U.S. (CO_2 sources based on NatCarb, 2010)

To approximate the amount of alkalinity available in the form of CKD, the amount of CKD generated per source has been calculated based on the CO_2 emissions of the source with a clinker to CO_2 production ratio of 1, and a CKD to clinker production ratio of 0.060 [19, 43]. Based on data from Huntzinger et al., the expected extent of reaction for CKD is approximately 77% [20b]. The amount of SS generated per source was calculated based on the CO_2 emissions of the source, a ratio of CO_2 emitted to steel produced of 0.64, and a ratio of steel produced to slag generated of 8.33 [18e, 44]. Based on previous work by Huijgen et al., the expected extent reacted for SS is 75% [18c]. To assess the potential for mineral carbonation using industrial alkalinity sources, the estimated total U.S. production of CKD, FA, and SS listed in Table 9.4 were used.

To determine the geographic relationship between CO_2 emissions sources and industrial alkalinity sources, the locations of industrial alkalinity sources have been mapped in relation to U.S. CO_2 emissions. Figure 9.4 provides the locations of stationary CO_2 emissions in the Continental U.S. overlain with circles representing CKD, FA, and SS production locations, scaled to the relative annual production.

Almost all locations where CKD, FA, or SS are produced are also producing CO_2 since FA is a byproduct of coal combustion and because the use of cement kilns, and iron and steel manufacturing processes are energy-intensive, thereby generating extensive CO_2 emissions.

9.4.3 Market Availability of Synthetic Aggregate Use

Data associated with the mined aggregate production in the U.S. including crushed stone or sand and gravel were compiled for 5269 U.S. sites from the 'Mineral Operations—Sand and gravel' and 'Mineral Operations—Crushed Stone' databases of the 2005 National Atlas (map layers compiled by the Minerals Information Team of the USGS) [45]. The mined aggregate volumes are illustrated in Fig. 9.5 by state, with the darker shades representing the higher production areas. In addition, the synthetic aggregate produced from the reaction of CO_2 (from point-source power plant emissions) with industrial alkalinity sources is presented in Fig. 9.5 and divided between slag, CKD, and fly ash sources. Given that the alkalinity source is the limiting resource for mineral carbonation, the location of the synthetic aggregate production was assumed to be that of the alkalinity source. Figure 9.5 further illustrates that the extent of synthetic aggregate production in total is an order of magnitude smaller than the mined aggregate volumes, and it can be assumed that most synthetic aggregate will find a market locally.

The extent to which synthetic aggregate production from mineral carbonation using industrial waste products (*i.e.*, alkalinity sources) could replace mined aggregate has also been investigated. The potential total U.S. synthetic aggregate production using FA, CKD, and SS has been estimated at 24 Mt/year, approximately 1.2% of total U.S. mined aggregate. In other words the available alkalinity from industrial byproduct waste is significantly limited. Table 9.5 shows a comparison

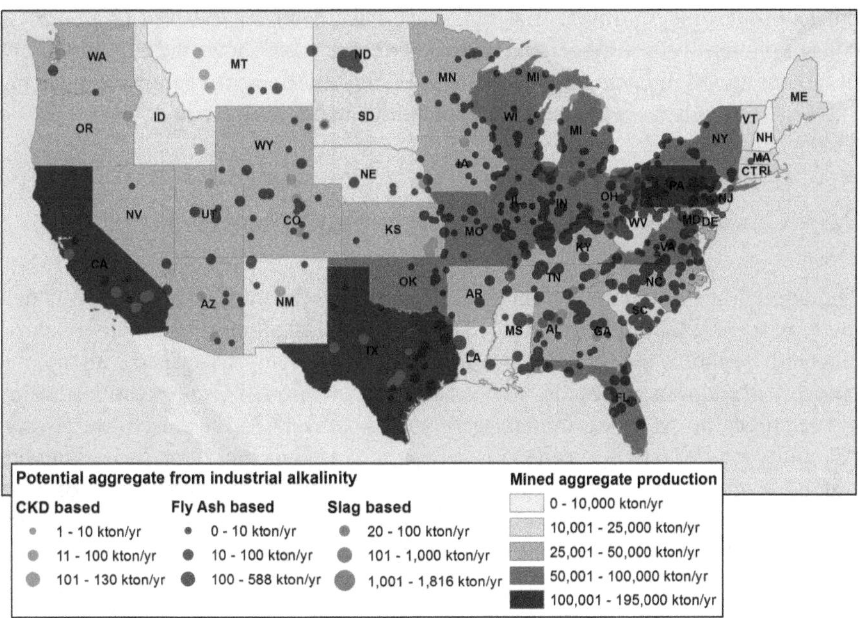

Fig. 9.5 Comparison of mined aggregate volumes to synthetic aggregate production from reaction of CO_2 emissions with industrial-based alkalinity sources in the Continental U.S.

Table 9.5 Top states of synthetic aggregate replacement of mined aggregate

State	Mined aggregate (kton/yr)	Synthetic aggregate (kton/yr)	% Synthetic of mined aggregate
North Dakota	14,700	1,837	12.5
Indiana	65,100	5,464	8.4
Ohio	73,400	3,521	4.8
Montana	31,300	1,045	3.3
South Carolina	24,400	505	2.1

Table 9.6 CO_2 emissions and mitigation potential of mined aggregate

Mined aggregate	CO_2 emissions (kg CO_2/ton mined)	Mitigated CO_2[a] (Mt CO_2/yr)
Crushed granite	3.6	0.09
Crushed limestone	4.5	0.12
Sand and gravel	23.9	0.62

[a]Assuming 1.2% of U.S. mined crushed stone, crushed limestone, or sand and gravel aggregate replaced with synthetic aggregate

of mined aggregate to synthetic aggregate for all of the states in which the synthetic aggregate may potentially replace more than 2% of the mined aggregate.

In addition to the CO_2 mitigation potential from synthetic aggregate production, it is interesting to also consider the CO_2 emissions associated with mining aggregate and the subsequent mitigation potential of displacing the mined aggregate with synthetic aggregate. The CO_2 emissions associated with the mining of crushed granite, crushed limestone, and industrial sand have been determined based on the direct consumption of energy associated mining [46]. Table 9.6 shows the CO_2 emissions per ton of mined material, and the potential CO_2 emissions mitigation by replacing 1.2% of U.S. mined aggregate production with synthetic aggregate.

9.5 Challenges of Mineral Carbonation Processes for CCS

The upper limit of the mitigation potential of mineral carbonation for CCS is dictated by the availability of alkalinity. The available alkalinity from mining natural silicate-based minerals is limited due to the energy intensity associated with mineral recovery in addition to the slow kinetics of the dissolution of divalent cations, which are required for achieving mineral carbonation conversion on a useful timescale. Alkalinity may also be sourced from industrial byproduct waste materials, but with limited potential. It has been estimated that CO_2 mitigation via carbonation with industrial byproducts is on the order of approximately 10 Mt CO_2/year at best, combining the CO_2 captured from the mineral carbonation process and the avoided CO_2 generated from aggregate mining practices. Although mineral carbonation for CCS may not serve to mitigate large volumes of CO_2, it may be one aspect comprising the portfolio of strategies, and has the additional benefit of simultaneous sequestration of other hazardous materials.

References

1. (a) Lackner KS, Butt DP, Wendt CH (1997) Progress on binding CO_2 in mineral substrates. Energy Convers Manag 38:259–264; (b) Sipilä J, Teir S, Zevenhoven R (2008) Carbon dioxide sequestration by mineral carbonation, Literature review update 2005–2007. Abo Akademi University, Turku; (c) Metz B (2005) IPCC special report on carbon dioxide capture and storage. Cambridge University Press, Cambridge, p 431
2. Gerdemann SJ, O'Connor WK, Dahlin DC, Penner LR, Rush H (2007) Ex situ aqueous mineral carbonation. Environ Sci Technol 41(7):2587–2593
3. Doney SC, Fabry VJ, Feely RA, Kleypas JA (2009) Ocean acidification: the other CO_2 problem. Ann Rev Mar Sci 1:169–192
4. Goff F, Lackner KS (1998) Carbon dioxide sequestering using ultramafic rocks. Environ Geosci 5(3):89–102
5. Rau GH, Caldeira K (1999) Enhanced carbonate dissolution: a means of sequestering waste CO_2 as ocean bicarbonate. Energy Convers Manag 40(17):1803–1813
6. House KZ, Schrag DP, Harvey CF, Lackner KS (2006) Permanent carbon dioxide storage in deep-sea sediments. Proc Natl Acad Sci USA 103(33):12291
7. Guyot F, Daval D, Dupraz S, Martinez I, Ménez B, Sissmann O (2011) CO_2 geological storage: the environmental mineralogy perspective. Comptes Rendus Geosci 343:246–259
8. Teir S, Eloneva S, Fogelholm CJ, Zevenhoven R (2006) Stability of calcium carbonate and magnesium carbonate in rainwater and nitric acid solutions. Energy Convers Manag 47(18–19):3059–3068
9. Defaud F, Martinez I, Shilobreeva S (2009) Experimental study of Mg-rich silicate carbonation at 400 and 500°C and 1 kbar. Chem Geol 265;79–87
10. Zevenhoven R, Fagerlund J (2010) Fixation of carbon dioxide into inorganic carbonates: the natural and artifical "weathering of silicates," in carbon dioxide as chemical feedstock. Wiley-VCH Verlag GmbH & Co, Weinheim
11. (a) Krevor SC, Lackner KS (2009) Enhancing process kinetics for mineral carbon sequestration. Energy Procedia 1(1):4867–4871; (b) Krevor SCM, Lackner KS (2011) Enhancing serpentine dissolution kinetics for mineral carbon dioxide sequestration. Int J Greenh Gas Con 5(4):1073–1080; (c) Prigiobbe V, Mazzotti M (2011) Dissolution of olivine in the presence of oxalate, citrate and CO_2 at 90°C and 120°C. Chem Eng Sci 66(24):6544–6554
12. Kelemen PB, Matter J, Streit EE, Rudge JF, Curry WB, Blusztajn J (2011) Rates and mechanisms of mineral carbonation in peridotite: natural processes and recipes for enhanced, in situ CO_2 capture and storage. Ann Rev Earth Planet Sci 39:545–576
13. Oelkers EH, Cole DR (2008) Carbon dioxide sequestration a solution to a global problem. Elements 4(5):305
14. Ityokumbul MT, Chander S, O'Connor WK, Dahlin DC, Gerdemann SJ (2001) Reactor design considerations in mineral sequestration of carbon dioxide. Pennsylvania State University, Energy and GeoEnvironmental Engineering, University Park, PA
15. Kharaka YK, Thordsen JJ, Kakouros E, Herkelrath WN (2005) Impacts of petroleum production on ground and surface waters: Results from the Osage-Skiatook Petroleum Environmental Research A site, Osage County, Oklahoma. Environmental Geosciences 12(2):127
16. (a) Soong Y, Fauth DL, Howard BH, Jones JR, Harrison DK, Goodman AL, Gray ML, Frommell EA (2006) CO_2 sequestration with brine solution and fly ashes. Energy Convers Manag 47(13–14):1676–1685; (b) Soong Y, Goodman AL, McCarthy-Jones JR, Baltrus JP (2004) Experimental and simulation studies on mineral trapping of CO_2 with brine. Energy Convers Manag 45(11–12):1845–1859
17. Constantz BR, Youngs A, Holland TC (2010) CO_2-sequestering formed building materials. U.S. Patent 7771684, Calera Corporation
18. (a) Stolaroff JK, Lowry GV, Keith DW (2005) Using CaO- and MgO-rich industrial waste streams for carbon sequestration. Energy Convers Manag 46(5):687–699; (b) Teir S, Eloneva S, Fogelholm CJ, Zevenhoven R (2007) Dissolution of steelmaking slags in acetic acid for precipitated calcium carbonate production. Energy 32(4):528–539; (c) Huijgen WJJ, Comans RNJ

(2005) Mineral CO_2 sequestration by steel slag carbonation. Environ Sci Technol 39(24):9676–9682; (d) Lekakh SN, Rawlins CH, Robertson DGC, Richards VL, Peaslee KD (2008) Kinetics of aqueous leaching and carbonization of steelmaking slag. Metall Mater Trans B 39(1):125–134; (e) Bonenfant D, Kharoune L, Sauvé S, Hausler R, Niquette P, Mimeault M, Kharoune M (2008) CO_2 sequestration potential of steel slags at ambient pressure and temperature. Ind Eng Chem Res 47(20):7610–7616

19. Huntzinger DN, Eatmon TD (2009) A life-cycle assessment of Portland cement manufacturing: comparing the traditional process with alternative technologies. J Clean Prod 17(7):668–675
20. (a) Huntzinger DN, Gierke JS, Kawatra SK, Eisele TC, Sutter LL (2009) Carbon dioxide sequestration in cement kiln dust through mineral carbonation. Environ Sci Technol 43(6):1986–1992; (b) Huntzinger DN, Gierke JS, Sutter LL, Kawatra SK, Eisele TC (2009) Mineral carbonation for carbon sequestration in cement kiln dust from waste piles. J Hazard Mater 168(1):31–37
21. (a) Rendek E, Ducom G, Germain P (2006) Carbon dioxide sequestration in municipal solid waste incinerator (MSWI) bottom ash. J Hazard Mater 128(1):73–79; (b) Rendek E, Ducom G, Germain P (2007) Influence of waste input and combustion technology on MSWI bottom ash quality. Waste Manag 27(10):1403–1407; (c) Toller S, Kärrman E, Gustafsson JP, Magnusson Y (2009) Environmental assessment of incinerator residue utilisation. Waste Manag 29(7):2071–2077
22. Hind AR, Bhargava SK, Grocott SC (1999) The surface chemistry of bayer process solids: a review. Colloids Surface A 146(1–3):359–374
23. (a) Fernàndez-Bertos M, Simons S, Hills C, Carey P (2004) A review of accelerated carbonation technology in the treatment of cement-based materials and sequestration of CO_2. J Hazard Mater 112(3):193–205; (b) Li X, Bertos MF, Hills CD, Carey PJ, Simon S (2007) Accelerated carbonation of municipal solid waste incineration fly ashes. Waste Manag 27(9):1200–1206
24. Van Oss HG, Padovani AC (2003) Cement manufacture and the environment, Part II: environmental challenges and opportunities. J Ind Ecol 7(1):93–126
25. Corish A, Coleman T (1995) Cement kiln dust. Concrete 29(5):40–42
26. American Coal Ash Association (2008) Coal Combustion Product (CCP) Production & Use Survey Report
27. Vassilev SV, Vassileva CG (2005) Methods for characterization of composition of fly ashes from coal-fired power stations: a critical overview. Energy Fuels 19(3):1084–1098
28. (a) Gunning PJ, Hills CD, Carey PJ (2010) Accelerated carbonation treatment of industrial wastes. Waste Manag 30(6):1081–1090; (b) Uliasz-Bochenczyk A, Mokrzycki E, Piotrowski Z, Pomykala R (2009) Estimation of CO_2 sequestration potential via mineral carbonation in fly ash from lignite combustion in Poland. Energy Procedia 1(1):4873–4879
29. Montes-Hernandez G, Pérez-Lúpez R, Renard F, Nieto JM, Charlet L (2009) Mineral sequestration of CO_2 by aqueous carbonation of coal combustion fly-ash. J Hazard Mater 161(2–3):1347–1354
30. Tawfic TA, Reddy KJ, Gloss SP, Drever JI (1995) Reaction of CO_2 with clean coal technology ash to reduce trace element mobility. Water Air Soil Pollut 84(3–4):385–398
31. Navarro C, Díaz M, Villa-García MA (2010) Physico-chemical characterization of steel slag. study of its behavior under simulated environmental conditions. Environ Sci Technol 44(14):5383–5388
32. van Oss HG (2003) Slag-iron and steel – minerals yearbook. United States Geological Survey (USGS)
33. Huijgen WJJ, Ruijg GJ, Comans RNJ, Witkamp GJ (2006) Energy consumption and net CO_2 sequestration of aqueous mineral carbonation. Ind Eng Chem Res 45(26):9184–9194
34. Johnston M, Clark MW, McMahon P, Ward N (2010) Alkalinity conversion of bauxite refinery residues by neutralization. J Hazard Mater 182(1–3):710–715
35. Agrawal A, Sahu KK, Pandey BD (2004) Solid waste management in non-ferrous industries in India. Resour Conserv Recycl 42(2):99–120
36. Khaitan S, Dzombak DA, Lowry GV (2009) Chemistry of the acid neutralization capacity of bauxite residue. Environ Eng Sci 26(5):873–881

37. Sahu RC, Patel RK, Ray BC (2010) Neutralization of red mud using CO_2 sequestration cycle. J Hazard Mater 179(1–3):28–34
38. (a) Arsova L, Haaren R, Goldstein N, Kaufman SM, Themelis NJ (2008) The state of garbage in America. Biocycle 49(12):22–24; (b) Wiles CC (1996) Municipal solid waste combustion ash: state-of-the-knowledge. J Hazard Mater 47(1–3):325–344
39. Prigiobbe V, Polettini A, Baciocchi R (2009) Gas-solid carbonation kinetics of air pollution control residues for CO_2 storage. Chem Eng J 148(2–3):270–278
40. Willett JC, Bolen WP (2011) Crushed stone and sand and gravel in the fourth quarter 2010: U.S. geological survey mineral industry surveys. United States Geological Survey (USGS)
41. Langer WH (1998) Natural aggregates of the conterminous united states: U.S. Geological Survey Bulletin 1954. United States Geological Survey (USGS)
42. Rubin ES Integrated environemental control model, version 6.2.4. http://www.cmu.edu/epp/iecm/index.html
43. (a) Jacott M, Reed C, Taylor A, Winfield M (2003) Energy use in the cement industry in north america: emissions, waste generation and pollution control, 1990–2001; May 30; (b) International ICF (2007) energy trends in selected manufacturing sectors: opportunities and challenges for environmentally preferable energy outcomes; U.S. Environmental Protection Agency: March 2007
44. Van Oss HG (2011) Slag, iron and steel–2009 [Advance Release] – 2009 minerals yearbook. United States Geological Survey (USGS)
45. (a) PHMSA, Pipeline Information Management Mapping Application. Pipeline and Hazardous Materials Safety Administration (PHMSA), U.S. Dept. of Transportation: 2005; (b) EIA (2007) About U.S. natural gas pipelines – transporting natural gas. Energy Information Administration, U.S. Department of Energy, Washington, DC, p 76
46. (a) Sector 21 (2010a) EC0721SP1: mining: subject series: product summary: products or services statistics: 2007; (b) sector 21 (2010b) ec0721sm1: mining: subject series: materials summary: selected supplies minerals received for preparation purchased machinery and fuels consumed by type by industry: 2007; (c) Sector 21 (2010c) ec0721i1: mining: industry series: detailed statistics by industry for the united states: 2007

Appendix

Appendix A: Basic Principles from Combustion Science

Understanding fossil fuel oxidation assists in determining the most effective and affordable CO_2 capture processes. Envisioning the application of a CO_2 capture technology to a hydrocarbon-based energy conversion process requires knowledge of the temperature, pressure, chemical environment, and CO_2 concentration for the specific application. Appendix A provides a brief overview of basic principles from combustion science, since advanced energy conversion processes may be developed that incorporate strategies that maximize CO_2 exhaust concentrations for more efficient and effective capture.

Mole and Mass Fractions and Partial Pressure Consider a multicomponent gas mixture of n_1 moles of species 1, n_2 moles of species 2, etc. One mole of any substance is equivalent to 6.02×10^{23} molecules (or particles) of that substance and is expressed as gram-moles (*i.e.*, mol), which is equal to the mass of the substance's molecular (or atomic) mass in grams. The term "molecular mass", although more correct, is often replaced by the term molecular weight and is referred to as such throughout the text. The mole fraction of species i, y_i, is defined as the fraction of the total number of moles in the mixture of type i. The mole fraction can be expressed as:

$$y_i \equiv \frac{n_i}{n_1 + n_2 + \ldots + n_i + \ldots} = \frac{n_i}{n_{TOTAL}}. \tag{A.1}$$

The mass fraction of species i, Y_i, can be expressed in a similar manner, in which the mass of species i is compared with the total mass of the mixture as:

$$y_i \equiv \frac{m_i}{m_1 + m_2 + \ldots + m_i + \ldots} = \frac{m_i}{m_{TOTAL}}. \tag{A.2}$$

By definition, the sum of all of the individual mole (or mass) fractions must be unity, *i.e.*,

$$\sum_i y_i = 1 \tag{A.3}$$

J. Wilcox, *Carbon Capture*,
DOI 10.1007/978-1-4614-2215-0, © Springer Science+Business Media, LLC 2012

$$\sum_i Y_i = 1. \tag{A.4}$$

Mole fractions and mass fractions can be converted easily from one to the other by multiplying by the ratio of the molecular weight (M) of a given species and the molecular weight of the total mixture, *i.e.*,

$$Y_i = y_i\left(\frac{M_i}{M_{mix}}\right) \tag{A.5}$$

$$y_i = Y_i\left(\frac{M_{mix}}{M_i}\right) \tag{A.6}$$

The molecular weight of a mixture, M_{mix}, may be calculated with knowledge of the mole or mass fraction of a given species as:

$$M_{mix} = \sum_i y_i M_i \tag{A.7}$$

$$M_{mix} = \frac{1}{\sum_i Y_i M_i} \tag{A.8}$$

The mole fraction of a given species may also be used to calculate the partial pressure of the ith species, p_i. The partial pressure of a given species is the pressure of that species if it was separated from the mixture at the same temperature and volume of the mixture. For a non-interacting (ideal) mixture of gases, the total mixture pressure is the sum of the individual partial pressures, *i.e.*,

$$p = \sum_i p_i \tag{A.9}$$

The partial pressure is related to the mixture composition and total pressure by the mole fraction as,

$$p_i = y_i p \tag{A.10}$$

Stoichiometry Through the expression of a general hydrocarbon fuel by C_xH_y, the general oxidation reaction may be expressed as,

$$C_xH_y + a(O_2 + 3.76N_2) \rightarrow xCO_2 + (y/2)H_2O + 3.76aN_2, \tag{A.11}$$

where $a = x + y/4$.

It is important to note that this is a stoichiometric expression and that in an excess oxidizing environment this expression will change, with O_2 on the right hand side of the equation as a product species.

The stoichiometric air-fuel ratio may be calculated by,

$$\left(\frac{A}{F}\right)_{stoic} = \left(\frac{m_{air}}{m_{fuel}}\right)_{stoic} = 4.76a\left(\frac{M_{air}}{M_{fuel}}\right) \tag{A.12}$$

where M_{air} and M_{fuel} are the molecular weights of the air and fuel, respectively. The stoichiometry is a quantitative relationship between the reactants and products of a given reaction. For example, the coefficients of each molecular formula in Eq. (A.11) are termed stoichiometric coefficients. The stoichiometric quantity of oxidizing agent (*e.g.*, air) is the amount of air required to burn a corresponding quantity of fuel to completion. If greater than the stoichiometric quantity is provided, the mixture is termed fuel-lean and if less than the stoichiometric quantity of oxidizing agent is provided the mixture is termed fuel-rich.

The *equivalence ratio*, Φ, is a term that describes whether a fuel-oxidizer mixture is rich, lean, or stoichiometric and is defined as:

$$\Phi = \frac{\left(^{A}/_{F}\right)_{stoic}}{\left(^{A}/_{F}\right)} \tag{A.13}$$

Based upon this definition, $\Phi < 1$ for fuel-lean mixtures, $\Phi > 1$ for fuel-rich mixtures, and $\Phi = 1$ for stoichiometric mixtures. In power plant steam generators and gas-fired turbines, the percent excess air is a more common term and is defined as,

$$\%\ excess\ air = \frac{1 - \Phi}{\Phi} \cdot 100\%. \tag{A.14}$$

For example, excess air of 10% means that 10% more than the stoichiometric oxidizer is used in the mixture for combustion. The oxidation of carbon to CO releases less energy than the complete oxidation to CO_2, due to the reservation of the chemical energy within the less stable CO molecule.

Example A.1 A propane (C_3H_8) gas-fired boiler operates with an oxygen concentration of 5 mole percent in the flue gas. Determine the operating air-fuel ratio and the equivalence ratio.

Solution

$$\text{Given}: \quad y_{O_2} = 0.05, \quad M_{propane} = 44.1 \text{ g/mol}$$

$$\text{Find}: \quad (A/F) \text{ and } \Phi.$$

The air-fuel ratio can be calculated using the mole fraction given for O_2 (0.05).

$$C_3H_8 + a(O_2 + 3.76N_2) \rightarrow 3CO_2 + 4H_2O + bO_2 + 3.76aN_2,$$

Such that a and b are related from the conservation of O atoms,

$$2a = 6 + 4 + 2b$$

or

$$b = a - 5$$

From the mole fraction definition, Eq. (A.1),

$$y_{O_2} = \frac{n_{O_2}}{n_{TOTAL}} = \frac{b}{3 + 4 + b + 3.76a}$$

Substituting the given O_2 mole fraction (0.05) and solving for a yields,

$$0.05 = \frac{a - 5}{2 + 4.76a}$$

or

$$a = 6.693$$

From the definition of air-fuel ratio, Eq. (A.12), the actual air-fuel ratio is,

$$\left(\frac{A}{F}\right) = \frac{4.76(6.693)(28.85)}{44.1} = 20.84$$

To find Φ, we must first determine $(A/F)_{stoich}$. From Eq. (A.11), $a = 5$,

$$\left(\frac{A}{F}\right)_{stoic} = \frac{4.76(5)(28.85)}{44.1} = 15.57$$

Using Eq. (A.13) to calculate Φ, yields,

$$\Phi = \frac{\left(A/F\right)_{stoic}}{\left(A/F\right)} = \frac{15.57}{20.84} = 0.75$$

This gas mixture is lean in fuel.

Example A.2 A natural gas-turbine engine operates at full capacity (5000 kW) at an equivalence ratio of 0.40 with an airflow rate of 17.1 kg/s. Assume the molecular formula of natural gas is $C_{1.16}H_{4.32}$ determine (a) the fuel mass and molar flow rate, (b) operating air-fuel ratio, (c) CO_2 and N_2 molar flow rates.

Solution

$$\text{Given:} \quad \Phi = 0.40 \quad M_{air} = 28.85 \text{ g/mol}$$

$$m_{air} = 17.1 \text{ kg/s} \quad M_{fuel} = 1.16\,(12.01) + 4.32\,(1.00) = 18.25 \text{ g/mol}$$

$$\text{Find}: \quad (A/F),\ \dot{m}_{fuel},\ \dot{n}_{fuel},\ \dot{n}_{CO_2},\ \dot{n}_{H_2O},\ \dot{n}_{O_2},\ \dot{n}_{N_2}$$

Recalling the definition of $(A/F)_{stoich}$ from Eq. (A.12) yields,

$$\left(\frac{A}{F}\right)_{stoic} = 4.76a\left(\frac{M_{air}}{M_{fuel}}\right),$$

with a = x + y/4 = 1.16 + 4.32/4 = 2.24, which yields,

$$\left({}^{A}\!/\!_{F}\right)_{stoic} = 4.76(2.24)\frac{28.85}{18.25} = 16.86,$$

and from Eq. (A.13) and the definition of equivalence ratio,

$$\left({}^{A}\!/\!_{F}\right) = \frac{\left({}^{A}\!/\!_{F}\right)_{stoic}}{\Phi} = \frac{16.86}{0.40} = 42.15$$

Since (A/F) is the ratio of the air to fuel flow rates,

$$\dot{m}_{fuel} = \frac{\dot{m}_{air}}{\left({}^{A}\!/\!_{F}\right)} = \frac{17.1\ \text{kg/s}}{42.15} = 0.41\ \text{kg/s}$$

Using the MW of the fuel (18.25 kg/kmol) converting from mass to molar flow rate,

$$\dot{n}_{fuel} = \frac{0.41\ \text{kg/s}}{18.25\ \text{kg/kmol}} \cdot 1000\frac{\text{mol}}{\text{kmol}} = 22.47\ \text{mol/s}$$

The molar flow rates of CO_2, H_2O, and N_2 can be determined based upon the stoichiometry from the general oxidation reaction,

$$C_{1.16}H_{4.32} + a\,(O_2 + 3.76N_2) \rightarrow 1.16CO_2 + 2.16H_2O + bO_2 + 3.76aN_2,$$

Such that a and b are related from the conservation of O atoms similar to Example A.1,

$$2a = 2.32 + 2.16 + 2b \quad \text{or} \quad b = a - 2.24$$

Using the definition of (A/F),

$$\left(\frac{A}{F}\right) = 4.76a\left(\frac{M_{air}}{M_{fuel}}\right),$$

Knowing that $(A/F) = 42.15$, solving for a and using substitution to determine b,

$$a = \left(\frac{42.15}{4.76}\right)\left(\frac{28.85}{18.25}\right) = 14$$

and $b = 14 - 2.24 = 11.76$, resulting in a new oxidation reaction expression:

$$C_{1.16}H_{4.32} + 14O_2 + 52.64N_2 \rightarrow 1.16CO_2 + 2.16H_2O + 11.76O_2 + 52.64N_2$$

The molar flow rate of the fuel $(\dot{n}_{fuel})$ multiplied by the stoichiometric ratios of fuel to CO_2, fuel to H_2O, and fuel to N_2 will provide the molar flow rates of each in the flue gas stream:

$$\dot{n}_{CO_2} = \dot{n}_{fuel} \cdot \left(\frac{1.16 \text{ mol } CO_2}{1 \text{ mol fuel}}\right) = \left(\frac{22.47 \text{ mol fuel}}{\text{s}}\right)\left(\frac{1.16 \text{ mol } CO_2}{1 \text{ mol fuel}}\right) = 26.07 \text{ mol } CO_2/\text{s}$$

$$\dot{n}_{H_2O} = \dot{n}_{fuel} \cdot \left(\frac{2.16 \text{ mol } H_2O}{1 \text{ mol fuel}}\right) = \left(\frac{22.47 \text{ mol fuel}}{\text{s}}\right)\left(\frac{2.16 \text{ mol } H_2O}{1 \text{ mol fuel}}\right) = 48.54 \text{ mol } H_2O/\text{s}$$

$$\dot{n}_{O_2} = \dot{n}_{fuel} \cdot \left(\frac{11.76 \text{ mol } O_2}{1 \text{ mol fuel}}\right) = \left(\frac{22.47 \text{ mol fuel}}{\text{s}}\right)\left(\frac{11.76 \text{ mol } O_2}{1 \text{ mol fuel}}\right) = 264.25 \text{ mol } O_2/\text{s}$$

$$\dot{n}_{N_2} = \dot{n}_{fuel} \cdot \left(\frac{52.64 \text{ mol } N_2}{1 \text{ mol fuel}}\right) = \left(\frac{22.47 \text{ mol fuel}}{\text{s}}\right)\left(\frac{52.64 \text{ mol } O_2}{1 \text{ mol fuel}}\right) = 1182.82 \text{ mol } N_2/\text{s}$$

The mass flow rate corresponding to each in the flue gas stream may be calculated by multiplying the molar flow rate by the molecular weight of each gas. Therefore, the mass flow rates for CO_2, H_2O, O_2 and N_2 are 1.15, 0.87, 8.46 and 33.12 kg/s, respectively.

Enthalpy and Heat of Combustion The definition of enthalpy as applied to combustion science is the sum of internal energy and PV work of an expanding fluid or solid on its surroundings. Absolute enthalpy, h is the reference state-defined enthalpy of a molecule, which is equal to the sum of the reference enthalpy, h_f° of the molecule at standard temperature and pressure (STP) conditions (298.15 K, 1 atm) and the sensible enthalpy change, Δh_s in going from STP to the conditions of interest. This relationship may be expressed for an ideal gas by,

$$h(T, p) = h_f^\circ(T_{ref}, p_{ref}) + \Delta h_s(T, p) \tag{A.15}$$

Table A.1 Heating values, heats of combustion and composition for various fuels [1, 2]

Fuel	HHV (MJ/kg)	Composition (dry, wt%)					
		C	H	O	S	N	Ash
Lignite coal	15.0	49.8	3.3	27.7	0.5	1.0	17.7
Anthracite coal	27.0	88.9	3.4	2.3	0.8	1.6	2.9
Natural gas	54.0	58.0	18.7	1.5	~0	21.6	0
Crude oil	42.0–45.3	88.0	8.2	0.5	3.0	0.1	0.2

Table A.2 Heating values and composition of common biomass fuels [1, 3]

Fuel	HHV (MJ/kg)	Moisture content (wt%)	Composition (wt%)					
			C	H	O	S	N	Ash
Wood	15.0	48.00	55.00	5.77	39.10	0.1	0.1	3.00
Grass/plants	17–21	29.8	44–49.2	5.0–6.0	39–43.4	0.01–0.2	0.3–1.2	6.9
Manure	9.56	44.0	7–54.1	0.8–6.7	5–48.3	0.3–1.2	2.33–4.7	42–43
Straw	14–16	10	54.7	6.0	38.9	0.07	0.3	4–5
Peat	17–22	70–90	56.7	6.0	35.9	0.2	1.7	4.9
Algae	18.15	31.9	53.8	7.4	30.9	0.5	7.5	6.1

The reference states of single atom-type substances are represented by their lowest energy states at standard conditions. For instance, the reference state for hydrogen is the diatomic species, H_2, since this gaseous state is the most stable form of hydrogen at STP. Correspondingly, the enthalpy of formation of H_2 is zero and the enthalpy of formation of H would be half the energy it takes to dissociate diatomic hydrogen into two corresponding atoms.

The enthalpy of reaction is synonymous to the enthalpy of combustion for combustion applications. The heat of combustion (Δh_{comb}) is numerically equivalent to the enthalpy of reaction, but opposite in sign. The heat of combustion provides knowledge of the energy release associated with a given reaction and is defined as,

$$\Delta h_{comb} = h_{prod} - h_{react} \tag{A.16}$$

where h_{prod} and h_{react} are the sum of product and reactant enthalpies, respectively.

The heat of combustion is also known as the heating value. The *higher heating value* (HHV) is defined as the heat of combustion with the inclusion of energy associated with water in a condensed phase (liquid). The energy required to evaporate water from the condensed liquid phase is approximately 285.8 kJ/mol. Therefore, including this energy into the heat of combustion provides an upper bound of the energy content within the chemical reaction. The alternative assumption of not including this energy is that the water within the fuel is all evaporated and that this energy is lost to the surroundings and cannot be recovered; this is known as the *lower heating value* (LHV). Examples of HHV and composition of various fossil and biomass fuels are listed in Tables A.1 and A.2, respectively. In the U.S., HHV is the standard measure used by the Department of Energy and others to quantify the energy content of fuels. In Europe and elsewhere, LHV has become the standard measure.

Appendix B: Common Gas Properties [4]

Chemical formula	Gas	Molecular weight	Boiling point at 0.1 Mpa (1 atm) (K)	Values at 0.1 Mpa, 289 K (1 atm, 15°C)			Specific heat at constant pressure at 0.1 Mpa (J/kg K)				Ratio of specific heats $k = c_p/c_v$ at 0.1 MPa				Molar Heat at Capacity at 339 K and 0.1 MPa (J/K mol)	Critical conditions	
				Specific gravity (Air = 1.00)	Density (kg/m^3)	Specific volume (m^3/kg)	233 K	289 K	339 K	422 K	233 K	289 K	339 K	422 K		Temperature (K)	Pressure (MPa)
CO	Carbon monoxide	28.01	81.89	0.967	1.1822	0.846	1042.6	1038.4	1042.6	1055.1	1.4	1.4	1.4	1.4	29.14	134.44	3.496
CO_2	Carbon dioxide	44.01	194.67	1.528	1.8678	0.535	791.3	841.6	891.8	1063.5	1.34	1.3	1.28	1.25	38.95	304.44	7.398
H_2	Hydrogen	2.016	20.22	0.070	0.0851	11.757	13918	14273	14412	14495	1.42	1.41	1.4	1.4	29.04	46.11[b]	2.255[b]
H_2O	Water (steam)	18.02	373.00	0.632[a]	0.5975[a]	1.674[a]	–	2076.8	–	2302.9	–	1.32[a]	–	1.31	33.80	647.22	21.974
H_2S	Hydrogen sulfide	34.08	211.33	1.175	1.4369	0.696	975.6	996.5	1017.4	1050.9	1.34	1.33	1.32	1.3	34.66	373.89	9.005
N_2	Nitrogen	28.02	77.44	0.967	1.1822	0.846	1042.6	1042.6	1042.6	1046.8	1.4	1.4	1.4	1.4	29.16	126.11	3.392
N_2O	Nitrous oxide	44.02	184.67	1.531	1.8710	0.534	–	879.3	–	–	–	1.3	–	–	40.39	310.00	7.267
NO	Nitric oxide	30.01	121.89	–	1.2687	0.788	1000.7	996.5	996.5	1000.7	1.38	1.39	1.39	1.38	29.78	179.44	6.591
O_2	Oxygen	32.00	90.22	1.105	1.3504	0.741	912.8	917.0	925.3	946.3	1.4	1.4	1.39	1.38	29.63	154.44	5.047
SO_2	Sulfur dioxide	64.06	263.00	2.254	2.7552	0.363	–	615.5	–	–	–	1.25	–	–	41.39	430.56	7.874
Air (dry)		28.97	78.70	1.000	1.2222	0.818	1004.2	1004.2	1008.3	1016.7	1.4	1.4	1.4	1.4	29.16	132.50	3.766
Flue Gas		~29	–	–	–	–	–	–	–	–	–	–	–	–	–	272.65	9.090

[a] At 289 K

[b] Approximate average for 373–589 K and 0.1–1.36 MPa (1–13.6 atm)

Appendix C: Generalized Gas Compressibility Charts (Courtesy of Dresser-Rand, Painted Post, New York)

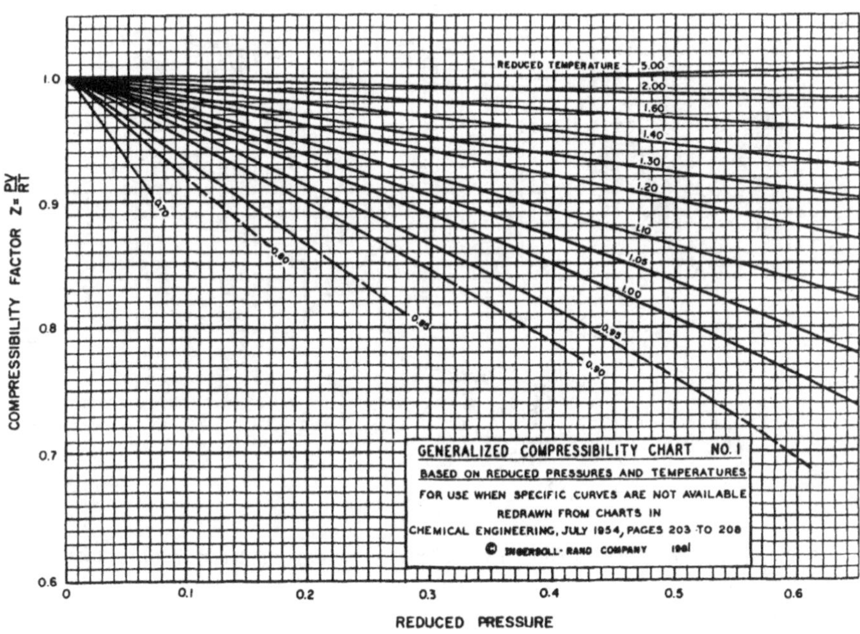

Fig. C.1 Generalized compressibility chart for low values of reduced pressure [4b]

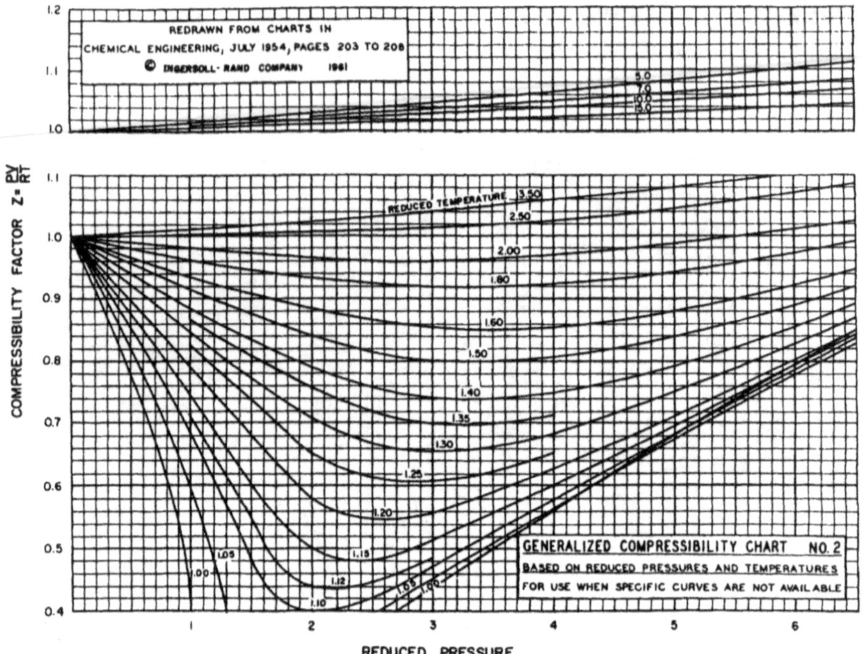

Fig. C.2 Generalized compressibility charts for medium values of reduced pressure [4b]

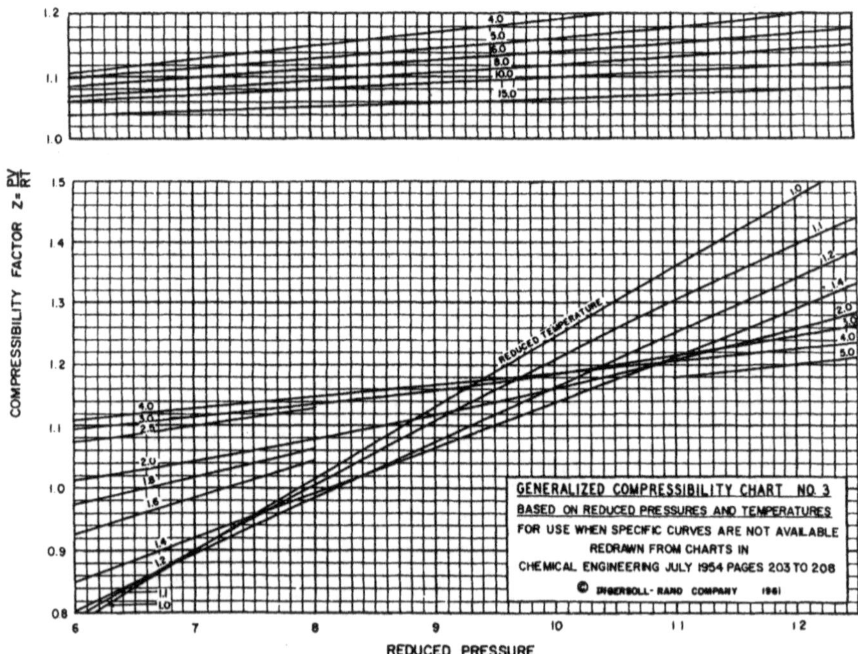

Fig. C.3 Generalized compressibility chart for moderately high values of reduced pressure [4b]

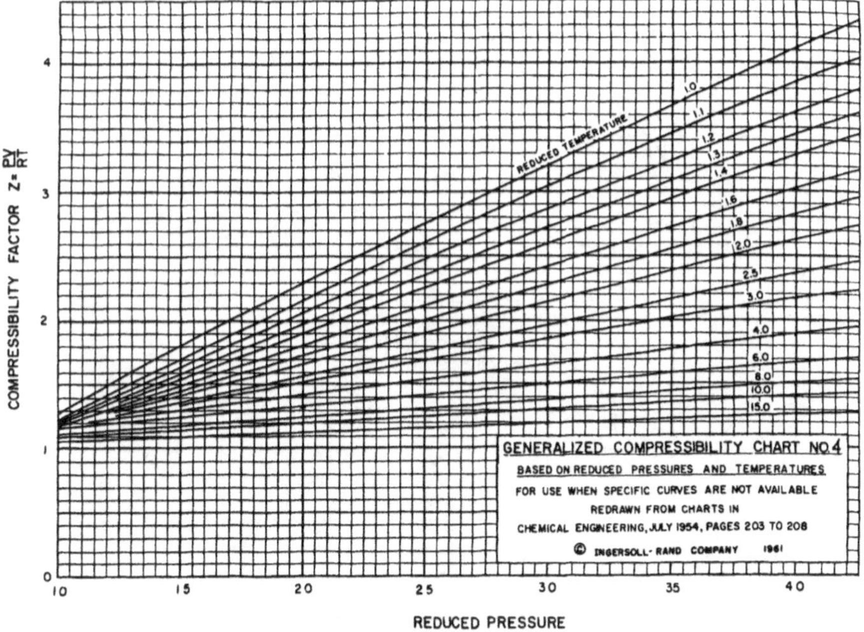

Fig. C.4 Generalized compressibility chart for very high values of reduced pressure [4b]

Appendix D: Quick Selection Method for Multistage Compressors (Courtesy of Dresser-Rand, Painted Post, New York)

The flow chart is provided to illustrate the graphical methods of rapidly selecting multistage compressors. Among the many purely graphical methods of rapidly selecting multistage compressors, this one was developed in 1965 by Don Hallock of the Elliott Company in, Jeannette, Pa. To use this method, the following quantities must be known:

1. W– weight flow, in lb/min or scfm (standard ft^3/min)
2. p_1– inlet pressure, in psia
3. R_p– pressure ratio (discharge psia/inlet psia)
4. t_1– inlet temperature, in °F
5. M– molecular weight
6. k– ratio of specific heats

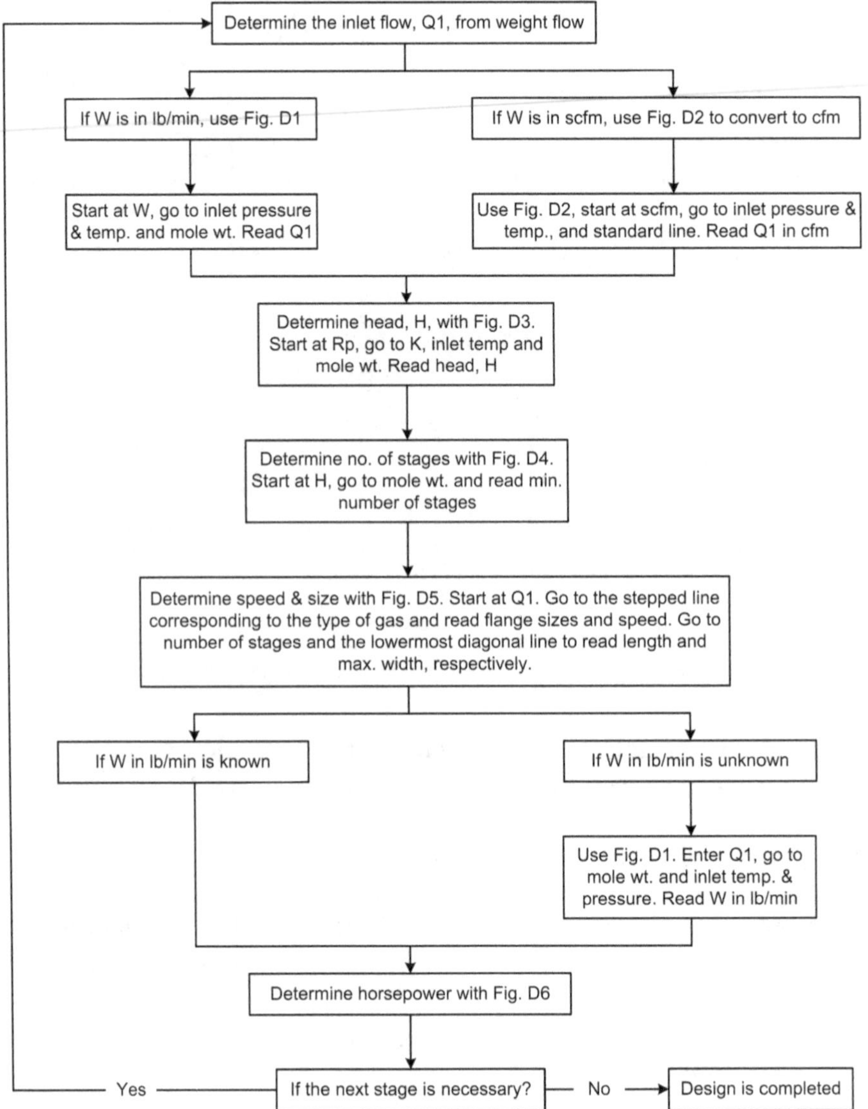

Determine the inlet flow, Q1, from weight flow
If W is in lb/min, use Fig. D1
If W is in scfm, use Fig. D2 to convert to cfm
Start at W, go to inlet pressure & temp. and mole wt. Read Q1
Use Fig. D2, start at scfm, go to inlet pressure & temp., and standard line. Read Q1 in cfm
Determine head, H, with Fig. D3. Start at Rp, go to K, inlet temp and mole wt. Read head, H
Determine no. of stages with Fig. D4. Start at H, go to mole wt. and read min. number of stages
Determine speed & size with Fig. D5. Start at Q1. Go to the stepped line corresponding to the type of gas and read flange sizes and speed. Go to number of stages and the lowermost diagonal line to read length and max. width, respectively.
If W in lb/min is known
If W in lb/min is unknown
Use Fig. D1. Enter Q1, go to mole wt. and inlet temp. & pressure. Read W in lb/min
Determine horsepower with Fig. D6
Yes
If the next stage is necessary?
No
Design is completed

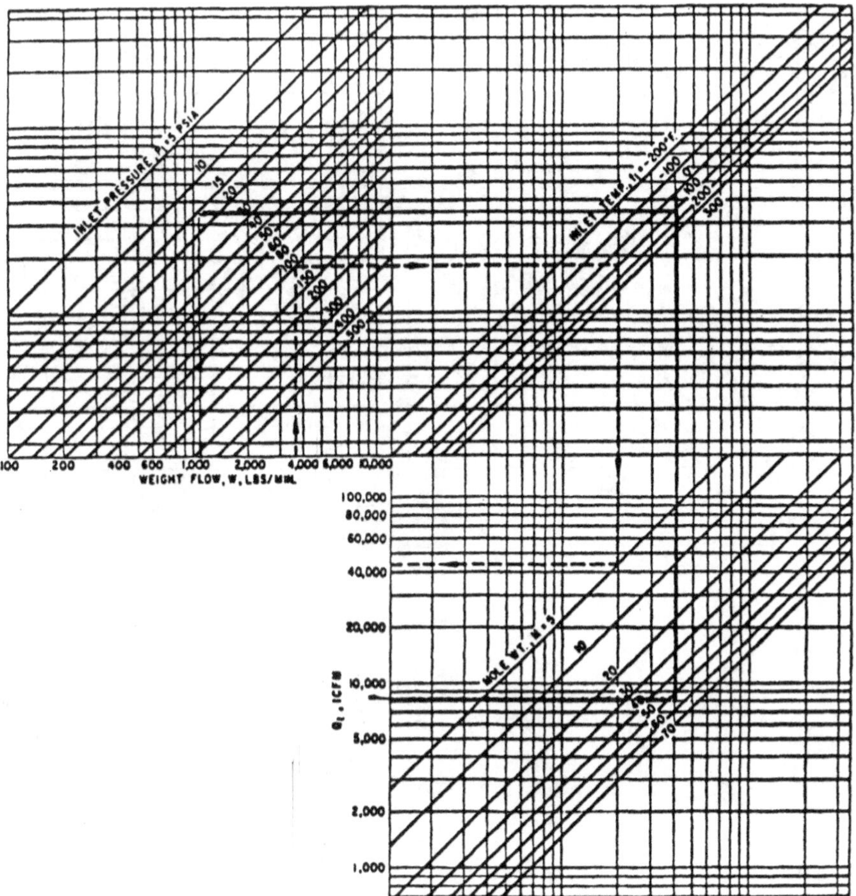

Fig. D.1 If the weight flow of gas W (lbs/min) is known, use this chart to find the inlet flow Q_1 (icfm) [4b]

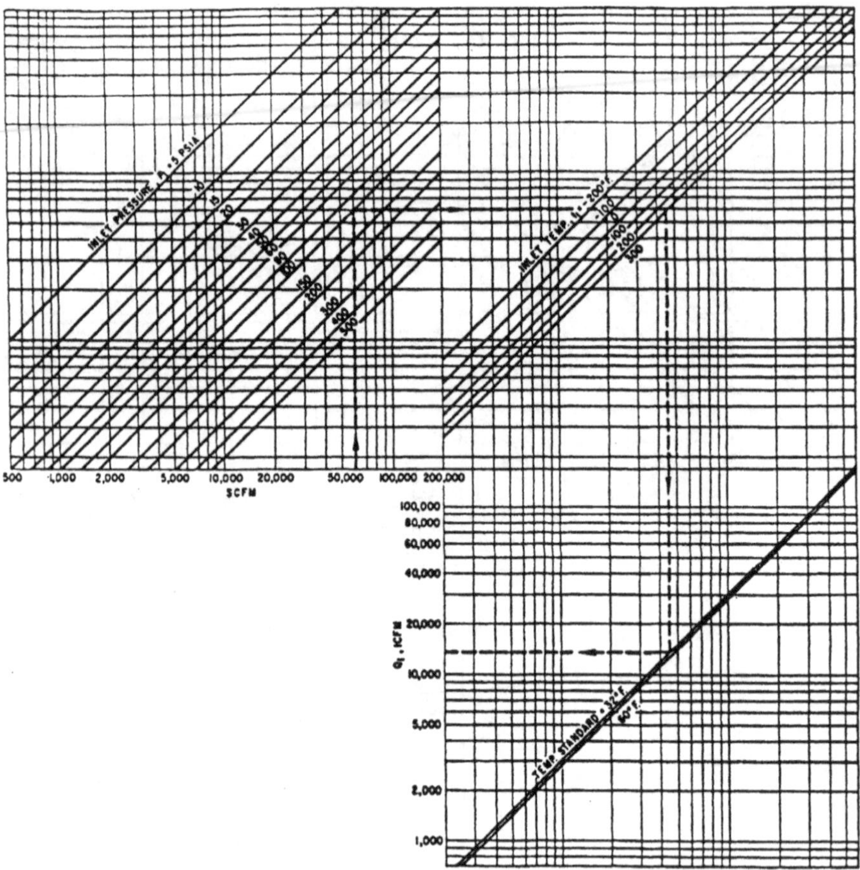

Fig. D.2 If the scfm value is known, use this chart to find the inlet flow Q_1 (icfm) [4b]

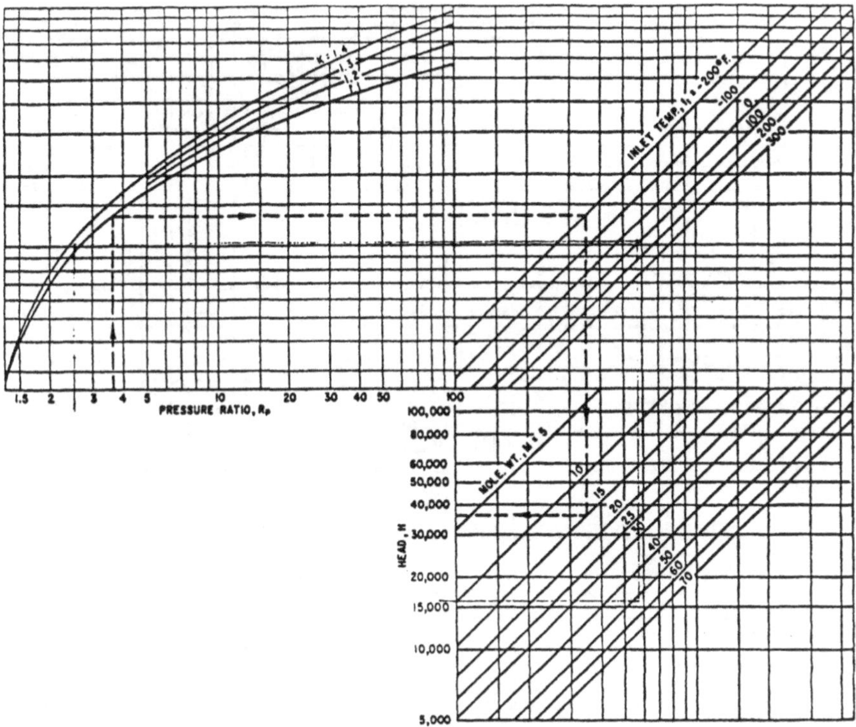

Fig. D.3 Enter this chart at R_p, the pressure ratio (discharge/inlet, psia), to find the head H [4b]

Fig. D.4 Enter this chart with the H value from Fig. D.3 to find the number of stages required

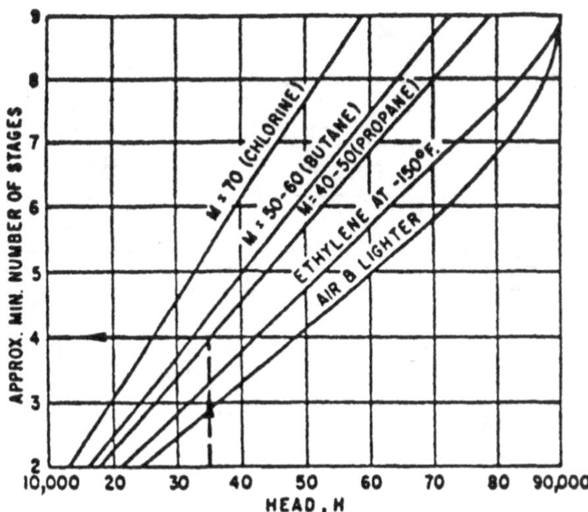

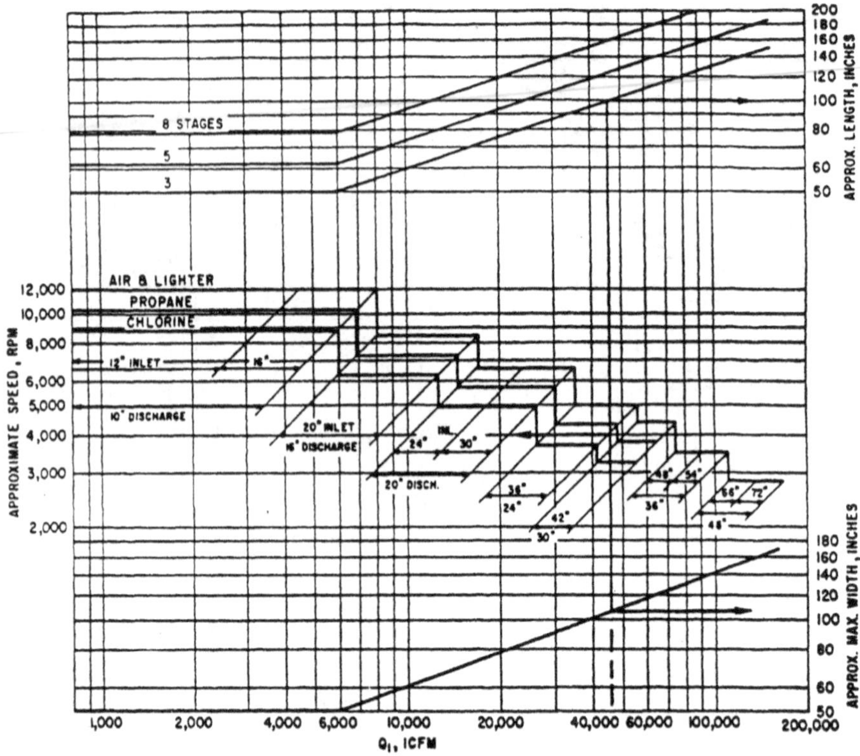

Fig. D.5 Enter this chart at the Q_1 value from Figs. D.1 or D.2 to find the speed, width, length and flange sizes [4b]

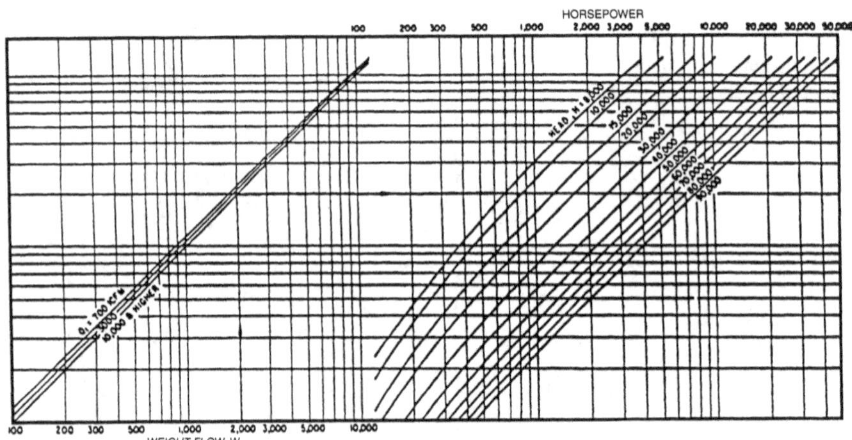

Fig. D.6 Enter this chart at the weight flow of gas W and proceed to find the compressor horsepower required [4b]

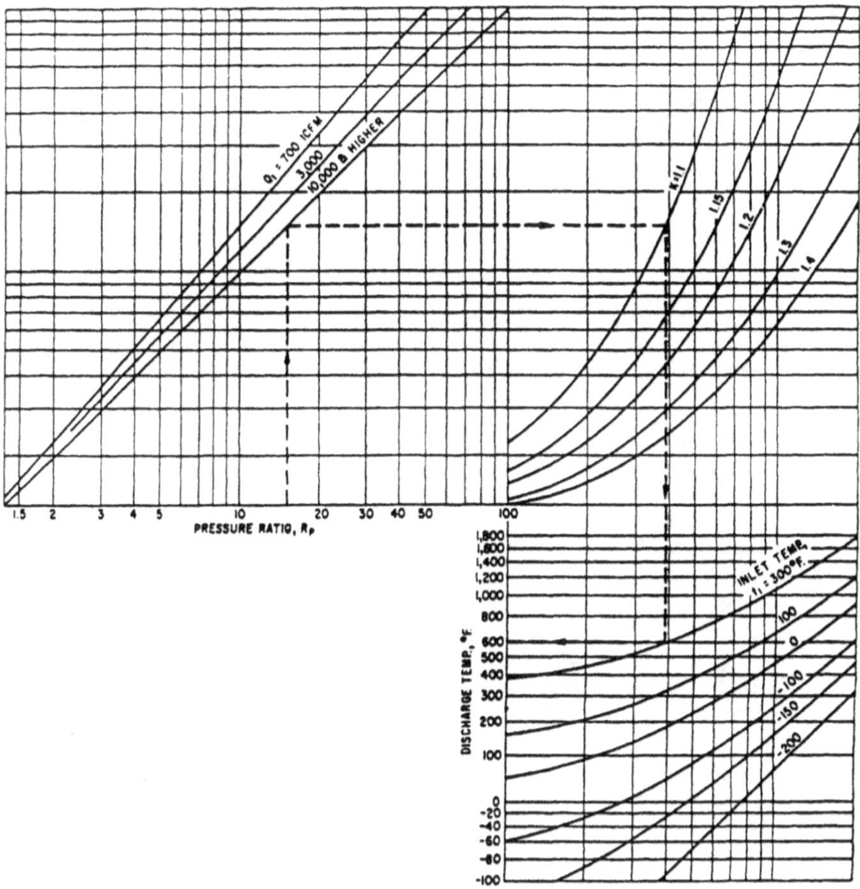

Fig. D.7 The discharge temperature can be found on this chart [4b]

Appendix E: Compressibility, Fugacity, and Solubility of CO_2 in Water

p (atm)	0°C			10°C			15°C		
	Z	f/p	$x_{CO_2} \times 10^3$	Z	f/p	$x_{CO_2} \times 10^3$	Z	f/p	$x_{CO_2} \times 10^3$
1	0.99263	0.99266	1.445	0.99359	0.99862	0.985	0.99400	0.99400	0.802
2	0.98527	0.98538	2.89	0.98719	0.98724	1.946	0.98800	0.98806	1.587
4	0.97055	0.97098	5.60	0.97440	0.97470	3.80	0.97602	0.97630	3.11
6	0.95576	0.95676	8.14	0.96156	0.96230	5.56	0.96401	0.96464	4.58
8	0.94082	0.94275	10.51	0.94862	0.95004	7.23	0.95193	0.95314	5.98
10	0.92567	0.92886	12.71	0.93554	0.93788	8.81	0.93971	0.94174	7.32
12	0.91022	0.91512	14.74	0.92224	0.92584	10.30	0.92732	0.93042	8.60
14	0.89440	0.90146	16.60	0.90868	0.91388	11.70	0.91470	0.91918	9.82
16	0.87814	0.88792	18.31	0.89480	0.90198	13.02	0.90181	0.90796	10.99
18	0.86136	0.87442	19.85	0.88054	0.89016	14.24	0.88860	0.89690	12.09
20	0.84400	0.86097	21.23	0.86585	0.87836	15.38	0.87501	0.88580	13.13
22	0.82596	0.84753	22.46	0.85068	0.86657	16.44	0.86101	0.87474	14.12
24	0.80719	0.83413	23.52	0.83495	0.85483	17.41	0.84654	0.86370	15.05
26	0.78760	0.82073	24.43	0.81863	0.84305	18.29	0.83156	0.85263	15.92
28	0.76712	0.80731	25.19	0.80166	0.83130	19.09	0.81601	0.84160	16.73
30	0.74567	0.79387	25.79	0.78397	0.81950	19.80	0.79985	0.83053	17.49
32	0.72319	0.78039	26.23	0.76552	0.80770	20.44	0.78303	0.81941	18.19
34	–	–	–	0.74624	0.79583	20.98	0.76549	0.80830	18.83
36	–	–	–	0.72608	0.78393	21.45	0.74720	0.79713	19.42

p (atm)	20°C			25°C			30°C		
	Z	f/p	$x_{CO_2} \times 10^3$	Z	f/p	$x_{CO_2} \times 10^3$	Z	f/p	$x_{CO_2} \times 10^3$
1	0.99437	0.99436	0.692	0.99470	0.99472	0.608	0.99529	0.99532	0.473
2	0.98874	0.98880	1.371	0.98941	0.98946	1.207	0.99060	0.99062	0.943
4	0.97750	0.97772	2.70	0.97885	0.97904	2.39	0.98121	0.98136	1.868
6	0.96624	0.96678	3.99	0.96827	0.96874	3.58	0.97183	0.97218	2.78
8	0.95491	0.95596	5.24	0.95764	0.95854	4.65	0.96241	0.96810	3.67
10	0.94348	0.94524	6.44	0.94691	0.94844	5.74	0.95293	0.95410	4.54
12	0.93190	0.93458	7.60	0.93607	0.93840	6.80	0.94337	0.94516	5.39
14	0.92013	0.92402	8.72	0.92506	0.92842	7.83	0.93368	0.93624	6.23
16	0.90812	0.91350	9.81	0.91385	0.91852	8.83	0.92385	0.92738	7.05
18	0.89584	0.90302	10.84	0.90240	0.90864	9.80	0.91384	0.91856	7.85
20	0.88324	0.89258	11.84	0.89069	0.89882	10.75	0.90364	0.90980	8.64
22	0.87029	0.88218	12.80	0.87866	0.88900	11.66	0.89320	0.90106	9.40
24	0.85693	0.87178	13.70	0.86629	0.87918	12.55	0.88250	0.89230	10.15
26	0.84312	0.86140	14.59	0.85353	0.86940	13.41	0.87152	0.88356	10.89
28	0.82883	0.85100	15.43	0.84035	0.85960	14.24	0.86022	0.87484	11.60
30	0.81401	0.84058	16.22	0.82672	0.84976	15.05	0.84858	0.86610	12.30
32	0.79862	0.83013	16.98	0.81259	0.83995	15.82	0.83657	0.85735	12.99
34	0.78262	0.81967	17.70	0.79793	0.83010	16.57	0.82415	0.84857	13.65
36	0.76596	0.80917	18.30	0.78270	0.82021	17.29	0.81132	0.83979	14.30

p (atm)	50°C			75°C			100°C		
	Z	f/p	$x_{CO_2} \times 10^3$	Z	f/p	$x_{CO_2} \times 10^3$	Z	f/p	$x_{CO_2} \times 10^3$
1	0.99603	0.99602	0.342	0.99696	0.99696	0.248	0.99766	0.99764	0.187
2	0.99206	0.99208	0.688	0.99393	0.99394	0.495	0.99532	0.99534	0.373
4	0.98414	0.98424	1.354	0.98788	0.98794	0.984	0.99066	0.99070	0.743
6	0.97623	0.97646	2.02	0.98184	0.98196	1.465	0.98601	0.98606	1.111
8	0.96831	0.96876	2.66	0.97581	0.97604	1.941	0.98137	0.98150	1.477
10	0.96035	0.96112	3.30	0.96976	0.97014	2.41	0.97678	0.97692	1.841
12	0.95234	0.95352	3.93	0.96369	0.96432	2.87	0.97207	0.97240	2.20
14	0.94426	0.94596	4.55	0.95759	0.95850	3.33	0.96740	0.96786	2.56
16	0.93608	0.93844	5.15	0.91450	0.95268	3.78	0.96272	0.96338	2.92
18	0.92779	0.93098	5.75	0.94525	0.94692	4.22	0.95801	0.95890	3.27
20	0.91938	0.9235	6.34	0.98899	0.94116	4.65	0.95327	0.95442	3.62
22	0.91081	0.91606	6.91	0.93266	0.93542	5.08	0.94849	0.94996	3.97
24	0.90208	0.90866	7.47	0.92625	0.92972	5.50	0.94367	0.94554	4.32
26	0.89316	0.90124	8.03	0.91973	0.92402	5.92	0.93880	0.94108	4.66
28	0.88403	0.89384	8.57	0.91312	0.91832	6.33	0.93388	0.93668	5.01
30	0.87467	0.88644	9.10	0.90638	0.91260	6.78	0.92889	0.93224	5.35
32	0.86507	0.87902	9.62	0.89952	0.90688	7.18	0.92383	0.92784	5.68
34	0.8552	0.87160	10.13	0.89253	0.90122	7.52	0.91871	0.92342	6.02
36	0.84505	0.86417	10.63	0.88538	0.89550	7.90	0.91350	0.91898	6.35

Appendix F: Parameters for Calculating Gas Diffusivity

Table F.1 Lennard-Jones potential parameters [5]

Substance		σ (Å)	$\varepsilon_{12}/k_B T$
Air		3.711	78.6
CH_4	Methane	3.758	148.6
CO	Carbon monoxide	3.690	91.7
CO_2	Carbon dioxide	3.941	195.2
C_2H_6	Ethane	4.443	215.7
H_2	Hydrogen	2.826	59.7
H_2O	Water	2.641	809.1
N_2	Nitrogen	3.798	71.4
O_2	Oxygen	3.467	106.7

Table F.2 The collision intergral Ω [5]

$k_B T/\varepsilon_{12}$	Ω	$k_B T/\varepsilon_{12}$	Ω	$k_B T/\varepsilon_{12}$	Ω
0.30	2.662	1.65	1.153	4.0	0.8836
0.40	2.318	1.75	1.128	4.2	0.8740
0.50	2.066	1.85	1.105	4.4	0.8652
0.60	1.877	1.95	1.084	4.6	0.8568
0.70	1.729	2.1	1.057	4.8	0.8492
0.80	1.612	2.3	1.062	5.0	0.8422
0.90	1.517	2.5	0.9996	7	0.7896
1.00	1.439	2.7	0.9770	9	0.7556
1.10	1.375	2.9	0.9576	20	0.6640
1.30	1.273	3.3	0.9256	60	0.5596
1.50	1.198	3.7	0.8998	100	0.5130
1.60	1.167	3.9	0.8888	300	0.4360

Appendix G: Kinetic and Thermodynamic parameters for CO_2 Absorption Reactions

G.1. CO_2–H_2O

The rate constant and equilibrium constant values as used in the model for CO_2–H_2O reactions [6]. The values are at infinite dilution and zero ionic strength. c_{H_2O} is the concentration of water in dilute solution (55.6 M)

Reaction	Kinetics	Equilibrium	References
$CO_2 + H_2O \underset{k_{-1}}{\overset{k_1,K_1}{\rightleftharpoons}} H_2CO_3$	*General expression*		Soli et al. [7]
	$k_1 = \frac{e^{(22.66-7799/T)}}{c_{H_2O}} M^{-1}s^{-1}$		
	$k_{-1} = e^{(30.15-8018/T)}s^{-1}$	$K_1 = \frac{k_1}{k_{-1}} M^{-1}$	
	At 30°C		
	$k_1 = 4.68 \times 10^{-2}s^{-1a}$	$K_1 = 1.15 \times 10^{-3}$	
	$k_{-1} = 40.65\ s^{-1}$	$\log K_1 = -2.94$	
$CO_2 + OH^- \underset{k_{-2}}{\overset{k_2}{\rightleftharpoons}} HCO_3^-$	*General expression*		Kucka et al. [8]
	$k_2 = 3.279 \times 10^{13} e^{-6613/T} M^{-1}s^{-1}$	$K_2 = \frac{k_2}{k_{-2}} M^{-1}$	
	$k_{-2} = \frac{k_2}{K_2} s^{-1}$		
	At 30°C		Pinsent et al. [9]
	$k_2 = 1.24 \times 10^4 M^{-1}s^{-1}$	$K_2 = 3.21 \times 10^7\ M^{-1}$	
	$k_{-2} = 3.86 \times 10^{-4}s^{-1b}$	$\log K_2 = 7.51$	
$H^+ + CO_3^{2-} \overset{K_3}{\longleftrightarrow} HCO_3^-$	*General expression*		Harned et al. [10]
	–	$\log\ K_3 = \frac{2902.39}{T} + 0.02379T - 6.4980$	
	At 30°C		
	–	$K_3 = 1.95 \times 10^{10} M^{-1}$	
		$\log K_3 = 10.29$	

Reaction	Kinetics	Equilibrium	References
$H^+ + HCO_3^- \overset{K_4}{\longleftrightarrow} H_2CO_3$	*General expression*		Edwards et al. [11]
	–	Calculated as $\frac{K_1}{K_1'}$ where $K_1' = \frac{c_{H^+} c_{HCO_3^-}}{c_{CO_2} c_{H_2O}}$ and $\log K_1' = \frac{-5251.43}{T} - 36.7816 \log T + 102.2685 - \log c_{H_2O}$	
	At 30°C		Harned et al. [10]
	–	$K_4 = 2.46 \times 10^3 M^{-1}$[c] $\log K_4 = 3.39$	
$H^+ + OH^- \overset{K_5}{\longleftrightarrow} H_2O$	*General expression*		Edwards et al. [11]
	–	$\log K_5 = \frac{5839.48}{T} + 22.4773 \log T - 61.2060 + \log c_{H_2O}$	
	At 30°C		Maeda et al. [12]
	–	$\log K_5 = 13.83$	
$MEA + H^+ \overset{K_6}{\longleftrightarrow} MEAH^+$	*General expression*		–
	–	–	
	At 30°C		Bates et al. [13]
	–	$K_6 = 2.24 \times 10^9 M^{-1}$ $\log K_6 = 9.35$	

[a]The rate constant for the reaction of CO_2 with H_2O is defined as the pseudo-first-order rate constant

[b]k_{-2} is calculated based on the principle of microscopic reversibility

[c]The protonation is defined as given in the equation; it is common to define this value differently, using the sum over the concentration of H_2CO_3 and dissolved CO_2 as 'carbonic acid'

G.2. CO_2-amine

Rate and equilibrium constants for the reactions of MEA with carbonic acid and bicarbonate, and protonation constant for the carbamate [6]

Reaction	Kinetics	Equilibrium		References
		via kinetics	via titrations	
$H_2CO_3 + RNH_2 \underset{k_{-14}}{\overset{k_{14}}{\rightleftharpoons}} RNHCOOH + H_2O$	$k_{14} = 9.16(1) \times 10^2 (M^{-1}s^{-1})$	$\log K_{14} = 5.25(1)$	$\log K_{14} = 5.63(2)$	McCann et al.
	$k_{-14} = 5.14(8) \times 10^{-3} (s^{-1})$			
$HCO_3^- + RNH_2 \underset{k_{-15}}{\overset{k_{15}}{\rightleftharpoons}} RNHCOO + H_2O$	$k_{15} = 1.05(2) \times 10^{-3} (M^{-1}s^{-1})$	$\log K_{15} = 1.15(1)$	$\log K_{15} = 1.54(2)$	McCann et al.
	$k_{-15} = 7.43(4) \times 10^{-5} (s^{-1})^a$			
$CO_2 + RNH_2 \underset{k_{-16}}{\overset{k_{16}}{\rightleftharpoons}} RNHCOOH$	$k_{16} = 6.11(3) \times 10^3 (M^{-1}s^{-1})^a$	$\log K_{16} = 2.31(1)$	$\log K_{16} = 2.69(2)$	McCann et al.
	$k_{-16} = 29.8(3) (s^{-1})$			
$RNHCOO^- + H^+ \overset{K_{17}}{\longleftrightarrow} RNHCOOH$	–	–	$\log K_{17} = 7.49(5)$	McCann et al.

[a] k_{-15} is computed as $k_{-15} = (k_{15}\, k_{-14}\, K_{17})/(k_{14}\, K_4)$; k_{-16} is computed as $k_{-16} = (k_{16}\, k_{-14}\, k_{-1})/(k_{14}\, k_1)$, based on the principle of microscopic reversibility

G.3. CO_2-PZ

The rate constant and equilibrium constant values as used in the model for CO_2–amine and amine protonation reactions [14]. The values are at infinite dilution and zero ionic strength. c_{H_2O} is the concentration of water in dilute solution (55.6 M)

Reaction	Kinetics	Equilibrium	References
$CO_2 + PZ \underset{k_{-7}}{\overset{k_7,K_7}{\rightleftharpoons}} PZCOO^- + H^+$	$k_7 = 5.8 \times 10^4 exp\left(-4200\left(\frac{1}{T} - \frac{1}{298}\right)\right)\left(M^{-1}s^{-1}\right)$	$\log K_7 = -3.75 + \frac{1570.4}{T}$	k_7 [15]
	$k_{-7} = \frac{k_7}{K_7}\left(M^{-1}s^{-1}\right)$	$+\log K'_1 - \log c_{H_2O}$	K_7 [16]
$CO_2 + PZCOO^- \underset{k_{-8}}{\overset{k_8,K_8}{\rightleftharpoons}} PZ(COO)_2^{2-} + H^+$	$k_8 = 5.95 \times 10^4 exp\left(-4270\left(\frac{1}{T} - \frac{1}{298}\right)\right)\left(M^{-1}s^{-1}\right)$	$\log K_8 = -1.587 + \frac{574.2}{T}$	k_8 [15]
	$k_{-8} = \frac{k_8}{K_8}\left(M^{-1}s^{-1}\right)$	$+\log K'_1 - \log c_{H_2O}$	K_8 [16]
$H + PZ \overset{K_9}{\longleftrightarrow} PZH^+$	–	$\log K_9 = 5.172 + \frac{1890}{T}$ $- \log c_{H_2O}$	Bishnoi et al. [17]
$H^+ + PZH^+ \overset{K_{10}}{\longleftrightarrow} PZH_2^{2+}$	–	$\log K_{10} = 5.3$	
$H^+ + PZCOO^- \overset{K_{11}}{\longleftrightarrow} HPZCOO$	–	$\log K_{11} = 4.354 + \frac{1517}{T}$	Ermatchkov et al. [16]
$CO_2 + AMP \underset{k_{-12}}{\overset{k_{12},K_{12}}{\rightleftharpoons}} AMPCOO^- + H^+$	$k_{12} = \exp\left(23.69 - \frac{5176.49}{T}\right)\left(M^{-1}s^{-1}\right)$	$\log K_{12} = -5.016 + \frac{1104.1}{T}$	k_{12} [18]
	$k_{-12} = \frac{k_{12}}{K_{12}}\left(M^{-1}s^{-1}\right)$	$+\log K'_1 - \log c_{H_2O}$	K_{12} [19]
$H^+ + AMP \overset{K_{13}}{\longleftrightarrow} AMPH^+$	–	$\log K_{13} = 0.9761 + \frac{2603}{T}$	Regressed from data in Littel et al. [20]

Appendix H: Design for Trayed Columns

Calculating the Number of Equilibrium Trays There are both graphical and algebraic methods for determining the number of equilibrium trays for a column, but only the algebraic approach will be considered. The Kremser method, which was modified by Edmister [21] is applicable for the absorption and stripping of dilute mixtures (*i.e.*, ≤ 12 vol%). With CO_2 as the solute, the fraction absorbed is given by:

$$\frac{A_{CO_2}^{N+1} - A_{CO_2}}{A_{CO_2}^{N+1} - 1} = \text{fraction of } CO_2 \text{ absorbed,} \tag{H.1}$$

while the fraction of CO_2 stripped is given by:

$$\frac{S_{CO_2}^{N+1} - S_{CO_2}}{S_{CO_2}^{N+1} - 1} = \text{fraction of } CO_2 \text{ stripped} \tag{H.2}$$

such that A_{CO_2} and S_{CO_2} are the solute absorption and stripping factors for CO_2, respectively, and N is the number of equilibrium stages required for the specified CO_2 fraction absorbed or stripped. The absorption and stripping factors are defined in terms of the entering liquid and gas flow rates into the column, denoted by L and V, respectively, and K_i defined as the vapor-liquid equilibrium ratio for CO_2 as follows:

$$A_{CO_2} = \frac{L}{K_{CO_2} V} \tag{H.3}$$

and

$$S_{CO_2} = \frac{K_{CO_2} V}{L} \tag{H.4}$$

The most common expressions for the equilibrium ratio are:

$$K_{CO_2} = \frac{p_O}{p}, \quad \text{for an ideal solution at subcritical temperature} \tag{H.5}$$

$$K_{CO_2} = \frac{\gamma_{i,L}^{\infty} p_O}{p}, \quad \text{for a non-ideal solution at subcritical temperature} \tag{H.6}$$

$$K_{CO_2} = \frac{H_{CO_2}}{p}, \quad \text{for a solute at supercritical temperature} \tag{H.7}$$

$$K_{CO_2} = \frac{p_O}{x_e p}, \quad \text{for a soluble solute at subcritical temperature} \tag{H.8}$$

such that p_O is the CO_2 vapor pressure, p is the gas pressure, $\gamma_{i,L}^{\infty}$ is the liquid activity coefficient at infinite dilution, H_{CO_2} is the Henry's law constant, and x_e is the liquid mole fraction of CO_2 at equilibrium conditions. The fraction of CO_2 absorbed or stripped may be determined given the number of equilibrium stages, in addition to the absorption and stripping factors.

Calculating the Stage Efficiency and Column Height The *stage efficiency* is required to determine the actual number of stages from the number of theoretical stages. The stage efficiency, E_0, is primarily dependent on the physical properties of the vapor and liquid streams. Stage efficiencies are typically determined using empirical correlations dependent on the solvent viscosity in addition to equilibrium ratio data. A general empirical correlation suggested by Peters et al. [22] is:

$$\log\ E_0 = 1.597 - 0.1991\log\frac{K_{CO_2}M_L\mu}{\rho/16.02} - 0.0896\left[\log\frac{K_{CO_2}M_L\mu}{\rho/16.02}\right]^2, \tag{H.9}$$

such that K_{CO_2} is the equilibrium ratio, M_L is the molecular mass of the liquid, μ is the viscosity of the solvent in units of cP (mPa s), and ρ is the density of the liquid in units of kg/m^3. The actual number of trays, N_{act} can be calculated from the stage efficiency, E_0, and number of equilibrium trays, N from:

$$N_{act} = \frac{N}{E_0} \tag{H.10}$$

It is important to note that this will provide an approximate estimate and that sophisticated commercial software programs exist for determining accurate stage efficiencies. It is also important to note that this correlation is appropriate for the case of discrete stages and is not be applicable for packed columns. The height of the column, H_c, may be calculated by:

$$H_c = (N_{act} - 1)\ H_s \tag{H.11}$$

such that N_{act} is the actual number of trays and H_s is the spacing between the trays. Typical tray spacings are either 0.46 or 0.61 m for large-diameter columns [22].

Calculating the Column Diameter Determining the column diameter requires knowledge of the vapor velocity at column flooding conditions, V_f, which may be calculated from:

$$V_f = C_{SB}\left(\frac{\sigma}{20}\right)^{0.2}\left(\frac{\rho_L - \rho_V}{\rho_V}\right)^{0.5} \tag{H.12}$$

such that C_{SB} is the Souders and Brown flooding conditions in units of m/s, σ is the surface tension in units of dyne/cm, and ρ_L and ρ_V are the liquid and vapor stream densities, respectively. The actual vapor velocity at the top and bottom of the column may be assumed as 70 to 90% of the vapor velocity at flooding conditions. The column cross-sectional area is calculated from:

$$A = \frac{Q}{u} \tag{H.13}$$

and

$$A_c = A + A_d \tag{H.14}$$

such that A is the column area, A_d is the downcomber area, A_c is the cross-sectional area of the column, Q is the vapor volumetric flow rate, and u is the gas velocity (either at the top or bottom of the column). The column diameter may then be calculated from $D_c = 4(A_c)^{1/2}/\pi$. The larger of the top or bottom diameter calculated may be used to represent the entire column diameter.

Appendix I: Figures Used in Determining Height and Width of a Packed Tower

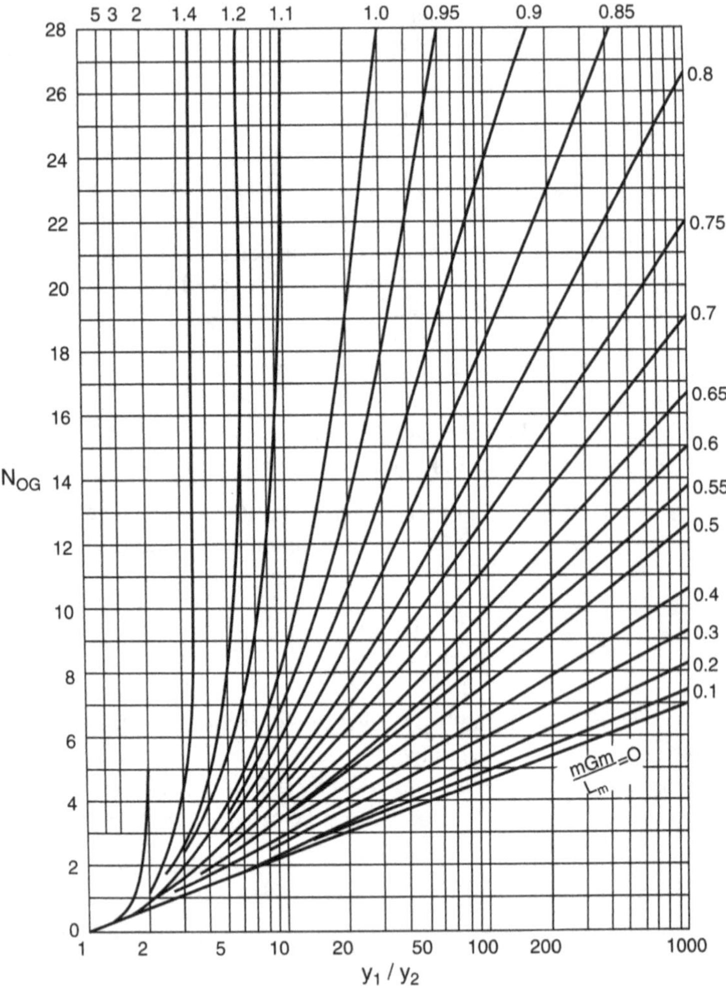

Fig. I.1 Number of transfer units N_{OG} as a function of y_1/y_2 and $m(V/L)_m$ [23]

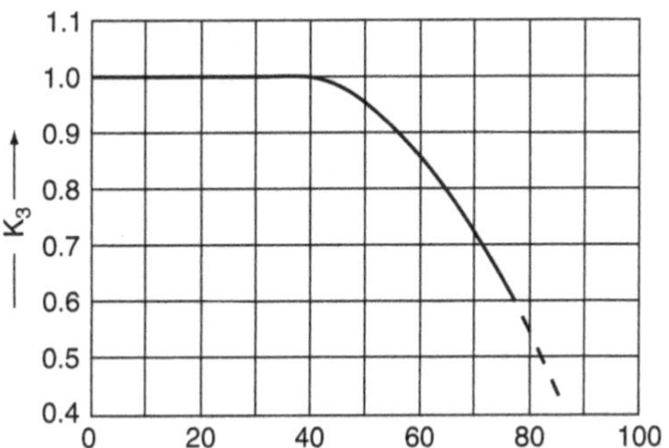

Fig. I.2 Percentage flooding corrections factor [23]

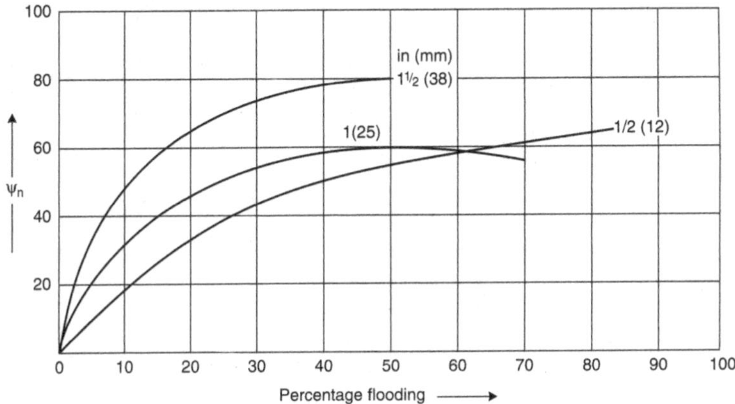

Fig. I.3 Factor for H_G for Berl saddles [23]

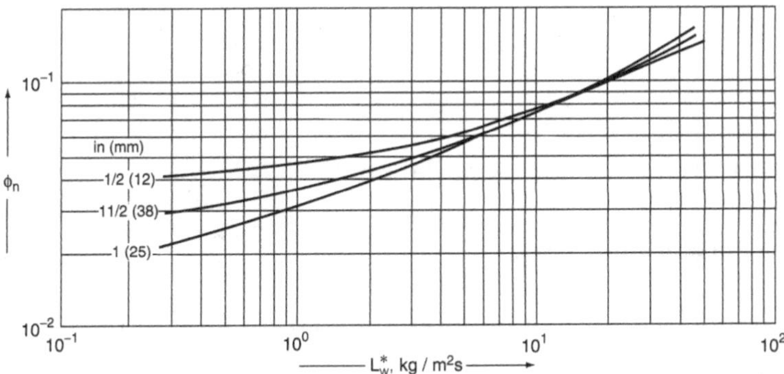

Fig. I.4 Factor for H_L for Berl saddles [23]

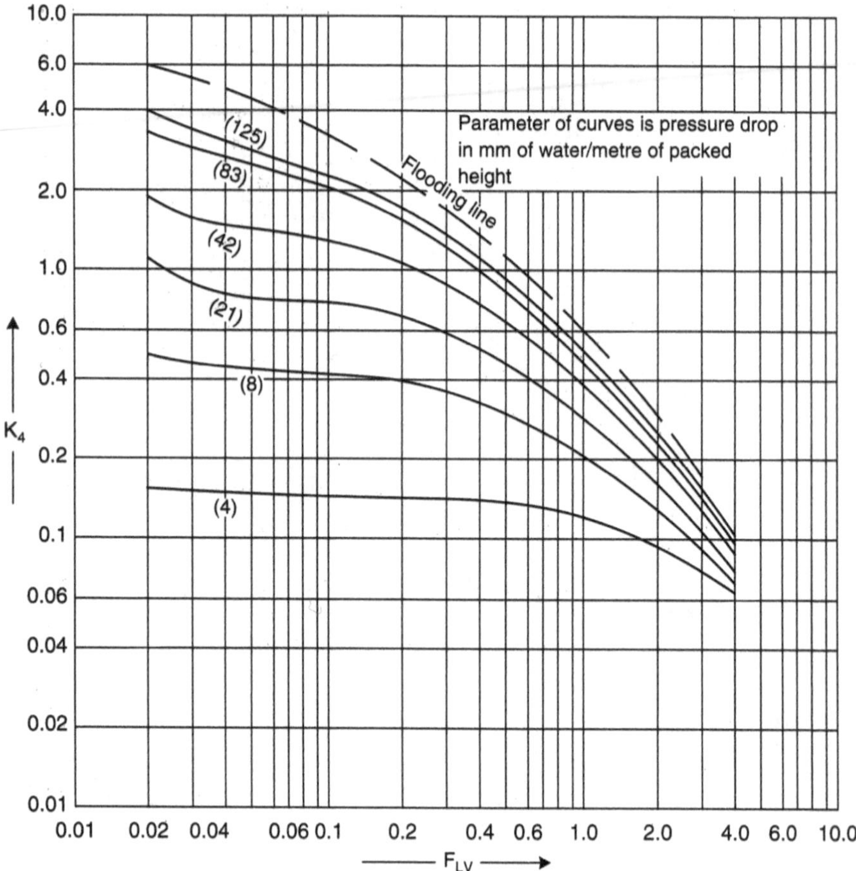

Fig. I.5 Generalized pressure drop correlation [23]

Appendix J: Vapor Pressure Parameters

Antoine Equation

$$\log p_O = A - \frac{B}{T + C} \tag{J.1}$$

Where

$$\mathrm{T} = [^\circ\mathrm{C}]$$

$$p_O = [\mathrm{mmHg}]$$

- The antoine Equation should never be used outside the stated temperature.
- The constants A, B, and C form a set. Never use one constant from one tabulation and the other constants from a different tabulation.

Constants for the Antoine Equation [24]

Formula	Name	A	B	C	T_{min} (°C)	T_{max} (°C)
Ar	Argon	6.84064	340.271	271.801	−189.37	−122.29
CO	Carbon monoxide	6.72527	295.228	268.243	−205.00	−140.23
CO_2	Carbon dioxide	7.58828	861.82	271.883	−56.57	31.04
H_2	Hydrogen	6.14858	80.948	277.532	−259.2	−239.97
H_2O	Water	8.05573	1723.64	233.076	0.01	373.98
H_2S	Hydrogen sulfide	7.11958	802.227	249.61	−85.47	100.38
N_2	Nitrogen	6.72531	285.573	270.087	−210.00	−147.05
N_2O	Nitrous oxide	6.81488	550.604	228.438	−90.85	36.42
NO	Nitric oxide	7.70299	473.03	249.866	−163.65	−93.00
O_2	Oxygen	6.83706	339.209	268.702	−218.8	−118.57
SO_2	Sulfur dioxide	7.33311	1013.46	237.647	−75.48	157.6

References

1. Kirk-Othmer (2007) Coal, 5th ed, vol 2. Wiley, Chichester, p 2762
2. Speight JG (2011) Combustion of hydrocarbons. Elsevier, Oxford, p 648
3. (a) Pisupati SV (2004) Fuel chemistry. In: Meyers RA (ed) Encyclopedia of physical science and technology, 3rd ed. Academic Press, London, pp 253–274; (b) Overend RP (2003) Heat, power and combined heat and power. National Energy Technology Laboratory (NETL), U.S. Department of Energy's (DOE), Amsterdam, pp 63–102; (c) O'Sullivan P (2005) On the value of lakes. The Lakes Handbook, p 3–24; (d) Kay B, Lal R, Kimble J, Follett R, Stewart B (1998) Soil structure and organic carbon: a review. Soil processes and the carbon cycle, p 198; (e) Morfit C (1848) Manures, their composition, preparation, and action upon soils: with the quantities to be applied. Being a field companion for the farmer. From the French of standard authorities. Lindsay and Blakiston, Philiadelphia
4. (a) Sass BM, Farzan H, Prabhakar R, Gerst J, Sminchak J, Bhargava M, Nestleroth B, Figueroa J (2009) Considerations for treating impurities in oxy-combustion flue gas prior to sequestration. Energy Procedia 1(1):535–542; (b) Bloch HP (2006) A practical guide to compressor technology. Wiley-Interscience, Hoboken
5. Lide DR (2008) CRC handbook of chemistry and physics. CRC Press, Boca Raton, p 2736
6. McCann N, Phan D, Wang X, Conway W, Burns R, Attalla M, Puxty G, Maeder M (2009) Kinetics and mechanism of carbamate formation from CO_2 (aq), carbonate species, and monoethanolamine in aqueous solution. J Phys Chem A 113(17):5022–5029
7. Soli AL, Byrne RH (2002) CO_2 system hydration and dehydration kinetics and the equilibrium CO_2/H_2CO_3 ratio in aqueous NaCl solution. Mar Chem 78(2–3):65–73
8. Kucka L, Müller I, Kenig EY, Górak A (2003) On the modeling and simulation of sour gas absorption by aqueous amine solutions. Chem Eng Sci 58:3571–3578
9. Pinsent BRW, Pearson L, Roughton FJW (1956) The kinetics of combination of carbon dioxide with hydroxide ions. T Faraday Soc 52:1512–1520
10. Harned HS, Scholes SR Jr (1941) The ionization constant of HCO_3^- from 0 to 50°C. J Am Chem Soc 63(6):1706–1709

11. Edwards T, Maurer G, Newman J, Prausnitz J (1978) Vapor liquid equilibria in multicomponent aqueous solutions of volatile weak electrolytes. AIChE J 24(6):966–976
12. Maeda M (1987) Bull Chem Soc Jpn 60:3233–3239
13. Bates RG, Pinching GD (1951) J Res Natl Bur Stand 46:349–352
14. Puxty G, Rowland R (2011) Modeling CO_2 mass transfer in amine mixtures: PZ-AMP and PZ-MDEA. Environ Sci Technol 45:2398–2405
15. Samata A, Bandyopadhyay SS (2007) Chem Eng Sci 62:7312–7319
16. Ermatchkov V, Kamps ÁPS, Speyer D, Maurer G (2006) Solubility of carbon dioxide in aqueous solutions of piperazine in the low gas loading region. J Chem Eng Data 51(5):1788–1796
17. Bishnoi S, Rochelle GT (2000) Absorption of carbon dioxide into aqueous piperazine: reaction kinetics, mass transfer and solubility. Chem Eng Sci 55(22):5531–5543
18. Saha AK, Bandyopadhyay SS (1995) Kinetics of absorption of CO_2 into aqueous solutions of 2-amino-2-methyl-1-propanol. Chem Eng Sci 50:3587–3598
19. Silkenbāumer D, Rumpf B, Lichtenthaler RN (1998) Ind Eng Chem Res 37:3133–3141
20. Littel RJ, Bos M, Knoop GJ (1990) Dissociation constants of some alkanolamines at 293, 303, 318, and 333 K. J Chem Eng Data 35:276–277
21. Edmister WC (1957) Absorption and stripping-factor functions for distillation calculation by manual-and digital-computer methods. Am Inst Chem Eng 3(2):165–171
22. Peters MS, Timmerhaus KD, West RE (2003) Plant design and economics for chemical engineers, 5th ed, p 988. McGraw-Hill, New York
23. Towler G, Sinnott RK (2009) This figure was published in Chemical engineering design principles, pp 754–851, Copyright Elsevier
24. Yaws CL, Narasimhan PK, Gabbula C (2009) Yaws' handbook of antoine coefficients for vapor pressure, 2nd ed. Knovel, New York. http://www.knovel.com/web/portal/browse/display?_EXT_KNOVEL_DISPLAY_bookid=1183&VerticalID=0. Accessed 17 August 2011

Glossary

2nd-law efficiency The ratio of the minimum amount of work or energy required to perform a task to the amount actually used.

Absolute pressure The pressure referenced against a perfect vacuum, equal to gauge pressure plus atmospheric pressure.

Absorption factor (A) The ratio of the slope of the operating line *L/V* to that of the vapor-liquid equilibrium ratio or *K*-value; $A = L/KV$. The value of *A* is normally made greater than unity to allow nearly complete removal of solute from the vapor phase.

Acid dissociation constant The equilibrium constant for dissociation in the context of acid-base reactions. It is a measurement of the acid strength in solution.

Activity coefficient The thermodynamic quantity accounting for deviations from ideal behavior in a chemical mixture.

Adsorbate An atom or molecule adsorbed onto a surface.

Adsorbent A porous material with a high surface area that adsorb adsorbates onto its surface.

Adsorption isotherm Represents the amount of adsorbate on an adsorbent as a function of pressure at a fixed temperature.

Adsorptive The component present in the fluid phase capable of being adsorbed and is generally referred to as a fluid molecule in the gas phase prior to adsorption.

Advection Transport associated with a fluid's bulk motion in a particular direction.

Alkalinity Ability of a given solution to neutralize acid.

Amphoteric A substance that can act as an acid or a base, *e.g.*, water, ammonia, amino acid, etc.

Anion An ion with more electrons than protons yielding a net negative charge.

J. Wilcox, *Carbon Capture*,
DOI 10.1007/978-1-4614-2215-0,

Anode Electrode that oxidizes a chemical species and delivers electrons to an external load.

Annual photon flux molar density (PFD) One of two primary components required to determinethe maximum annual growth of biomass (*i.e.*, CH_2O in the simplest case) derived from microalgae, the other being the energy content of the biomass produced.

Arrhenius expression Rate constant expression; $k = Ae^{-\frac{E_a}{RT}}$, where A is the pre-exponential factor, E_a is the activation energy, T is the temperature, and R is the gas constant.

Axial flow compressor Rotating airfoil-based compressors in which the working fluid flows parallel to the axis of rotation.

Band gap The energy gap between the top of the filled valence band and bottom of the empty conduction band.

Biomass accumulation efficiency Efficiency associated with the reduction of biomass due to the energy expenditure required of the cellular functions of the microalgae.

Biomass energy content Energy (kJ/g) contained within biomass from algae.

Biomimetics The implementation of the biological methods found in nature to the problems of engineering systems or modern technology. In the context of CO_2 capture, applications involving carbonic anhydrase are termed biomimetic since carbonic anhydrase is a natural enzyme that is found in the red blood cells of mammals.

Break point The point when the concentration of the fluid leaving a sorbent bed increases as unadsorbed solute begins to emerge.

Breakthrough curve The movement of the mass-transfer zone through a bed.

Breakthrough time or breakthrough point (break point) The time at which the leading point of the mass transfer zone just reaches the end of the sorbent bed. This is also known as the breakthrough point.

Brønsted acid An acid that is able to lose or donate a hydrogen cation or proton (H^+).

Bubble-point curve The temperature-concentration profile obtained from temperatures at which the first bubble of vapor is formed from a vaporizing liquid at a given concentration.

Calcination A thermal treatment process in which solid materials are heated to high temperature in air or oxygen. The most common application is the decomposition of limestone (calcium carbonate) to lime (calcium oxide) and CO_2.

Capillary condensation A process occurring in porous solids, in which multilayer adsorption from a gas proceeds to the point at which pore spaces are filled with liquid and separated from the gas phase by a meniscus. Capillary condensation is often accompanied by hysteresis.

Capillary fibers Hollow fibers having diameter greater than 500 μm.

Carbamate bond The bond between a carbon and nitrogen atom. In the context of carbon capture, the carbon atom is from the CO_2 molecule and the nitrogen atom is from an amine.

Carbonation conversion The percentage of calcium or magnesium that react to yield carbonate minerals.

Catalyst Reagent that participates in the reaction by increasing the rate of the reaction without being consumed by the reaction itself.

Cathode Electrode that reduces a chemical species and acquires electrons from an external load.

Cation An ion with fewer electrons than protons yielding a net positive charge.

Centrifugal compressor A dynamic compressor that depends on the transfer of energy from a rotating impeller to the air.

Chemical equilibrium The state at which the forward reaction rate and the reverse reaction rate are equal, resulting in no net change in the concentrations of reactants and products with time.

Chemisorption The adsorption mechanism resulting from a chemical bond formation between adsorbent and the adsorbate.

Clathrate Chemical substance consisting of a lattice of one type of molecule that traps and contains a second type of molecule.

Compression efficiency The ratio of the theoretical power to the actual power required for compression.

Compression ratio The ratio between the absolute discharge pressure and the absolute inlet pressure.

Compression stroke The phase of a compression in which the piston moves from the bottom end to the top end of the cylinder to compress the gas trapped inside of the cylinder.

Concentration polarization Formation of concentration gradients within each fluid on either side of a membrane.

Contact angle The angle between two of the interfaces at the three-phase line of contact. For example, the angle at which a liquid or vapor interface meets a solid surface.

Convection Transport associated with combined molecular diffusion and advection (*i.e.*, bulk diffusion).

Critical temperature (T_c) The temperature above which a gas is not able to liquefy, no matter how much pressure is applied.

Cryogenic distillation Separation process in which gas components are separated according to the differences in their boiling temperature. It takes place either at low temperatures or high pressures.

Current density Current flowing through a conductor per unit area.

Delivery stroke (exhaust stroke) The phase of compression in which the discharge valve is opened to release the overpressurized gas trapped in the cylinder after the piston reaches to a certain point.

Delta loading The net theoretical removal capacity of a sorbent bed; also referred to as the working capacity.

Dew-point curve The temperature-concentration profile obtained from temperatures at which the first drop of liquid is formed from a condensing gas at a given concentration.

Diffusivity The rate of transfer of the solute in a given fluid under the driving force of a concentration (or pressure) gradient.

Dipole moment The measurement of molecular polarity defined as the product of the magnitude of charge and distance between the charges.

Dissolution Process in which minerals, rocks or soils are dissolved by acidic solutions such as acid rain, volcanic eruptions, etc.

Electrolyte Substance that is sandwiched between two electrodes and transports ions (not electrons) within an electrochemical cell.

Elementary reaction A chemical reaction in which its chemical species react directly to form products in a single step and pass through a single transition state.

Energetic efficiency The recoverable energy contained in the product formed from catalytic CO_2 reduction.

Enhancement factor The ratio of the average rate of absorption into an agitated liquid in the presence of reaction to the average rate of absorption without reaction.

Equilibrium constant The ratio between the concentration of products and reactants at equilibrium.

Equilibrium stage A hypothetical stage where the equilibrium between the liquid and vapor phase occurs.

Equivalence ratio The ratio of the stoichiometric air-fuel ratio to the actual air-fuel ratio for a given mixture.

Excess adsorption The additional amount of fluid particles present as a consequence of adsorption, *i.e.*, the additional amount present over the gas phase at the same temperature, pressure, and volume.

Excess free volume The volume existing in glassy polymers that remains frozen in the polymer matrix due to the inability of the polymer chains to rotate.

Exchange current density Reflects the rates of the backward and forward reactions, with the net current zero at equilibrium.

Expansion stroke The phase of a compression cycle in which both the discharge and inlet valves are closed after the delivery stroke and the piston is being forced to move to expand the gas.

Faradaic efficiency The ratio of the number of electrons going to a desired product to the total number of electrons transferred in the circuit.

Feed Inlet gas stream that is exposed to the membrane surface for separation.

Fermi level Energy level with a 50% probability of being occupied by an electron. Energy levels lower in energy than the Fermi level are occupied with electrons, whereas energy levels higher than the Fermi level empty.

Flooding velocity The gas velocity inside the column that causes liquid holdup or suspension at a given stage. The liquid can rapidly accumulate on higher gas flow rates. A column should be operated at flow rates below the flooding point. The flooding velocity depends strongly on the type and size of packing as well as the liquid mass velocity.

Flux The amount of fluid that flows through a given cross-sectional area per unit time.

Free volume The volume of the polymer that is not occupied by the polymer molecules themselves.

Fuel-lean mixture A fuel-oxidizer mixture that has significantly *less* fuel in comparison to the stoichiometric combustion ratio.

Fuel-rich mixture A fuel-oxidizer mixture that has significantly *more* fuel in comparison to the stoichiometric combustion ratio.

Fugacity The pressure dependence of the Gibbs free energy of a real gas. It has units of pressure and contains all the information associated with the nonideality of the gas. For an ideal gas the fugacity equals the pressure.

Full-spectrum solar energy The full-spectrum solar energy represents the total solar irradiance incident on the algae cultivation system, and is affected by cloud coverage, aerosols, ozone, and other gases in the atmosphere.

Gauge pressure The pressure (which is zero) referenced against ambient air pressure, so it is equal to the difference between the absolute pressure and atmospheric pressure.

Glass transition temperature, (T_g) Temperature at which the reversible transition from glassy states into rubbery states in amorphous materials occurs.

Global reaction The overall reaction consisting of a series of elementary reactions.

Hardpan soil Dense layer of soil found below the uppermost topsoil layer. Water drainage is limited due to the high clay content in hardpan soil, thereby hindering root growth.

Heat of adsorption The heat or energy associated with adsorbing an adsorbate onto an adsorbent surface. Also termed enthalpy change.

Height of a mass-transfer unit (HTU) A measure of the separation effectiveness of the packing material in the separation process. The larger mass-transfer coefficient results in a smaller value of HTU. HTU has units of length.

Heterogeneous (reaction) Reaction in which the reactants are in different phases and/or catalysts are used.

Higher heating value (HHV) The amount of heat produced by the complete combustion of a unit quantity of fuel of which all of the products of combustion are brought back to the original precombustion temperature including the condensation of any water vapor produced.

Homogeneous (reaction) Reaction in which the reactants are in the same phase.

Hydrate Water-containing substance. In the context of a clathrate hydrate, water in its solid phase (ice) traps gas molecules within its caged configuration. When the gas is methane, this is termed a methane hydrate.

Hydrolysis Chemical weathering process affecting silicate and carbonate minerals in which water ionizes slightly and reacts with silicate minerals.

Hydronium ion (H_3O^+) The aqueous cation that forms from water in the presence of hydrogen ions. It is very acidic with a pKa of -1.7. Its acidity is the standard used to evaluate the strength of an acid in water.

Ideal gas Gas composed of a set of randomly-moving, non-interacting particles that obeys the equation of state $pV = nRT$.

Ionic liquid A salt typically having a high viscosity and exhibiting a low vapor pressure.

Isobaric process A change of a system in which *pressure* remains constant.

Isothermal process A change of a system in which *temperature* remains constant.

Knudsen diffusion The diffusion mechanism occurring in a pore with a narrow diameter (2–50 nm) on the order of molecular or several molecular distances and is due to the collision of molecules with the pore wall.

Ladder polymers Type of polymer composed of aromatic or heterocycles. Their structure is comprised of double-stranded chains linked by hydrogen bonds or chemical bonds at regular intervals.

Length of the unused bed (LUB) The additional section of the sorbent bed required to account for the spreading of the concentration front or the bed length remaining at breakthrough.

Lewis acid An acid that acts as an electron acceptor.

Liquid-mole fraction The amount of liquid (in moles) divided by the total number of moles in the system.

Lower heating value (LHV) As opposed to the higher heating value, the water component of a combustion process is in the vapor state after combustion. It is determined by subtracting the heat of vaporization of water from the higher heating value.

Macropore Pore with width greater than 50 nm.

Mass-transfer coefficient The coefficient of the driving force (*e.g.*, concentration or pressure gradient) for mass transfer.

Mass-transfer zone (MTZ) The region in which most of the mass transfer (or adsorption) takes place as the concentration wave moves through the bed. The shape of the mass-transfer zone depends on the adsorption isotherm, flow rate, and diffusion characteristics.

Mean free path The average distance taken by a moving particle between successive collisions.

Membrane module Membrane housing, *i.e.*, plate-and-frame, spiral-wound, hollow fiber, tubular, and monolith.

Membrane selectivity (α) The ratio of permeabilities for a binary mixture.

Mesopore Pore of intermediate size (*e.g.*, 2–50 nm).

Micropore Pore with width less than 2.0 nm.

Minimum fluidization velocity The minimum flow rate that allows particles to become fluidized.

Molarity The unit of solution concentration; usually present in mol/L (*i.e.*, 1 M = 1 mol/L).

Molecular diffusion The net transport of molecules that results from their molecular motion in the absence of turbulent mixing. It occurs when the concentration (or pressure) gradient of a particular gas in a mixture differs from its equilibrium value.

Molecular sieving Separation of fluid particles according to the difference in their molecular dimensions.

Natural weathering Alteration of the properties of the rocks, soils and minerals due to their exposure to Earth's atmosphere, biological, chemical and physical processes.

Non-arable Land which is unsuitable for farming due to unfavorable conditions, such as limited water and harsh climate.

Number of mass-transfer units (NTU) A measure of the difficulty of the separation process. NTU is equivalent to the number of theoretical stages. The purity of the product from the separation process corresponds to the NTU. A higher NTU is necessary for a very high purity product.

Nutraceutical Pertaining to nutrition for medicinal purposes.

Overpotential The potential difference between the applied electrode potential and the standard potential.

Oxyfuel combustion The process of burning a fuel using pure oxygen in place of air as the primary oxidant.

Partition function (*K*-value) The concentration ratio of vapor to liquid at a given temperature and pressure.

Permeability, (P_A) A membrane's ability to permeate a particular gas, which is an intrinsic property of a membrane and is the product of the diffusivity and solubility, $D_A S_A$.

Permeance Property that defines the volume of gas flowing through the membrane per unit area, per time.

Permeate Portion of the feed that is able to pass through the membrane.

pH A measure of the acidity or basicity of an aqueous solution. Solutions are considered acidic with a pH less than 7 and basic (or alkaline) with a pH greater than 7.

Photoinhibition Reemission of photons that results in cellular damage due to heat release.

Photon transmission efficiency The configuration of the growth set up and subsequent losses in incident solar energy.

Photon utilization efficiency Accounts for the reduction in full photon absorption due to suboptimal growth conditions of the microalgal culture.

Photosynthetic efficiency The ratio of chemical energy obtained from photosynthesis and the total sunlight energy used in photosynthesis.

Physisorption The adsorption mechanism associated with intermolecular forces, *i.e.*, van der Waals forces. Physisorption does not involve chemical bonding, and purely involves charge interactions.

pKa The logarithm value of the acid dissociation constant (K_a). The larger the value of pK_a, the smaller the extent of dissociation.

Poiseuille flow The bulk flow of a viscous fluid in a channel.

Polarization The process of inducing a charge.

Polarizibility The ease of distortion of the electron cloud of an atom or molecule by an external electric field such as the presence of a nearby ion or dipole.

Polytropic cycle An expansion or compression of a gas in which the quantity pV^n is held constant, where p and V are the pressure and volume of the gas and n is a constant.

Postcombustion capture The process that removes CO_2 after combustion of the fuel and prior to its compression, transportation and storage.

Precombustion capture The process that removes CO_2 before combustion, typically to produce hydrogen from the shift conversion.

Primary amine Derivative of ammonia (NH_3) in which one hydrogen atom is replaced by a carbon-based side chain. The empirical formula for a primary amine is R_1-NH_2.

Protonation The addition of a proton (hydrogen ion) to an atom, molecule or ion, normally to generate a cation.

Quadrupole moment The distribution of either electric charge or magnetization equivalent to two dipoles that point in opposite directions.

Rank (of coal) The classification of coal usually determined by the content of volatile materials and the chemical structure. Rank is correlated with the energy content with anthracite the highest ranked coal and lignite the lowest ranked coal.

Reciprocating compressor Type of compressor that uses a piston driven by a crankshaft to increase gas pressure.

Reciprocating positive displacement The increase in the pressure of a gas is achieved by reducing its volume. In such operations, a piston within a cylinder is used as the compressing and displacing element.

Reduction-oxidation (redox) Reactions in which a chemical species either loses (oxidized) or gains (reduced) electrons.

Relative humidity The ratio of the partial pressure of water in the atmosphere at a given temperature to the vapor pressure of pure water at the same temperature. It is often expressed as a percentage.

Relative pressure The ratio between the equilibrium and the saturation pressure of a given adsorbate at the temperature of adsorption.

Retentate (residue) The stream that does not pass through the membrane during membrane separation processes.

Reynolds Number Dimensionless number represented by the ratio of inertial forces to viscous forces.

Saturation (vapor) pressure The pressure of a gas (vapor) in thermodynamic equilibrium with its condensed phase in a closed system. Also known as vapor pressure.

Schmidt Number Dimensionless number that is the ratio of momentum diffusivity and mass diffusivity.

Secondary amine Derivative of ammonia (NH_3) in which two hydrogen atoms are replaced by carbon-based side chains. The empirical formula for a primary amine is R_1-NH-R_2.

Separation factor (selectivity) Membrane's ability to effectively separate gases and can be obtained from the ratio of the component permeabilities.

Sherwood Number Dimensionless number that is the ratio of convective to diffusive mass transport.

Skin A dense and relatively thin surface layer of asymmetric membrane.

Solute Substance dissolved in a given solvent.

Sorbent The material used to sorb (adsorb/desorb/absorb) liquids or gases.

Sorption The general terminology referred to both adsorption and desorption processes.

Specific heat The amount of heat per unit mass required to raise the temperature by one degree Celsius.

Stage cut The ratio of the permeate flow to feed flow yielding the fraction of feed transported through membrane.

Stage efficiency A correction factor applied to the number of ideal stages to obtain the number of actual stages in a column and is sometimes referred to as the plate efficiency.

Steric hindrance The resistance that occurs when the size of a molecule prevents chemical reactions that are observed in related smaller molecules in confined spaces.

Stripping factor (*S*) The ratio of the slope of the vapor-liquid equilibrium ratio (or K-value) to that of the operating line, *i.e.*, L/V; $S = 1/A$ or KV/L. Generally, conditions are chosen to make S greater than unity.

Superficial velocity The volumetric fluid flow rate divided by the cross-sectional area of the bed. (in reference to packed-bed systems)

Supermicropore Pore with width between 0.7 to 2.0 nm.

Surface tension The ability of the surface of a liquid to resist an external force.

Termolecular reaction A single reaction step involving the simultaneous collision of any combination of three molecules, ions, or atoms. This type of reaction is rare in solutions and gas mixtures.

Thermal expansion coefficient Coefficient describing the extent of the change in the size of the object with the change in temperature.

Tortuosity The deviation from the linear path of a particle moving from point A to point B in a three-dimensional pore system.

Ultramicropore Pore with width less than 0.7 nm.

Vapor pressure The pressure of a vapor, which is in equilibrium with its condensed phase. Also known as saturation (vapor) pressure.

Void fraction A measure of the void space within a given material, calculated as the fraction of the void volume over the total volume of the porous material.

Volatility Propensity of a substance to vaporize.

Voltage potential A measure of the electronic energy; measured with respect to a reference electrode, *e.g.*, Ag/AgCl.

Working capacity The net theoretical removal capacity of an adsorption bed; also referred to as delta loading.

Zwitterion A molecule with a positive and negative electrical charge at different locations within the molecule.

Index

J. Wilcox, *Carbon Capture*,
DOI 10.1007/978-1-4614-2215-0,